METHODS IN CELL BIOLOGY

VOLUME XVIII

Chromatin and Chromosomal Protein Research. III

Methods in Cell Biology

Series Editor: **DAVID M. PRESCOTT**
DEPARTMENT OF MOLECULAR, CELLULAR
AND DEVELOPMENTAL BIOLOGY
UNIVERSITY OF COLORADO
BOULDER, COLORADO

Methods in Cell Biology

VOLUME XVIII

*Chromatin and Chromosomal
Protein Research. III*

Edited by

GARY STEIN and JANET STEIN

DEPARTMENT OF BIOCHEMISTRY AND
MOLECULAR BIOLOGY
UNIVERSITY OF FLORIDA
GAINESVILLE, FLORIDA

LEWIS J. KLEINSMITH

DIVISION OF BIOLOGICAL SCIENCES
UNIVERSITY OF MICHIGAN
ANN ARBOR, MICHIGAN

1978

ACADEMIC PRESS • New York   San Francisco   London
A Subsidiary of Harcourt Brace Jovanovich, Publishers

ACADEMIC PRESS, INC.
111 Fifth Avenue, New York, New York 10003

United Kingdom Edition published by
ACADEMIC PRESS, INC. (LONDON) LTD.
24/28 Oval Road, London NW1

LIBRARY OF CONGRESS CATALOG CARD NUMBER: 64–14220

ISBN 0–12–564118–4

PRINTED IN THE UNITED STATES OF AMERICA

CONTENTS

Part A.   Chromatin Fractionation. II

1. *Fractionation of Chromatin into Template-Active and Template-Inactive Portions*
Michael Savage and James Bonner

2. *Fractionation of Chromatin by Buoyant Density-Gradient Sedimentation in Metrizamide*
G. D. Birnie

3. *Dissection of the Eukaryotic Chromosome with Deoxyribonucleases*
Richard Axel

9. Radioimmunoassay of Nonhistone Proteins
Michael E. Cohen, Lewis J. Kleinsmith, and A. Rees Midgley

10. Immunofluorescent Techniques in the Analysis of Chromosomal Proteins
L. M. Silver, C. E. C. Wu, and S. C. R. Elgin

Part C. Sequencing of Histones

11. Peptide Mapping and Amino Acid Sequencing of Histones
Robert J. DeLange

Part D.   Physical Properties of DNA–Nuclear Protein Complexes

LIST OF CONTRIBUTORS

Numbers in parentheses indicate the pages on which the authors' contributions begin.

ALLEN T. ANSEVIN, Department of Physics, University of Texas System Cancer Center, M. D. Anderson Hospital and Tumor Institute, Texas Medical Center, Houston, Texas (397)

RICHARD AXEL, Institute of Cancer Research and Department of Pathology, College of Physicians and Surgeons, Columbia University, New York, New York (41)

J. P. BALDWIN, Biophysics Laboratories, Portsmouth Polytechnic, Portsmouth, United Kingdom (295)

MICHAEL W. BERNS, Developmental and Cell Biology, University of California, Irvine, Irvine, California (277)

G. D. BIRNIE, Wolfson Laboratory for Molecular Pathology, The Beatson Institute for Cancer Research, Glasgow, Scotland (23)

DAVID BLOCH, Botany Department and Cell Research Institute, The University of Texas at Austin, Austin, Texas and the Cell Biology Section, National Center for Toxicological Research, Jefferson, Arkansas (247)

JAMES BONNER, Division of Biology, California Institute of Technology, Pasadena, California (1)

E. M. BRADBURY, Biophysics Laboratories, Portsmouth Polytechnic, Portsmouth, United Kingdom (295)

EUGENE CHIN, Department of Zoology, The University of Texas at Austin, Austin, Texas (247)

F. CHYTIL, Departments of Biochemistry and Medicine, Vanderbilt University School of Medicine, Nashville, Tennessee (123)

MICHAEL E. COHEN, Division of Biological Sciences, The University of Michigan, Ann Arbor, Michigan (143)

R. D. COLE, Department of Biochemistry, University of California, Berkeley, Berkeley, California (189)

ROBERT J. DELANGE, Department o. Biological Chemistry, UCLA School of Medicine, Los Angeles, California (169)

S. C. R. ELGIN, Department of Biochemistry and Molecular Biology, Harvard University, Cambridge, Massachusetts (151)

GERALD D. FASMAN, Graduate Department of Biochemistry, Brandeis University, Waltham, Massachusetts (327)

CHI-TSEH FU, Botany Department and Cell Research Institute, The University of Texas at Austin, Austin, Texas and the Cell Biology Section, National Center for Toxicological Research, Jefferson, Arkansas (247)

EDMOND J. GABBAY, Departments of Chemistry and Biochemistry, University of Florida, Gainesville, Florida (351)

TORU HIGASHINAKAGAWA, Mitsubishi Kasei Institute for Life Sciences, Tokyo, Japan (55)

R. P. HJELM, JR., Biophysics Laboratories, Portsmouth Polytechnic, Portsmouth, United Kingdom (295)

M. W. HSIANG, Department of Biochemistry, University of California, Berkeley, Berkeley, California (189)

LEWIS J. KLEINSMITH, Division of Biological Sciences, The University of Michigan, Ann Arbor, Michigan (143)

ROGER D. KORNBERG, Department of Biological Chemistry, Harvard Medical School, Boston, Massachusetts (429)

HSUEH JEI LI, Division of Cell and Molecular Biology, State University of New York at Buffalo, Buffalo, New York (385)

A. REES MIDGLEY, Reproductive Endocrinology Program, The University of Michigan, Ann Arbor, Michigan (143)

ADA L. OLINS, The University of Tennessee–Oak Ridge Graduate School of Biomedical Sciences and The Biology Division, Oak Ridge National Laboratory, Oak Ridge, Tennessee (61)

RONALD H. REEDER, Department of Embryology, Carnegie Institution of Washington, Baltimore, Maryland (55)

RANDOLPH L. RILL, Department of Chemistry and Institute of Molecular Biophysics, Florida State University, Tallahassee, Florida (69)

HANS RIS, Department of Zoology, University of Wisconsin, Madison, Wisconsin (229)

DENNIS E. ROARK, Departments of Biology and Physics, Maharishi International University, Fairfield, Iowa (417)

MICHAEL SAVAGE, Division of Biology, California Institute of Technology, Pasadena, California (1)

BARBARA RAMSAY SHAW, Department of Chemistry, Duke University, Durham, North Carolina (69)

L. M. SILVER, The Committee on Higher Degrees in Biophysics, Harvard University, Cambridge, Massachusetts (151)

B. DAVID STOLLAR, Departments of Biochemistry and Pharmacology, Tufts University School of Medicine, Boston, Massachusetts (105)

JEAN O. THOMAS, Department of Biochemistry, University of Cambridge, Cambridge, England (429)

KENSAL E. VAN HOLDE, Department of Biochemistry and Biophysics, Oregon State University, Corvallis, Oregon (69)

HARVEY L. WAHN, Department of Embryology, Carnegie Institution of Washington, Baltimore, Maryland (55)

W. DAVID WILSON, Department of Chemistry, Georgia State University, Atlanta, Georgia (351)

C. E. C. WU, Department of Biology, Harvard University, Cambridge, Massachusetts (151)

PREFACE

During the past several years considerable attention has been focused on examining the regulation of gene expression in eukaryotic cells with emphasis on the involvement of chromatin and chromosomal proteins. The rapid progress that has been made in this area can be largely attributed to development and implementation of new, high-resolution techniques and technologies. Our increased ability to probe the eukaryotic genome has far-reaching implications, and it is reasonable to anticipate that future progress in this field will be even more dramatic.

We are attempting to present, in four volumes of *Methods In Cell Biology,* a collection of biochemical, biophysical, and histochemical procedures that constitute the principal tools for studying eukaryotic gene expression. Contained in the first volume (Volume 16) are methods for isolation of nuclei, preparation and fractionation of chromatin, fractionation and characterization of histones and nonhistone chromosomal proteins, and approaches for examining the nuclear-cytoplasmic exchange of macromolecules. The second volume (Volume 17) deals with further methods for fractionation and characterization of chromosomal proteins, including DNA affinity techniques. Also contained in this volume are methods for isolation and fraction of chromatin, nucleoli, and chromosomes. This volume (Volume 18) focuses on approaches for examination of physical properties of chromatin as well as chromatin fractionation, and immunological and sequence analysis of chromosomal proteins. In the fourth volume (Volume 19) enzymic components of nuclear proteins, chromatin transcription, and chromatin reconstitution are described. Volume 19 also contains a section on methods for studying histone gene expression.

In compiling these four volumes we have attempted to be as inclusive as possible. However, the field is in a state of rapid growth, prohibiting us from being complete in our coverage.

The format generally followed includes a brief survey of the area, a presentation of specific techniques with emphasis on rationales for various steps, and a consideration of potential pitfalls. The articles also contain discussions of applications for the procedures. We hope that the collection of techniques presented in these volumes will be helpful to workers in the area of chromatin and chromosomal protein research, as well as to those who are just entering the field.

We want to express our sincere appreciation to the numerous investigators who have contributed to these volumes. Additionally, we are indebted to Bonnie Cooper, Linda Green, Leslie Banks-Ginn, and the staff at Academic Press for their editorial assistance.

GARY S. STEIN
JANET L. STEIN
LEWIS J. KLEINSMITH

Part A. Chromatin Fractionation. 11

Chapter 1

Fractionation of Chromatin into Template-Active and Template-Inactive Portions

MICHAEL SAVAGE AND JAMES BONNER

Division of Biology,
California Institute of Technology,
Pasadena, California

I. Introduction

In recent years several approaches have been made to separate transcriptionally repressed from transcriptionally active regions of chromatin. The earlier fractionation methods devised to separate the transcriptionally active from inactive regions were based on autoradiographic and cytological observations with isolated cells and nuclei (*1*). In these studies the transcriptionally inert regions are seen as highly condensed material (termed heterochromatin) while those regions active in RNA transcription (as judged by incorporation of radioactive RNA precursors) are visualized as extended diffuse fibers (called euchromatin). Other microscopic observations also indicate that chromatin is in an extended fibrous state when transcriptionally active. These observations include puffing of the *Drosophila* salivary gland chromosomes (*2*), the extended conformation of balbiani rings (*3*), as well as the extended configuration of rRNA genes in the act of producing ribosomal RNA (*4,5*).

The term chromatin fractionation states implicitly that one or more forms of chromatin are to be separated from one or more other forms. This goal has been approached in several different ways, all depending on the different structural and chemical properties of the varied kinds of chromatin. The first report concerning such fractionation involved the separation, by differential centrifugation, of euchromatin (less condensed)

1

from heterochromatin (more condensed) (6). Frenster's demonstration that the condensed heterochromatin was more easily sedimented than the extended euchromatin set the groundwork for the fractionation of sheared chromatin by sucrose density gradients. This approach, introduced by Chalkley and Jensen (7), has been used widely in studies of chromatin fractionation. We (8) have based our fractionation on the fact that expressed, template-active sequences are more readily attacked by nucleases than are nonexpressed sequences. McCarthy and his colleagues have made use of the fact that the DNA of expressed sequences melts at a lower temperature than does the DNA nonexpressed sequences. This procedure recovers the template-active DNA in denatured form. McCarthy *et al.* (9) have also based a fractionation on the fact that the template-active portions of the genome bind RNA polymerase molecules while the nonexpressed portions of the genome do not [confirmed by Marushige and Bonner (10)]. Finally, sheared chromatin has been separated into fractions by ion-exchange chromatography (11, 12).

The regions which are active in RNA synthesis probably are interspersed between inactive, nontranscribed regions. To liberate template-active (expressed) sequences from inactive (nonexpressed) sequences, chromatin must be sheared or clipped into fragments smaller than the length of the whole template-active sequence. Statistical calculations by Davidson *et al.* (13) show that on the average the fragments liberated must be one-third or less of the length of the expressed sequence. The average length of the transcriptional unit in rat liver chromatin has been calculated by Davidson *et al.* (13) to be 6500 ± 500 base pairs. Thus the shearing method chosen for use in chromatin fractionation should produce fragments on the order of 2000 base pairs or less, with minimal alteration of the structural aspects of the chromatin. Such shearing or clipping may be done in a variety of ways which include sonication, hydrodynamic shear, and nuclease clipping. Although several investigators have used sonication for shearing of chromatin, this does not appear to be a recommended procedure. Sonication of chromatin to the size (a few hundred to a few thousand base pairs) required for separation of template-active from template-inactive fragments results in the liberation of histones from the fragments, and artifactually increases template activity (Marushige and Bonner, unpublished). The effect of mechanical shear on chromatin structure has also been examined in detail. Noll *et al.* (14) have shown that chromatin prepared by methods involving mechanical shear has staphylococcal nuclease and trypsin digestion patterns markedly different from those obtained with native chromatin or with chromatin sheared by brief incubations with nuclease. Doenecke and McCarthy (15) using restriction modification methylases, have analyzed the movement of histones along the DNA strand during chromatin fragmentation. They find that histones move laterally along the DNA fiber during the mechanical shearing process. In conclusion, caution must be exercised to avoid artifacts arising due to rearrangement of chromosomal proteins during the preparation and shearing of chromatin prior to fractionation.

II. Criteria for Fractionation

A. Template Activity

Template activity in support of DNA-dependent RNA polymerase is an obvious candidate as a criterion for fractionation of template-active from template-inactive chromatin. The expressed portion of the genome must by definition be capable of transcription, and this has generally been found. Template activity, the rate of transcription of native chromatin by a fixed amount of RNA polymerase relative to the rate of transcription of purified homologous DNA by the same amount of RNA polymerase, has been measured for several kinds of chromatin (16). The template activity correlates well with the metabolic activity of the tissue or cell type being examined. For example, rat liver chromatin, a tissue active in RNA synthesis, has a template activity of 18–20%, while chicken erythrocyte chromatin which is little transcribed has a template activity of 2%. Regardless of the fractionation method used, the amount of chromatin isolated in the template-active fraction should reflect the template activity of the isolated native chromatin. With a few exceptions (17), all methods of chromatin fractionation yield fractions more active and less active in RNA synthesis. The template-inactive fraction generally possesses some capacity to support DNA-dependent RNA synthesis. It appears that template activity of whole chromatin (as defined above) is the same whether the *Escherichia coli* or homologous RNA polymerase is used (18 and Van den Broek, personal communication). However, the template activity of the putatively template-active fraction may be higher than that measured with *E. coli* RNA polymerase by as much as a factor of 30 when homologous form II RNA polymerase is used (21). This drastic difference may reflect differences in recognition elements in template-active chromatin, or it may merely indicate that the template-active fraction, as isolated, has a large proportion of single-stranded DNA, since eukaryotic RNA polymerase II prefers denatured to native DNA as a substrate.

B.   Nascent RNA

The template-active fraction would be expected to bear nascent RNA. Chromatin pulse labeled with, for example, [^{3}H]uridine *in vivo*, should yield an active chromatin fraction bearing all the nascent RNA. This criterion has been used by several investigators. Thus, Billing and Bonner (*19*) found that the fraction of chromatin most readily attacked by nuclease (shown by other criteria, discussed below, to be the expressed fraction) bears the great bulk of the nascent RNA. The same is true of chromatin fractions separated by sucrose density gradient centrifugation (*18,20,21*). It is also true for the active and inactive fractions of chromatin as separated by ECTHAM–cellulose chromatography (*11,12,22*). This criterion is inconclusive, however, because the nascent RNA transcripts are very sensitive to clipping by cellular ribonucleases and by the hydrodynamic forces required to shear chromatin (see above). Therefore it is important to establish that the cofractionating nascent RNA is intimately associated with the template-active fraction, since the fractionation of free RNA would be coincident with the template-active chromatin in most fractionation schemes.

C.   Endogenous RNA Polymerase

It might be expected that RNA polymerase would be found associated with the expressed portion of the genome. RNA polymerase molecules are clearly visible in the electron micrographs of template-active chromatin (*4*) and are not obvious in electron micrographs of template-inactive portions of chromatin. More compelling, it has been shown for the case of rat liver chromatin that all of the endogenous RNA polymerase activity is found in the putative template-active fraction as separated from the template-inactive by nuclease digestion (*10*). The same is true for chromatin fractionated by sucrose density gradient centrifugation (*9*). B. J. McCarthy (personal communication) has used RNA polymerase to purify the template-active portion of chromatin in a slightly different way. A large excess of *E. coli* RNA polymerase is added to sheared chromatin. The *E. coli* RNA polymerase binds to the template-active portion of chromatin, not to the template-inactive portion. The fragments containing a large number of bound RNA polymerase molecules are separated from the fragments that do not contain such RNA polymerase by exclusion chromatography.

In summary, ability to bind RNA polymerase at physiological ionic strength and/or presence of RNA polymerase in a fraction forms one additional criterion for a meaningful fractionation of chromatin into expressed and nonexpressed regions.

D. Sequence Complexity

Fractionation with respect to DNA sequences may be demonstrated in two ways. First, the inactive heterochromatin has been shown (23) to be enriched in the highly repetitive satellite DNAs. These satellite DNA sequences differ in base composition, and therefore band at a different buoyant density than the bulk of the nuclear DNA. Thus the finding that the satellite DNA is localized in the inactive regions indicates that fractionation has occurred (24).

The second, and more conclusive, type of sequence fractionation involves those sequences actively involved in transcription. These sequences should be preferentially localized in the template-active fraction. The expressed portion of the genome would be expected to include a subset of the single copy sequences of the entire genome. These sequences would be expected to be hybridizable by messenger RNA, and to be hybridizable to nuclear RNA as well. A study of the sequence structure of the template-active fraction of chromatin as compared to that of the template-inactive fraction of chromatin has been carried out by Gottesfeld *et al.* for rat liver chromatin (25). In this case it is shown, as will be discussed in more detail below, that the template-active fraction of rat liver chromatin as prepared by nuclease digestion contains 10% of the single copy sequences, and that these sequences are to a large degree different from the single copy sequences contained in the template-active portion of rat brain chromatin. Gottesfeld *et al.* (26) have also shown that template-active single copy DNA hybridizes to a much greater extent to whole cell RNA than does the single copy DNA of the template-inactive fraction of rat liver chromatin. Studies of the sequence structure of the putative template-active fraction as compared with whole genomal DNA provide a powerful tool for the verification of whether or not a fractionation has been achieved.

E. Use of Probes

Recently Hawk *et al.* (27) have investigated the validity of ECTHAM–cellulose (11) and glycerol (18) gradient fractionation using mouse cells infected with the Moloney strain of murine leukemia virus.

In vivo these cells produce abundant RNA homologous to the Moloney type C leukemia virus but not RNA homologous to the type B mouse mammary tumor virus or globin RNA. Fractionation of chromatin into expressed and nonexpressed fractions and hybridization with cDNA copies of the viral and globin RNAs showed that the sequences (genes) for production of type C and type B viral products were equally distributed between the two fractions of chromatin. They also found a random distribution of globin sequences in the template-active and inactive fractions. Thus although fractions differing in physical properties (suggesting valid fractionation) were obtained, they were unable to separate active from inactive genes by these methods.

III.   Methods of Fractionation

A.   Separation of Euchromatin and Heterochromatin

Separation of metabolically active euchromatin from metabolically inactive heterochromatin has been described by Frenster et al. (6). They lysed lymphocyte nuclei, briefly sonicated the nuclei, and then separated heavier from lighter fractions by differential centrifugation. That chromatin which pellets at 100 g is classified as heterochromatin and that which does not pellet at 3000 g, but does pellet at 78,000 g, is classified as euchromatin. The latter constitutes about 20% of the total. If the nuclear RNA had been labeled in life then the specific activity of the lighter fraction exceeded that of the heavy fraction by a factor of 3.

This method of fractionation has been applied to the fractionation of chromatins of rat liver and rat Novikoff ascites cells by DeBellis et al. (17). These investigators used template activity in support of RNA synthesis by E. coli RNA polymerase as their criterion. By this assay, euchromatin and heterochromatin of both sources possessed identical activities.

B.   Fractionation by Sucrose Density Gradient Centrifugation

Following the early lead of Frenster (6,28), many investigators have separated sheared chromatin by sucrose density gradient centrifugation (29). Chromatin is sheared either in the VirTis blender or by other means, such as sonication, loaded on a linear or isokinetic sucrose gradient, and centrifuged. Provided only that a suitably small size of fragment has been achieved (10,000 base pairs or less of DNA), two fractions result—a heavy fraction and a light fraction. The light fraction makes up about 10–20% of the total DNA. The first thorough study of this fractionation is that of Chalkley and Jensen (7). These workers showed that the heavy fraction can be converted to the light fraction by treatment with 4 M urea and that this conversion is irreversible after removal of the urea. They concluded that the heavy fraction consists of fragments similar to those of the light fraction, but linked by interfragment bands perhaps due to histone H1. Chalkley and Jensen found that the light fraction has a somewhat higher template activity than the heavy fraction (about a factor of two). The template activity of the heavy fraction, after conversion to light fraction, is the same as that of material originally light, according to Chalkley and Jensen (7).

Typical of reports on the fractionation of chromatin by sucrose density gradient centrifugation is that of Berkowitz and Doty (20). These workers investigated chick embryo reticulocyte chromatin, which was sheared by sonication and applied to a sucrose density gradient; a slowly sedimenting fraction then was separated from a rapidly sedimenting fraction. The slowly sedimenting or light fraction constituted 12% of the total. The template activity of the slowly sedimenting fraction was twice that of the rapidly sedimenting fraction (determined with E. coli RNA polymerase) and, in addition, the concentration of nascent globin RNA transcripts as determined by hybridization of nascent mRNA to globin cDNA was 5 times more abundant in the transcripts of the light fraction than among the transcripts of the heavy fraction. The chromatin had been sheared to such a degree that the light fraction DNA was on the average 700 base pairs long while the template-inactive or heavy fraction DNA was on the average 1500 base pairs long. Similar results have been found by Rodriques and Becker (30) in the fractionation of rat liver chromatin by glycerol gradient centrifugation. They find similar histone composition in the light and heavy fractions, although more histone was associated with the heavy fraction. The nonhistone proteins show marked differences with unique species in the euchromatin and heterochromatin fractions.

In 1975 Doenecke and McCarthy described similar results with *Drosophila* chromatin sheared either to 2 kb (kilobase) or to 0.6 kb and fractionated on a sucrose gradient (31,32). This yields about 30% of the total chromatin as a light or slowly sedimenting fraction. This slowing sedimenting fraction possessed essentially all of the template activity for support of RNA synthesis by E. coli polymerase and also bore the bulk of the nascent RNA of the whole chromatin (9). Murphy et al. working with mouse myeloma chromatin sheared by VirTis shearing (or by the French press) and fractionated on sucrose or glycerol gradients find 11% of the total DNA in the light or slowly sedimenting fraction (18). This slowly sedimenting fraction has the bulk of the template activity as determined by transcription of the homologous RNA polymerase and also bears the bulk of the nascent RNA labeled *in vivo*.

Interestingly, many investigators of chromatin fractionation do not provide chemical analyses of either their starting material or the derived fractions. The data of Table I illustrate the histone/DNA and nonhistone chromosomal protein/DNA ratio for calf thymus chromatin and its fractions as separated by sucrose density gradient centrifugation and similar data for chick embryo reticulocyte chromatin. In both cases histone is marginally enriched in the heavy (template less active) fraction. Nonhistone proteins

TABLE I

CHEMICAL COMPOSITIONS OF SHEARED CHROMATIN AND THEIR SUBFRACTIONS AS
SEPARATED BY SUCROSE DENSITY-GRADIENT CENTRIFUGATION

	Fraction	Histone/DNA	Nonhistone/DNA
Calf thymus chromatin (7)	Whole chromatin	0.95	0.33
	Light fraction	0.91	0.32
	Heavy fraction	1.03	0.31
Chick embryo reticulocyte chromatin (20)	Whole chromatin	1.20	0.28
	Light fraction	1.04	0.55
	Heavy fraction	1.20	0.20

are marginally enriched in the light fraction in the case of reticulocyte chromatin, and this finding is reflected in other cases in the literature.

Henner *et al.* (21) have used sucrose density gradient centrifugation to separate the two fractions of sheared calf thymus chromatin. They have made the interesting discovery that although the difference in template activity between light and heavy fractions is only 2-fold when measured with *E. coli* RNA polymerase, it is 30-fold when measured with homologous RNA polymerase II.

Our group (unpublished observations) has fractionated VirTis-sheared rat liver chromatin (DNA size = 6 kb) on isokinetic sucrose gradients. This results in a light peak ($\sim 12\%$ of total chromatin) and a pelleted heavy fraction. The sequence complexity of the light peak DNA has been examined by reassociation kinetics. Our results suggest that the complexity of single copy sequences in the DNA of the light peak chromatin is the same as for whole genomal DNA (Fig. 1). This result casts doubt on the validity of chromatin thus fractionated.

In summary, it may be said of separation of sheared chromatin into fractions by sucrose density gradient centrifugation, that a fractionation is achieved and that the amount or proportion of the chromosomal DNA in the light fraction approximates the template activity of the particular chromatin used. It is clear also that in general the light fraction has a slightly higher template activity (ca. 2 times) in support of RNA synthesis than does the heavy fraction. In no case has it been clearly shown that the DNA of the light fraction contains the DNA sequences which are expressed in the particular cells or tissues from which the chromatin is obtained. The fact that heavy can be transformed to light (7) by treatment by 4 *M* urea suggests that the two fractions differ in some physical (packing) way. The results of Fig. 1 suggest no difference in sequence complexity between the two fractions.

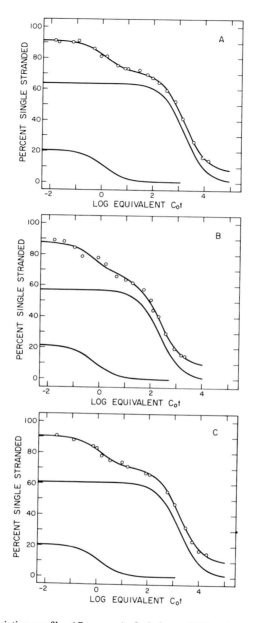

FIG. 1. Reassociation profiles (C_0t curves) of whole rat DNA and of DNA purified from the "light fraction" of the sucrose gradient. In both cases the DNA was subsequently sheared to a single-strand length of 350 bases, denatured, and allowed to reassociate in 0.12 M phosphate buffer at 60°. Samples were removed periodically and single-stranded separated from double-stranded DNA by hydroxyapatite chromatography. (A) Whole rat DNA. (B) Template-active DNA prepared by method of Gottesfeld et al. (33) (C) Sucrose "light" fraction DNA (A) and (C) are identical.

C.   Fractionation of Chromatin by Ion-Exchange Chromatography (ECTHAM Cellulose)

ECTHAM–cellulose chromatography of chromatin was developed by Reeck et al. (11). ECTHAM–cellulose is a weak ion exchanger with ionized groups of low pK. It is suitable for the fractionation of chromatin by solutions of low ionic strength. For this procedure the chromatin is first sheared. Reeck et al. (11, 12) have used sonication and have reduced the chromatin to a DNA fragment size of about 1000 base pairs. The sheared material is applied to the ECTHAM–cellulose at pH 7.2 and the column developed by pH gradient up to pH 9. The increasing pH is effected by the developing agent, 10 mM Tris base–10 mM NaCl. The chromatin fragments apparently consist of a large variety of materials of slightly different pKs. The most spectacular differences in biological properties are between the fractions which elute first and those which elute last, i.e., at the least and most basic pH. The fractions which elute first are higher melting than whole chromatin; those which elute last are lower melting than whole chromatin. Reeck et al. (12) have also reported differences in the distribution of histone and nonhistone proteins in ECTHAM–cellulose fractionated chromatin. The early-eluting fraction is nearly devoid of template activity in support of RNA synthesis by E. coli polymerase (34). The late-eluting fractions have about half the template activity of protein-free DNA. Similarly the RNA polymerase initiation sites are concentrated in the late-eluting fractions. Physical measurements suggest that the DNA of the early-eluting fraction is much less extended than the DNA of the late-eluting fraction.

Again, although ECTHAM–cellulose chromatography provides a way to clearly separate fractions of chromatin of quite different properties, particularly with respect to ability to serve as template for RNA polymerase, no work has yet been reported which firmly associates the late-eluting fractions with that part of the genome which codes for cellular RNA. In addition, the ECTHAM–cellulose fractionation method would appear not to be a good one for large-scale preparation of template-active and template-inactive fractions of chromatin.

D.   Thermal Elution Chromatography

McConaughy and McCarthy (35) have developed a procedure for the fractionation of chromatin by thermal chromatography. The chromatin is first sheared with a VirTis homogenizer and the sheared chromatin applied to hydroxyapatite in 0.12 M phosphate buffer in a jacketed column at 60°. The temperature of the column is then gradually increased. A minor fraction of the DNA, about 2% in the case of chicken erythrocyte chromatin and about 10% in the case of liver chromatin, is eluted at 70°–75°. The remainder is eluted at temperatures ranging from about 80° to 95°. McConaughy and McCarthy show by hybridization of whole cell RNA from chick erythrocytes with the several DNA fractions that whole cell RNA hybridizes to the extent of 30% with the lower melting fraction and to only to a very small proportion of the DNA from the higher melting fraction. This method has not been developed extensively, although a variant of it has been reported by Markov *et al.* (*36*). These investigators fractionate chromatin by thermal precipitation of chromatin in phosphate buffer. In this procedure, the chromatin remaining soluble at 98° in 80 mM phosphate buffer corresponds to the template-active sequences as determined by DNA–RNA hybridization analysis. It appears to be suitable for the large-scale fractionation of chromatin into expressed and nonexpressed portions of the genome.

E. Phase Partition

Turner and Hancock (*37*) have fractionated chromatin by partition between two aqueous phases. The chromatin, in this case from cultured mouse cells, was first isolated and then sheared to a single-stranded size of approximately 600 bases. The two phases consisted of polyethylene glycol (Carbowax 6000) and dextran T500 in aqueous solution. The volume ratio of the two phases was 1:1. As in two-phase partition separations the chromatin was agitated mechanically for 5 minutes in a mixture of the two phases and the two phases were then separated by centrifugation. Each phase was next mixed with an equal amount of the other, etc., until 50 or more steps had been carried out.

The DNA of mouse chromatin separated into two major fractions. Nonhistone chromosomal protein (enriched in the template-active fraction of chromatin) is concentrated in the polyethylene glycol phase while DNA not associated with nonhistone chromosomal proteins is concentrated in the dextran phase. Pulse-labeled nascent RNA is concentrated in the polyethylene glycol phase with the nonhistone chromosomal proteins. This method, after further development, could prove to be a suitable one for the separation of template-active from template-inactive chromatin.

F. Fractionation by DNase II

We have used DNase II (*10*) which makes double-stranded clips to reduce the length of unsheared (purified) rat liver chromatin. The chromatin is dissolved in sodium acetate buffer (pH 6.6) at 10 A_{260} units/ml. DNase II

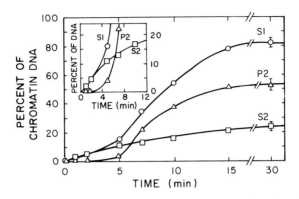

FIG. 2. Time course of separation of chromatin (rat liver) into fractions during incubation with DNase II and subsequent differential centrifugation. For identification of the fractions see Fig. 3. Chromatin was fractionated as described in text. The amount of nucleic acid in each fraction was determined by absorption at 260 nm in alkaline solution.

is added in the amount of 100 units/ml and the mixture incubated at 24°. Aliquots of the reaction mixture are removed from time to time and adjusted to pH 8 with 50 mM Tris buffer (pH 11.0) (which stops action of the enzyme), centrifuged at 15,000 g for 15 minutes, and the amounts of DNA in pellet

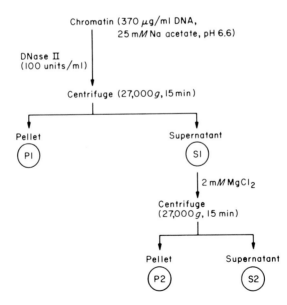

FIG. 3. Fractionation scheme for the separation of unsheared chromatin (rat liver) into fractions by successive treatment with DNase II, differential centrifugation, precipitation with magnesium chloride, and a second differential centrifugation.

TABLE II

CHEMICAL COMPOSITION OF RAT LIVER CHROMATIN AND ITS SUBFRACTIONS
AS PRODUCED BY NUCLEASE CLIPPING AND Mg^{2+} PRECIPITATION OF
HISTONE-RICH FRAGMENTS[a]

| | | Composition relative to DTA (w/w) | |
	Histone	Nonhistone	RNA
Unfractionated chromatin	1.06	0.65	0.05
Pl	1.15	0.58	0.05
LL S (nubodies)	1.03	0.05	0.0
S2	0.61	1.60	0.25
S2 Subfractions			
3–5 S	0.24	0.60	—
14 S	0.72	1.35	0.3–0.4
18.7 S	0.54	3.2	0.3–0.7

[a] From Gottesfeld and Bonner (37a).

and supernatant determined by absorption at 260 nm. The chromatin is increasingly solubilized with time as shown in Fig. 2. The pelleted fraction is known as Pl and the supernatant fraction as S1. To the supernatant 2 mM MgCl is added. This precipitates most histone-complexed DNA. The fractionation scheme is summarized in Fig. 3. The chemical compositions of the several fractions are different from one another. Thus, S2 is enriched and P1 impoverished in nonhistone chromosomal proteins. S2 is also enriched in RNA. The chemical compositions of the fractions are shown in Table II.

The fractionation described above is evidently based on differential sensitivity to nuclease attack of the different chromatin fractions. This is made clear by the sizes of the resulting DNA fractions. Thus after 5 minutes of nuclease digestion (Fig. 2) the single-stranded DNA size of the S2 DNA is approximately 350 bases while that of deproteinized Pl DNA is approximately 10 times as large. After 30 minutes of digestion S2 DNA is to a considerable part rendered acid-soluble while Pl DNA is still 200 or more base pairs in length. This difference in susceptibility to nuclease may result from differences in the chemical composition and structure of the two fractions. This will be returned to below.

That S2 DNA is the template-active fraction of chromatin DNA was first indicated by Billing and Bonner (19) who pulse-labeled the RNA of rat ascites cells, extracted the chromatin, and showed that DNase digestion released the nascent labeled RNA to fraction S2 with kinetics similar to those of the formation of S2 itself. Gottesfeld et al. (26) came to the same conclusion in experiments in which labeled tracer S2 DNA or Pl DNA was hybri-

TABLE III
Kinetic Components of Rat DNA and DNA Isolated from the Chromatin Fractions

DNA sample	% total chromatin DNA	Kinetic component	Fraction of DNA in component	$C_0t_{1/2}$ obs[a]	Estimated $C_0t_{1/2}$ for pure component[a]	Average kinetic complexity relative to E. coli[b] (base pairs)	Repetition frequency[c]
Unfractionated chromatin DNA	100	Very fast	0.083				
		Intermediate fast	0.150	0.63	0.095	1.2×10^5	2600
		Slow	0.094	32	3.01	3.8×10^6	50
		Nonrepetitive	0.627	2008	1259	1.6×10^9	1
		rms = 3.17%[f]					
Fraction S2 DNA (5-minute nuclease exposure)	11.3 ± 3.9[d]	Very fast	0.109				
		Intermediate fast	0.165	0.15	0.025	3.2×10^4	1200
		Slow	0.055	2.76	0.15	1.9×10^5	70
		Nonrepetitive	0.628	235	148	1.9×10^8	1
		rms = 1.89%[f]					
Fraction P1 DNA (5-minute nuclease exposure)	84.6 ± 4.8[d]	Very fast	0.047				
		Intermediate fast	0.131	0.54	0.071	9.0×10^4	2600
		Slow	0.175	33	5.8	7.3×10^6	40
		Nonrepetitive	0.608	1706	1037	1.3×10^9	1
		rms = 4.34%[f]					

Fraction P1 DNA (15-minute nuclease exposure)	23.9 ± 2.9[e]				3700
Very fast	0.060				
Intermediate fast	0.164	0.11	0.018	2.2 × 10⁴	
Slow	0.126	14.8	1.86	2.4 × 10⁶	30
Nonrepetitive	0.340	593	202	2.6 × 10⁸	1

rms = 3.95%[f]

[a] Calculated from a computer analysis (38) of the data of Fig. 3A–D of Gottesfeld et al. (25).

[b] The complexity of E. coli is 4.3 × 10⁶ base pairs (39). The $C_0't_{1/2}$ observed for E. coli DNA is 4.1. The difference in base composition of rat (41% GC) and E. coli DNA (50% GC) slows the rate of renaturation of rat DNA by a factor of 0.83 (40) relative to E. coli. This assumes that the components of each DNA studied are of the same average base composition as unfractionated rat DNA.

[c] Repetition frequency is calculated by dividing the chemical or analytic complexity of each component by the average kinetic complexity observed for that component. Chemical complexity is the total number of nucleotide pairs in a given component; it is obtained from the following expression: $CC = G(fC)(fD)$, where CC is the chemical complexity, G is the genome size of the rat [taken to be 2.1 × 10⁹ base pairs per haploid equivalent of DNA for rodents (41)], fC is the fraction of total chromatin DNA in a given chromatin sample, and fD is the fraction of DNA in the kinetic components.

[d] Data of Gottesfeld et al. (26).

[e] Data of Gottesfeld et al. (42).

[f] Root mean square (rms) deviation of computer analysis (38). After Gottesfeld et al. (25).

dized by a vast excess of whole cell RNA. S2 DNA is hybridized to whole cell RNA 5–7 times more extensively than P1 DNA. Also S2 DNA contains a vast majority of the expressed sequences. It must also contain nonexpressed sequences since only 60% of rat liver S2 DNA is hybridized by whole cell RNA.

That S2 DNA consists of a subset of the single copy sequences of whole rat liver DNA is shown by its self-reannealing profile (Fig. 1). The single copy sequences of S2 DNA reanneal approximately 10 times more rapidly than those of whole genomal DNA, and they therefore contain only 10% of the total single copy complexity. The S2 DNA after 5 minutes of digestion contains only 10% of whole genomal DNA. This 10% of genomal mass is a specific subset of sequences. The same is true of the middle repetitive DNA. S2 contains about 10% of the middle repetitive DNA mass and contains 10% of the middle repetitive complexity as shown in Table III.

That the S2 fraction is a specific subset of sequences is shown again by the fact that the S2 sequences of liver chromatin are shared only to the extent of about 30% with the S2 sequences of brain chromatin (25).

We conclude that mild nuclease digestion preferentially clips out a portion of the genome which includes the transcribed sequences. The physical basis of the preferential digestion appears to lie in the fact that the nonexpressed portion of the genome is organized in v-bodies (nucleosomes). DNA packaged as nucleosomes is different from B form DNA in several ways. It exhibits a packing ratio of 7 (43), is greatly stabilized against melting, and exhibits altered circular dichroism (CD) and optical rotatory dispersion (ORD) spectra (42). Nucleosomal DNA is not available for transcription by added RNA polymerase. Transcriptionally active DNA, on the other hand, exhibits properties similar to those of B form DNA even though it contains histone proteins. It is only slightly stabilized against melting, possesses CD and ORD spectra similar to those of B form DNA, and is actively transcribed by added RNA polymerase (8).

IV. Structure of Template-Active Chromatin

Cytological (1) and biochemical (6) data suggest that the actively transcribed fraction of chromatin is in a more extended state than are the condensed, inactive regions. Thus, if separation of the two can be achieved, one might expect to find structural as well as chemical differences between the active fraction and the inactive (repressed) region.

Present biochemical (44–46) and physical (47) studies on the structure of isolated chromatin have culminated in a generally accepted model of chromatin structure (48). This model suggests that the primary repeating structural element in chromatin consists of about 200 base pairs of DNA complexed with two each of the major histones H2A, H2B, H3, and H4 (49). The histones and DNA are associated in such a way as to fold the DNA in an approximately 7:1 packing ratio (43).

This substructure has been resolved in the electron microscope (47, 50) and by digestion of chromatin with endonucleases (44) and subsequent analysis of the digestion products on agarose or polyacrylamide gels. Digestion of rat liver chromatin with staphylococcal nuclease (44) indicates that as much as 85% of the chromatin DNA is complexed in these subunits. Nearly all the studies to date have been done on unfractionated chromatin. Thus the pattern observed reflects mainly that of the inactive regions, which represent 80–90% of the chromatin.

From the electron micrographs of active ribosomal genes, McKnight (51) has found that the length of the coding sequence is nearly the same as expected for B form DNA, while the adjacent inactive regions are organized in the now familiar "beads on a string" subunit structures described above.

The template-active chromatin, as isolated in our group by limited digestion with DNase II, has a CD spectrum very similar to that of B form (extended) DNA. The DNA of the inactive chromatin, however, is in a different conformation, one more similar to C form DNA, suggesting that it is in supercoiled subunit structures. Thus the packing ratio of DNA in the active fraction would appear to be approximately 1 while in the inactive fraction it is 7 (43).

Are the subunit structures present in the active chromatin? This question has been approached in different ways. One indirect approach to see if subunits are present in template-active DNA is to hybridize the mRNA (or its cDNA) to the subunit DNA of whole chromatin to see if the expressed RNA sequences are represented in subunit DNA. This approach has been taken by Lacey and Axel (52) Kuo et al. (53), and Reeves and Jones (54). In these reports a cDNA (complementary DNA) copy of poly(A)$^+$ mRNA was found to hybridize to isolated subunit DNA with the same kinetics and to the same extent as to whole cell DNA. This result implies that subunits are present in the transcriptionally active fraction of chromatin and that the DNA in these subunits is protected from nuclease attack.

A more direct way to determine the structural organization of the template-active regions is first to isolate the template-active chromatin, then to digest it with endonucleases, and then analyze the digestion products. Using this approach, members of our group (33) have analyzed the subunit structure of the template-active fraction of rat liver chromatin. Chromatin was fractionated with DNase II (as described above) and separated into template-active and inactive regions. The nuclease resistant complexes in these fractions were then analyzed as to chemical composition and sedimentation behavior. The sedimentation profiles of the template-active and template-

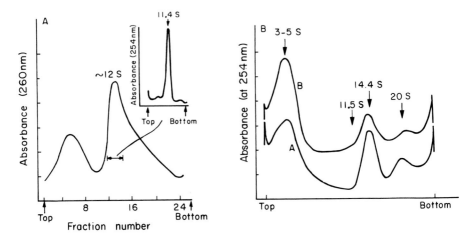

FIG. 4. Sucrose sedimentation patterns on isokinetic sucrose gradients of DNase II digest of: (A) template-inactive chromatin (Pl) and (B) template-active chromatin (S2). After Gottesfeld *et al.* (*33*).

inactive subunits are compared in Fig. 4. It is clear that the template-active chromatin subunits have higher sedimentation values (14 S and 20 S) than the subunits isolated from the inactive regions (11 S) (*25,55*). These results are in agreement with those of Paul and Duerksen (*56*) who report the production of 14 S particles upon autodigestion of mouse TCT chromatin by a Ca^{2+}/Mg^{2+}-activated endogenous endonuclease. These investigators do not see the 20 S component.

Table II compares the chemical compositions of the 14 S subunits and the 11 S subunits of inactive chromatin. The 14 S subunits of template-active chromatin possess all the major histone species present in the 11 S subunits of inactive chromatin, although at a lower histone/DNA ratio. The most striking properties of the template-active subunits is their high content of nonhistone protein and RNA. The nonhistones present in the 14 S peak are a heterogeneous group, and may reflect a varied population of subunits with similar sedimentation properties.

As reported by Paul and Duerksen (*56*), digestion of the 14 S subunit with staphylococcal nuclease reduces its sedimentation value to 11 S. In addition, treatment of the 14 S subunits with ribonuclease also reduces its sedimentation value to 11 S. We may therefore speculate that the DNA–histone core of the 14 S subunit is similar or identical to that of the 11 S subunits of inactive chromatin, but that the specific association of RNA and/or nonhistone proteins confers to this particle its higher sedimentation value. Thus the subunit organization of active chromatin is similar to that of inactive chromatin, consisting of alternating nuclease-sensitive and nuclease-resistant regions, with an average repeating subunit containing 200 base pairs of DNA. About 50% of the DNA of template-active chromatin is found in nuclease-resistant regions.

The template-active chromatin is nonetheless, as we have seen above, different from template-inactive chromatin. The former is soluble in $2 \, mM \, Mg^{2+}$ the latter is not. The former is more rapidly attacked by nuclease than the latter. The DNA of the template-active fraction is not stabilized against melting to the degree that the DNA of template-inactive chromatin is.

It is important to realize, therefore, that the 14 S nuclease-resistant 200 base pair subunit need not exist in the same structural conformation as the 11 S 200 base pair nuclease-resistant particle; i.e., the subunit may be associated with histones in such a way as to confer nuclease resistance without packing the DNA in a 7:1 packing ratio. The presence of negatively charged nonhistone chromosomal proteins and RNA may maintain the more extended conformation of the DNA–histone complex found in the actively transcribed regions. The altered ionic environment of template-active chromatin may shift the conformation of its subunit from a packed to an extended conformation and vice versa (57).

The presence of alternating nuclease-resistant and nuclease-sensitive regions in template-active chromatin is also indicated by staphyloccocal nuclease digestion of the extrachromosomal rRNA genes of *Tetrahymena pyriformis* (58,59). In these studies the extrachromosomal nucleolar ribosomal DNA (rDNA) of *Tetrahymena* was labeled with [³H]thymidine during its preferential replication during refeeding after prolonged starvation. The bulk of chromatin DNA was labeled during exponential growth with either ¹⁴C or ³²P. After preferential labeling of the rDNA the chromatin was digested with staphylococcal nuclease and the purified DNA fragments analyzed on gels. These studies demonstrate that the nuclease-sensitive regions of the actively transcribed rDNA are degraded by staphylococcal nuclease digestion to the same size and with the same kinetics as are those of total chromatin DNA. This is interpreted to mean that the subunits of the active regions have the same periodicity and length as the subunits in the inactive region. This agrees well with work described above. Thus Gottesfeld *et al.* (33) found the length of the DNA from the template-active, nuclease-resistant particle to be of the same length as the protected DNA fragment of inactive chromatin. It appears that the secondary structure of actively transcribed chromatin is similar to that of inactive-repressed chromatin in that both fractions consist of alternating nuclease-sensitive and nuclease-resistant regions. However, the structure of the DNA in the nuclease-resistant nucleoprotein particles of template-active chromatin appears to be quite different from the structure of DNA in the particles of template-inactive DNA.

ACKNOWLEDGMENT

This work was supported by the National Institutes of Health, USPHS (GM 13762).

REFERENCES

1. Littau, V. C., Allfrey, V. G., Frenster, J. H., and Mirsky, A. E., *Proc. Natl. Acad. Sci. U.S.A.* **52**, 93 (1964).
2. Berendes, H. D., *Int. Rev. Cytol.* **35**, 61 (1973).
3. Daneholt, B., and Hosick, H., *Cold Spring Harbor Symp. Quant. Biol.* **38**, 629 (1973).
4. Hamkalo, B. A., Miller, O. L., and Bakken, A. H., *Cold Spring Harbor Symp. Quant. Biol.* **38**, 915 (1973).
5. McKnight, S. L., and Miller, O. L., Jr., *Cell* **8**, 305 (1976).
6. Frenster, J. H., Allfrey, V.G., and Mirsky, A. E., *Proc. Natl. Acad. Sci. U.S.A.* **50**, 1026 (1963).
7. Chalkley, R., and Jensen, R., *Biochemistry* **7**, 4380 (1968).
8. Bonner, J., Gottesfeld, J. G., Garrard, W., Billing, R., and Uphouse, L., in "Methods in Enzymology," Vol. 40: Hormone Actions, Part E, Nuclear Structure and Function (B. W. O. Malley and J. G. Hardman, eds.), p. 97. Academic Press, New York, 1975.
9. McCarthy, B. J., Nistllura, J. T., Doenecke, D., Nasser, A. S., and Johnson, C. B., *Cold Spring Harbor Symp. Quant. Biol.* **38**, 763 (1973).
10. Marushige, K., and Bonner, J., *Proc. Natl. Acad. Sci. U.S.A.* **68**, 2941 (1971).
11. Reeck, G. R., Simpson, R. T., and Sober, H. A., *Proc. Natl. Acad. Sci. U.S.A.* **69**, 2317 (1972).
12. Reeck, G. R., Simpson, R. T., and Sober, H. A., *Eur. J. Biochem.* **49**, 407 (1974).
13. Davidson, N., Pearson, W., Gottesfeld, G., and Bonner, J., *Biochemistry* **15**, 2481 (1976).
14. Noll, M., Thomas, J. O., and Kornberg, R. D., *Science* **187**, 1203 (1975).
15. Doenecke, D., and McCarthy, B. J., *Eur. J. Biochem.* **64**, 405 (1976).
16. Bonner, J., Dahmus, M. E., Fambrough, P., Huang, R. C., Marushige, K., and Tuan, K., *Science* **159**, 47 (1968).
17. DeBellis, R., Benjamin, W., and Gellhorn, A., *Biochem. Biophys. Res. Commun.* **36**, 166 (1969).
18. Murphy, E. C., Jr., Hall, S. H., Shepard, J. H., and Weiser, R. S., *Biochemistry* **12**, 3843 (1973).
19. Billing, R. J., and Bonner J., *Biochim. Biophys. Acta* **281**, 453 (1972).
20. Berkowitz, E. M., and Doty, P., *Proc. Natl. Acad. Sci. U.S.A.* **72**, 3328 (1975).
21. Henner, D., Kelly, R. I., and Furth, J. J., *Biochemistry* **14**, 4764 (1975).
22. Simpson, R. T., and Polacow, I., *Biochem. Biophys. Res. Commun.* **55**, 1078 (1973).
23. Yasmineh, W. H., and Yunis, J. J. *Biochem. Biophys. Res. Commun.* **36**, 779 (1969).
24. Doerksen, J. D., and McCarthy, B. J., *Biochemistry* **10**, 1471 (1971).
25. Gottesfeld, J. M., Bagi, G., Berg, B., and Bonner, J., *Biochemistry* **15**, 2472 (1976).
26. Gottesfeld, J. M., Garrard, W. T., Bagi, G., Wilson, R. F., and Bonner, J., *Proc. Natl. Acad. Sci. U.S.A.* **71**, 2193 (1974).
27. Hawk, R. S., Anisowicz, A., Silverman, A. Y., Parks, W. P., and Scolnick, E. M., *Cell* **4**, 321 (1975).
28. Frenster, J. H., *Nature (London)* **206**, 680 (1965).
29. Nasser, D. S. and McCarthy, B. J. in "Methods in Enzymology," Vol. 40: Hormone Action, Part E, Nuclear Structure and Function (B. W. O'Malley and J. G. Hardman, eds.), p. 93. Academic Press, New York, 1975.
30. Rodrigues, L. V., and Becker, F. F., *Arch. Biochem. Biophys.* **173**, 438 (1976).
31. Doenecke, D., and McCarthy, B. J., *Biochemistry* **14**, 1366 (1975).

32. Doenecke, D., and McCarthy, B. J., *Biochemistry* **14**, 1373 (1975).
33. Gottesfeld, J. M., Murphy, R. J., and Bonner, J., *Proc. Natl. Acad. Sci. U.S.A.* **72**, 4404 (1975).
34. Simpson, R. T., *Proc. Natl. Acad. Sci. U.S.A.* **71**, 2740 (1974).
35. McConaughy, B. L., and McCarthy, B. J., *Biochemistry* **11**, 998 (1972).
36. Markov, G. G., Ivanov, I. G., and Pashev, I. G., *J. Mol. Biol.* **93**, 405 (1975).
37. Turner, G., and Hancock, R., *Biochem. Biophys. Res. Commun.* **58**, 437 (1974).
37a. Gottesfeld, J. M., and Bonner, J., *in* "The Molecular Biology of the Mammalian Genetic Apparatus," Vol. I. Isolation and Properties of the Expressed Portion of the Mammalian Genome (P. Ts'o, ed.), pp. 381–397. Elsevier/North-Holland Biomedical Press Amsterdam (1977).
38. Britten, R. J., Graham, B. E., Neufeld, B. R., in "Methods in Enzymology," Vol. 39: Hormone Action Part D, Isolated Cells, Tissues, and Organ Systems (J. G. Hardman and B. W. O'Malley, eds.), p. 363. Academic Press, New York, 1974.
39. Cairns, J., *Cold Spring Harbor Symp. Quant. Biol.* **28**, 43 (1963).
40. Wetmur, J. B., and Davidson, N., *J. Mol. Biol.* **31**, 349 (1968).
41. Lewin, B., *in* "Gene Expression: Eukaryotic Chromosomes," p. 7. Wiley, New York, 1974.
42. Gottesfeld, J. M., Bonner, J., Radda, G. K., and Walker, I. O., *Biochemistry* **13**, 2937 (1974).
43. Griffith, J. D., *Science* **187**, 1202 (1975).
44. Noll, M., *Nature (London)* **251**, 249 (1974).
45. Hewish, D. R., and Burgoyne, L. A., *Biochem. Biophys. Res. Commun.* **52**, 504 (1973).
46. Simpson, R. T., and Whitlock, J. P., Jr., *Nucleic Acids Res.* **3**, 117 (1976).
47. Olins, A. L., and Olins, D. E., *Science* **183**, 330 (1974).
48. Kornberg, R. D., *Science* **184**, 868 (1974).
49. Thomas, J. O., and Kornberg, R. D., *Proc. Natl. Acad. Sci. U.S.A.* **72**, 2626 (1975).
50. Woodcock, C. L. F., *J. Cell Biol.* **59**, 368a (1973).
51. McKnight, S. L., and Miller, O. L., Jr., *Cell* **8**, 305 (1976).
52. Lacey, E., and Exel, R., *Proc. Natl. Acad. Sci. U.S.A.* **72**, 3978 (1975).
53. Kuo, M. R., Sahasrabuddhe, C. G., and Saunders, G. F., *Proc. Natl. Acad. Sci. U.S.A.* **73**, 1572 (1976).
54. Reeves, R., and Jones, A., *Nature (London)* **260**, 495 (1976).
55. Hemminki, K., *Nucleic Acids Res.* **3**, 1499 (1976).
56. Paul, I. J., and Duerksen, J. D., *Biochim. Biophys. Res. Commun.* **68**, 97 (1976). 76).
57. Li, H. J., *Nucleic Acids Res.* **2**, 1275 (1975).
58. Piper, P. W., Celis, J., Kaltoft, K., Leer, J. C., Nielsen O. F., and Westergaard, O., *Nucleic Acids Res.* **3**, 493 (1976).
59. Mathis, D. J., and Gorovsky, M. A., *Biochemistry* **15**, 750 (1976).

Chapter 2

Fractionation of Chromatin by Buoyant Density-Gradient Sedimentation in Metrizamide

G. D. BIRNIE

Wolfson Laboratory for Molecular Pathology,
The Beatson Institute for Cancer Research,
Glasgow, Scotland

I. Advantages of a Nonionic Buoyant-Density Solute

Most of the methods which have been used in attempts to fractionate chromatin have been designed to take advantage of the differences observed *in vivo* between heterochromatin and euchromatin, namely, the former is granulated and highly condensed, while the latter appears to exist mainly as uncondensed, extended fibrils. Differential centrifugation, rate-zonal sedimentation through a stabilizing density gradient, and gel-exclusion chromatography are methods which should be readily applicable to the separation of two species of particle which are as different as heterochromatin and euchromatin appear to be *in vivo*. However, although chromatin has been fractionated by these methods, they have not been particularly useful. For one thing, none have yielded fractions which have been proved unequivocally to be enriched in active chromatin. For another, the reproducibility of fractionations is poor, and extremely dependent on the ionic composition of the chromatin solution or suspension. More recent information about the effect of ionic strength on the conformation and degree of aggregation of chromatin (*1,2*) and on the distribution of chromosomal proteins on the DNA (*3–6*) have gone a long way to explaining the earlier results [for review, see (*7*)].

Fractionations by these methods depend on a number of parameters including the size, shape, and density of the particles, while fractionation by isopycnic sedimentation depends solely upon differences in density. There is good evidence that the proteins are not evenly distributed along the DNA in chromatin (*1,6,8–10*) and, consequently, isopycnic sedimen-

23

tation in a suitable medium would appear to offer a more reproducible method of fractionating chromatin. Moreover, this method should not be dependent on fragment size so long as the chromatin has been sheared enough that each fragment consists mainly of one or other type of chromatin. However, chromatin has properties which have made it difficult to find a suitable medium for isopycnic sedimentation experiments. Buffers of high ionic strength readily dissociate DNA–protein complexes (*11*) so that buoyant density-gradient centrifugation in, for example, gradients of CsCl cannot be used to fractionate chromatin unless the chromatin is first fixed with formaldehyde (*12*). The reaction between formaldehyde and chromatin is not readily reversible, and chromatin fractions isolated in this way are usually of little use for further study. Until recently there have been no unreactive, nonionic gradient solutes which were generally suitable for the isopycnic sedimentation of unfixed chromatin. Because chromatin is grossly dehydrated in sucrose solutions it becomes too dense to band isopycnically in sucrose gradients, although it will band in dense glucose–sucrose solutions (*13*). However, these solutions are so viscous that sheared chromatin sediments too slowly and they can be used only for very high-molecular-weight material. Chromatin will separate into a series of bands containing different proportions of protein when sedimented to equilibrium in gradients of chloral hydrate (*14*), but this compound is not particularly suitable since it reacts with proteins and denatures DNA.

Recently it has been realized that certain X-ray contrast media and their derivatives possess many of the attributes required of a density-gradient solute suitable for the isopycnic sedimentation of unfixed nucleoproteins. Of those tried so far, metrizamide is the only one which is totally nonionic in aqueous solutions, and it is far and away the one which has been most widely exploited, having been used for experiments involving the isopycnic banding of macromolecules, macromolecular complexes, cell organelles, whole cells, and viruses. This chapter discusses the use of metrizamide for the isopycnic sedimentation of unfixed chromatin, and describes those properties of metrizamide which are important in that regard. For more detailed descriptions of the properties and uses of metrizamide gradients, reference should be made to recent reviews (*15,16*).

II.   Properties of Metrizamide

A.   Physicochemical Properties

Metrizamide (Nyegaard and Co. A/S, Oslo, Norway), the structure of which is shown in Fig. 1, is a tri-iodinated benzamido derivative of glucose;

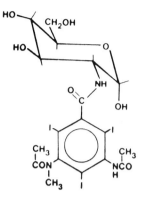

Fɪɢ. 1. Metrizamide: 2-(3-acetamido-5-N-methylacetamido-2,4,6-tri-iodobenzamido)-2-deoxy-ᴅ-glucose.

its molecular weight is 789. Metrizamide is an off-white powder which is stable at room temperature when kept dry and in the dark. It is soluble in water and in aqueous buffer solutions up to a concentration of 85% (w/v). Dilute solutions are easily prepared in the usual way; more concentrated ones [more than 20% (w/v)] are most easily prepared by adding small portions of the powder to the solvent over a period of hours with constant stirring. Preparation and storage of metrizamide solutions require some care; heating solutions above 50°C leads to release of free iodine, as does extended exposure to light, particularly direct sunlight and fluorescent light. Solutions must also be protected from bacterial contamination; their pH must be maintained within the range 2.5–12.5, otherwise the glycosidic linkage is hydrolyzed. Tautomerism of the glucose occurs at pH values greater than 8, but this does not seem to affect the properties of metrizamide as a buoyant density-gradient solute. Within these limits, solutions of metrizamide are perfectly stable and can be stored indefinitely at −20°C.

Detailed comparisons of the densities, viscosities, and osmolarities of solutions of metrizamide, sucrose, and Ficoll have been published (15). From a practical standpoint, the important differences between metrizamide solutions and isodense solutions of the others are these. First, the concentration of metrizamide is much lower than that of either sucrose or Ficoll. Second, the viscosity of the metrizamide solution is much lower. Third, the osmolarity is less than that of a sucrose solution, and less than that of a Ficoll solution of density 1.11 gm/cm³ or greater; moreover, the relationship between concentration and osmolarity of metrizamide solutions is approximately linear whereas that of Ficoll is an exponential func-

TABLE I

CONCENTRATION, REFRACTIVE INDEX, AND DENSITY OF AQUEOUS
METRIZAMIDE[a]

Concentration (gm/100 ml solution)	Refractive index at 20°C	Density (gm/cm³)	
		at 20°C	at 5°C
0	1.3320	0.998	1.000
10	1.3478	1.053	1.056
20	1.3640	1.107	1.110
30	1.3800	1.161	1.165
40	1.3964	1.216	1.221
50	1.4124	1.270	1.276
60	1.4288	1.324	1.331
70	1.4451	1.379	1.386
80	1.4612	1.433	1.441

[a] From data supplied by Nyegaard and Co. A/S.

tion of concentration. In addition, the maximum density obtainable with metrizamide in solution (1.46 gm/cm³) is much greater than that with Ficoll (1.19 gm/cm³) or sucrose (1.27 gm/cm³).

Table I shows the relationship between the refractive index, the concentration (w/v), and the density of each of a series of metrizamide solutions. Graphs drawn from these data enable the concentration and density of a metrizamide solution to be determined from its refractive index, which can be measured in a suitable refractometer with as little as 20 μl of the solution. The refractive index varies linearly with concentration and density; consequently, density at 20°C (ρ_{20}) can be calculated from refractive index at 20°C (η_{20}) using the equation:

$$\rho_{20} = 3.350\,\eta_{20} - 3.462 \tag{1}$$

In most cases, centrifugation is done at low temperature (say, 5°C) while refractive indices are measured at room temperature (say, 20°C). The relevant equation relating density at 5°C (ρ_s) and refractive index at 20°C (η_{20}) is:

$$\rho_s = 3.453\,\eta_{20} - 3.601 \tag{2}$$

The data in Table I, and Eqs. (1) and (2), relate to metrizamide solutions in distilled water, and appropriate corrections must be made for any buffer or salt in the solution. For most purposes such low concentrations of buffer or salt are used in metrizamide gradients that no correction is necessary unless very accurate estimates of buoyant density are wanted. If required, the refractive index of the metrizamide solution can be corrected as follows:

$$\eta(\text{corr}) = \eta(\text{obs}) - [\eta(\text{salt}) - \eta(\text{water})] \qquad (3)$$

Under conditions in which metrizamide is stable, it is chemically inert. It does not interact with any biological materials, with the sole exception of uncomplexed proteins—though not with proteins in nucleoprotein complexes (17, 18). For example, nucleoproteins are not disrupted by prolonged exposure to metrizamide; less protein is lost from, and RNA polymerase activity is better preserved in, nuclei recovered from metrizamide solutions than from concentrated sucrose (19); and metrizamide does not impair the infectivity of many viruses (20) or the viability of many mammalian cells (21). Although the activity of some enzymes is inhibited by metrizamide, the inhibition is completely reversed by the removal, or even adequate dilution, of the compound (16, 18, 22). The dense complexes formed between metrizamide and proteins are completely dissociable by removal of the metrizamide (18).

B. Formation of Gradients of Metrizamide

Metrizamide gradients can be formed with any of the gradient-producing devices which have been described or, most simply, by layering a series of metrizamide solutions of different concentrations in the centrifuge tube. With the latter method, diffusion results in the formation of smooth, reproducible gradients in 18 to 24 hours. Preformed gradients must, of course, be used for rate-zonal centrifugation; for isopycnic centrifugation of high-molecular-weight chromatin they have the advantage that centrifugation times are relatively short. Their disadvantage is that the volume of sample is quite severely limited. Alternatively, gradients of metrizamide can be formed *in situ* by centrifugation, in which case much larger sample volumes can be used. However, the viscosity of metrizamide solutions, which is high compared with, for example, that of CsCl solutions, means that the formation of an equilibrium gradient is a relatively slow process. For example, with a 30% (w/v) metrizamide solution ($\rho_s = 1.16 \, \text{gm/cm}^3$) centrifuged at 5°C in an M.S.E. (MSE Scientific Instruments Ltd., Crawley, Sussex, England) 10 × 10 aluminium fixed-angle rotor at 35,000 rpm, equilibrium is obtained only after 68 hours. More concentrated solutions require much longer times and/or higher speeds before equilibrium is achieved. Consequently, rarely, if ever, is it possible to use gradients of metrizamide which are truly equilibrium gradients.

The final shape of a gradient obtained by centrifugation of a metrizamide solution, whether in a preformed gradient or not, is virtually impossible to forecast except empirically. The factors which determine the gradient profile include speed, time, and temperature of centrifugation, initial density of the solution, as well as the type (fixed-angle or swing out) and physical dimensions of the rotor used. It is not possible to consider the interrelated effects of all of these parameters on gradient profiles in this article, and reference must be made to others in which examples of changes in profile consequent on changes in one or more of these parameters have been described (23–25). However, a few general rules can be given to serve as a guide for initial experiments. First, gradients are formed very slowly in swing-out rotors except where dilute solutions are centrifuged in small rotors at high speed. This means that preformed gradients can be used in swing-out rotors except where dilute solutions are centrifuged in small gradient shape during centrifugation. Second, gradients (though not equilibrium gradients) are formed quite rapidly in fixed-angle rotors, particularly if the length of the column in the direction of the centrifugal field is kept as short as possible. Thus, rotors in which the tube is held at a shallow angle relative to the axis of the rotor are particularly useful, especially if the gradient occupies half, or less, of the total volume of the tube. Third, the temperature of centrifugation should be as high, and the initial concentration of metrizamide as low, as is compatible with the stability of the particles and the fractionation being sought.

At first sight, the slow rate of formation of gradients and the dependence of the gradient profile on so many different parameters appear to be grave disadvantages as far as the use of metrizamide as a buoyant density-gradient material is concerned. However, in practice, these actually add considerably to the versatility of metrizamide gradients as a means for fractionating particles on the basis of buoyant density. After the first few hours of centrifugation, changes in gradient profile are relatively slow even in fixed-angle rotors, and low-molecular-weight macromolecules (e.g., DNA of molecular weight 5×10^5) can be banded isopycnically quite satisfactorily in metrizamide gradients formed *in situ*. Thus, although the dependence of the gradient profile on so many parameters means that considerable care is required in choosing centrifugation conditions, it also means that judicious selection of the rotor, of time, speed, and temperature of centrifugation, and of initial density of solution, almost always enables the shape of the gradient to be manipulated to optimize resolution over any particular range of densities.

Under the conditions normally used for centrifugation, metrizamide is chemically inert and does not react with the plastics from which centrifuge tubes are usually made. Both polypropylene and polycarbonate tubes are perfectly satisfactory in swing-out rotors. However, polycarbonate tubes often crack and leak when centrifuged in fixed-angle rotors for prolonged periods while containing solutions of metrizamide. Tubes of polypropylene survive one or more periods of centrifugation, but haircracks are produced which cause the tubes to fail in later experiments. The most likely cause is the relatively large difference in density between the top and bottom of a metrizamide gradient (usually between 0.24 gm/cm³ and 0.5 gm/cm³) which is too great for the more rigid polypropylene and polycarbonate tubes to accommodate easily. No such problem arises with, for example, CsCl gradients—in that case, the difference in densities is usually no more than 0.1 gm/cm³. In contrast, tubes of polyallomer, which are much more flexible, are completely unaffected and can be used safely many times over.

C. Fractionation and Analysis of Metrizamide Gradients

Although metrizamide gradients can be fractionated by any one of the usual methods, in practice we have found that, because of their viscosity, the upwards displacement method using the M.S.E. top-unloading device is the most convenient (26). A fine-bore tube is passed through the gradient to the bottom of the centrifuge tube, and a dense fluid is introduced to the bottom of the gradient either by gravity flow or by pumping. The gradient is displaced upwards and into a cone-shaped collecting cap sealed on the top of the centrifuge tube, from which it passes via a fine-bore plastic tube to the tubes in which the fractions are being collected. A burette is used as the reservoir for the displacing fluid since the volume of each fraction cut can be determined as the gradient is displaced simply by observing the change in the volume in the burette, a much simpler and more accurate procedure than either drop-counting or collecting fractions in graduated tubes. While it is possible to displace the gradients with a concentrated solution of metrizamide, by far the most convenient displacing fluid is Fluorochemical FC43 (3M Company). This fluid is inert chemically, immiscible with aqueous solutions, nonviscous, and very dense ($\rho \cong$ 1.9 gm/cm³). Though expensive, it is easily recovered after the gradient has been fractionated, and thus can be reused many times. This method of fractionating gradients of metrizamide (and other materials) is simple, accurate, and cheap.

The density of each fraction is calculated from its refractive index, which can be measured with 20 μl of solution, using Eqs. (1) or (2). Radioactivity is measured by liquid scintillation spectrometry using a scintillation fluid which is miscible with aqueous metrizamide; one suitable fluid consists of 1 volume of Triton X-100 mixed with 2 volumes of toluene containing 5 gm of PPO and 0.3 gm of POPOP per liter. The extent to which metrizamide quenches ³H and ¹⁴C is not serious. For example, 250 μl of 60% (w/v) metrizamide in a mixture consisting of 10 ml of Triton X-100 toluene-based scintillant and 1 ml of water quenches the radioactive counting rate of

[³H]DNA by 15%, and that of [¹⁴C]DNA by 10%. Quench correction curves can readily be constructed so enabling double-labeled counting of, for example, [³H]RNA and [¹⁴C]DNA to be done easily even when the samples contain a high concentration of metrizamide.

Direct spectrophotometric measurement of nucleic acids at 260 nm or proteins at 280 nm in fractions from metrizamide gradients is not possible because of the large molar extinction of metrizamide in the ultraviolet below about 300 nm (λ_{max} 242 nm). However, there are a number of ways in which this problem can be circumvented. For example, metrizamide is soluble in dilute acids, such as 0.5 M HClO$_4$ or trichloroacetic acid, and so DNA and RNA can be acid-precipitated in the cold, washed with cold dilute acid to remove metrizamide, and redissolved for estimation by the diphenyl-amine method and the orcinol reaction, respectively. It is better to use these chemical reactions rather than ultraviolet extinction measurements to estimate DNA and RNA in redissolved precipitates because removal of the last traces of metrizamide (essential because of its high molar extinction at 260 nm) requires extensive washing of the precipitate. Moreover, each fraction usually contains such a small quantity of nucleic acid it is advisable to add 100 to 200 μg of protein such as bovine serum albumin to each fraction to act as carrier before acid precipitation.

The concentration of a protein in a metrizamide gradient fraction can be determined in a number of ways, depending on the protein involved and/or the degree of accuracy required. In some cases, advantage can be taken of specific extinction at wavelengths at which metrizamide does not absorb light—for example, hemoglobin can be estimated directly by measuring extinction at 410 nm, even in the presence of high concentrations of metriz-amide. If other acid-precipitable materials are absent, the distribution of protein through a gradient can be determined by measuring turbidity at 450 nm after mixing each fraction with 1 ml of cold 0.5 M HClO$_4$. Precise measurements of protein concentrations can be obtained by the sodium dodecyl sulfate–amidoschwartz method of Schaffner and Weissmann (27) even when large amounts of other acid-precipitable materials are present. Enzymic activity is not impaired even by prolonged (e.g., 72 hours) exposure to metrizamide.

The recovery of macromolecules from metrizamide solutions can usually be accomplished without difficulty. Larger particles are most easily recovered by diluting the metrizamide solution with buffer, then centrifuging for long enough to pellet them. Metrizamide is soluble in 70% aqueous ethanol, and consequently nucleic acids and nucleoproteins can be precipitated by adding a one-ninth volume of 30% (w/v) NaCl and 2 volumes of ethanol. The precipitate can be collected after storage at −20 °C for at least 4 hours by centrifugation at 10,000 g for 10 minutes. Washing of the precipitate with

70% aqueous ethanol followed by 95% aqueous ethanol removes the residual metrizamide. Gel filtration, ultrafiltration, and dialysis (though this is slow) can also be used to separate metrizamide from nucleoproteins and macromolecules.

D.   Buoyant Densities in Metrizamide

All biological macromolecules and particles so far tested band isopycnically in metrizamide gradients, not because metrizamide solutions cover a wider range of densities than solutions of other materials but because metrizamide is virtually unhydrated in aqueous solution. In this, metrizamide differs from virtually all other buoyant-density solutes. In consequence, macromolecules and particles are not dehydrated even in concentrated solutions of metrizamide so that they band at densities corresponding to those of the fully hydrated particles. Thus, the buoyant densities of all biological materials, with the sole exception of proteins, are much lower in metrizamide than in other buoyant density-gradient media. This effect is particularly notable in the case of nucleic acids which band at extremely low densities in metrizamide gradients (Table II). Proteins are

TABLE II

BUOYANT DENSITIES IN METRIZAMIDE

	Buoyant density (gm/cm^3)
DNA	
Native mouse, Na$^+$ salt	1.118
Native mouse, Cs$^+$ salt	1.179
Denatured mouse, Na$^+$ salt	1.147
RNA	
Mouse ribosomal, Na$^+$ salt	1.170
Mouse Hn, Na$^+$ salt	1.168
Protein	
Urease	1.28
Catalase	1.27
α-Casein	1.24
Nucleoproteins	
Histone–DNA complex	1.168[a]
Chromatin (unsheared)	1.200[b]
Chromatin (sheared)	1.185[c]
	1.245[d]

[a–d] Ratios of protein-to-DNA in nucleoproteins were: (a) 1.1; (b) 1.5; (c) 1.3; (d) 1.8.

exceptional since they are normally hydrated to a very low extent (0.2–0.4 gm water/gm) and are not significantly dehydrated even in concentrated CsCl solutions. Because of their low hydration, proteins have high buoyant densities in metrizamide (Table II). As expected, nucleoproteins band at densities intermediate between those of nucleic acids and proteins (Table II), and can easily be separated from both of the latter because of the wide range of densities obtainable in metrizamide gradients. Moreover, the buoyant density of a nucleoprotein is proportional to the ratio of protein to nucleic acid in the complex (15, 17, 28), thus allowing buoyant density-gradient centrifugation in metrizamide to be used to fractionate mixtures of nucleoprotein particles according to the relative amounts of protein and nucleic acid in each species. RNA being denser than DNA in metrizamide also means that ribonucleoproteins can be separated from deoxyribonucleoproteins, though in this case the small difference between the buoyant densities of DNA and RNA causes there to be considerable overlap between ribonucleoproteins of a lower protein-to-RNA ratio and deoxyribonucleoproteins of higher protein-to-DNA ratio (Section III,B).

The extent to which macromolecules and particles are hydrated in aqueous solution is dependent upon the ions present in the solution. The high degree of hydration of DNA and RNA in water means that the buoyant density of these molecules is particularly sensitive to both the concentration and the species of anions and cations in the density gradient (16, 29). Consequently, it is important to control carefully the ionic composition of the density gradient in which a fractionation of nucleoprotein is to be attempted, particularly in cases where reproducibility is important. Particular care has to be taken when divalent cations, such as Mg^{2+}, are included in the density gradient, since these can bind specifically to certain particles causing marked changes in hydration and, consequently, in buoyant density. Although such specific binding of Mg^{2+} has been used to enhance certain fractionations (30), it must be emphasized that, in these circumstances, the fractionation is very sensitive to small changes in Mg^{2+} concentration. Thus, not only is it necessary to define the ionic composition of the metrizamide solutions, but it is also important to remember to control carefully the ionic composition of the solution of nucleoproteins to be fractionated.

III.   Fractionation of Chromatin

A.   Preparation of Chromatin

The method that we have used to prepare soluble chromatin from mouse cells grown in tissue culture and mouse liver has been described in detail

CHART 1

PREPARATION OF CHROMATIN FROM NUCLEI (7)

Nuclei

| homogenized in 30 volumes of 0.14 M
| NaCl–5 mM EDTA–50 mM Tris-HCl
| (pH 7.5)

Fine suspension

| stirred at 0°C for 20 minutes;
| centrifuged at 15,000 g
| for 10 minutes

Pellet

| homogenized in same volume of
| NaCl–EDTA–Tris;
| centrifuged at 15,000 g
| for 10 minutes

Pellet

| previous steps repeated twice

Pellet

| homogenized in same volume of
| 50 mM Tris-HCl, pH 7.5;
| centrifuged at 15,000 g
| for 10 minutes

Pellet

| homogenized in 1 mM EDTA–1 mM HEPES-
| NaOH (pH 7.0)
| *or* 1 mM EDTA–1 mM Tris-HCl (pH 7.5);
| dialyzed against same buffer
| for 2 hours at 5°C

SOLUBLE CHROMATIN

elsewhere (7), and a summary only is presented here (Chart 1). Chromatin prepared in this way can be banded isopycnically in preformed gradients of metrizamide or in gradients formed *in situ*. These preparations are also suitable for experiments in which the chromatin is sheared, for example by treatment with nucleases or by sonication, prior to centrifugation. Although the fractionation obtained is, to a large extent, dependent on the method used to prepare the chromatin, the techniques used to fractionate the chromatin are, of course, independent of how it is prepared.

B. Buoyant Density-Gradient Centrifugation

Although high-molecular-weight chromatin can be sedimented to its isopycnic point quite satisfactorily and rapidly in preformed 20–60% (w/v) metrizamide gradients, this method is not generally suitable for sheared chromatin because of the very long time of centrifugation required to ensure that the equilibrium positions have been reached by all species of particles. Moreover, the amount of chromatin which can be fractionated on a preformed gradient is limited. Consequently, most often it is preferable to mix the chromatin solution with a solution of metrizamide and form the density gradient by centrifugation in a fixed-angle rotor. The results of one such experiment in which this was done are illustrated by Fig. 2. The chromatin was prepared as described in Chart 1, from LS-cells (a substrain of mouse L-cells adapted to grow in suspension culture) which had been grown for 4 days in $[2\text{-}^{14}C]$thymidine, and pulse labeled with $[^3H]$ uridine for 15 minutes. The chromatin solution, containing 1 mg of DNA/ ml, was sheared to a small extent by ultrasonication for 5 seconds with a Branson Sonifier while immersed in an ice and water mixture. The DNA extracted from chromatin treated in this way was very heterogenous in molecular weight, ranging from about 2×10^5 to 3×10^6 and above, with

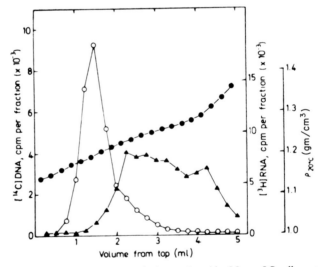

FIG. 2. Isopycnic banding of chromatin in metrizamide. Mouse LS-cells were grown for 4 days in $[2\text{-}^{14}C]$thymidine (0.05 μCi/ml, 59 μCi/mol) and then pulse labeled for 15 minutes with $[^3H]$uridine (10 μCi/ml, 24 Ci/mmol). Soluble chromatin was prepared and centrifuged to equilibrium in 41.5% (3/v) metrizamide as described in Section III; O——O, $[^{14}C]$ DNA; ▲——▲, $[^3H]$RNA; ●——●, density calculated from refractive index. From Rickwood and Birnie (7).

a peak at 5×10^5. The chromatin solution (1.5 ml) was mixed with 3.6 ml of a 70% (w/v) metrizamide solution in the same buffer as used to dissolve the chromatin [1 mM EDTA–1 mM HEPES-NaOH (pH 7.0)], together with sufficient of this buffer (approximately 0.9 ml) to reduce the final concentration of metrizamide to 41.5% (w/v). Ice-cold solutions were used throughout and, since the mixture was viscous at this low temperature, it was mixed thoroughly with a Vortex mixer. The refractive index of the mixture was measured to determine the concentration of metrizamide, and necessary adjustments were made by appropriate additions of buffer or stock metrizamide solution; care was taken to mix the solution thoroughly after each addition had been made. Five ml of the solution were then put in a 10-ml polyallomer centrifuge tube, the tube was filled with light paraffin oil (to prevent collapse during subsequent centrifugation) and centrifuged in an M.S.E. 10 × 10 ml aluminum fixed-angle rotor at 30,000 rpm (72,000 g_{max}) for 44 hours at 2°C. The rotor was allowed to come to a stop without braking, the tube was removed carefully to avoid disturbing the gradient, and the gradient was fractionated by upwards displacement with Fluorochemical FC43 (Section II, C). Fractions (0.25 ml each) were collected in glass tubes immersed in ice water; the refractive index of each fraction was measured and its density calculated (Section II, A), and the concentration of ^{14}C and ^{3}H in each was determined by liquid scintillation spectometry (Section II,C).

The results obtained from a similar experiment with highly sheared chromatin are illustrated in Fig. 3. In this case the chromatin was prepared from mouse-liver nuclei, and it was sheared by ultrasonication with the Branson Sonifier for a total of 2 minutes. The 0.25-in. tip was used at power setting 4 (3.5 A), and the solution was immersed in an ice and water mixture; sonication was done in 15-second bursts alternating with 15- to 30-second cooling periods. The molecular weight of the DNA extracted from chromatin treated in this way was about 2.5×10^5. Since the concentrations of DNA, RNA, and protein were to be determined by chemical analysis, a larger quantity of chromatin was required for this experiment. Consequently 4 ml of chromatin (containing 4 mg of DNA) were mixed with 9.5 ml of 70% (w/v) metrizamide and the concentration of metrizamide again adjusted precisely to 41.5% (w/v) by addition of buffer (about 2.5 ml). Three 10-ml polyallomer tubes, each containing 5-ml portions of this mixture, were centrifuged and the gradients fractionated exactly as described for the previous experiment. The density of each fraction was determined from its refractive index, and the fractions from the three gradients which corresponded in density were pooled for the analyses. In point of fact, since the initial densities of the solutions in the three centrifuge tubes were identical, and since care had been taken that each contained precisely the same volume

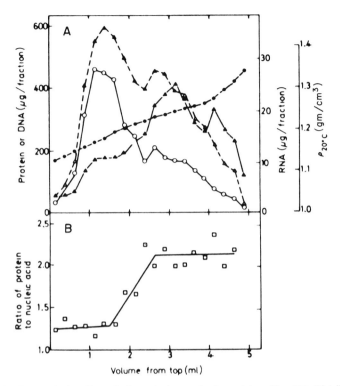

FIG. 3. Isopycnic banding of sheared chromatin in metrizamide: (A) Distribution of DNA, RNA, and protein; (B) ratio of protein to DNA + RNA. Soluble chromatin was prepared from mouse livers, sheared by sonication, and centrifuged to equilibrium in 41.5% (w/v) metrizamide as described by Section III; O——O, DNA; ▲——▲, RNA; ▲– – –▲, protein; □——□, weight ratio of protein to DNA + RNA; ●——●, density calculated from refractive index. From Rickwood and Birnie (7).

of mixture, the density of the same fractions from all three gradients corresponded very closely. One sample was taken from each pooled gradient fraction for determination of protein by the procedure of Schaffner and Weissmann (27) and a second for the measurement of the concentration of DNA and RNA after precipitation with 0.5 M $HClO_4$ (Section II,C).

Instead of pooling gradient fractions, a higher capacity rotor could have been used with appropriate adjustments to speed and time of centrifugation. For example, a similar fractionation could have been obtained by centrifuging 15-ml portions in an M.S.E. 8 × 35 ml titanium fixed-angle rotor at 27,000 rpm.

As well as illustrating the methodology of isopycnic sedimentation of chromatin in metrizamide gradients, these two experiments substantiate some points regarding the fractionation of chromatin by this means which were eluded to previously. First, the nucleoproteins are banded isopycni-cally in gradients formed *in situ* even though it is clear from the shapes of the gradients that equilibrium gradients of metrizamide have not yet been formed. Second, the centrifugation conditions chosen have resulted in the formation of a relatively shallow gradient over the density range of most immediate interest in these experiments. Centrifugation for longer, or at a higher speed, would have increased the density at the bottom of the gradient, and decreased that at the top. A similar effect would have resulted from a decrease in the initial concentration of metrizamide to, say, 30% (w/v). Under these conditions low-density material (e.g., uncomplexed DNA or RNA) would be completely separated from the nucleoproteins with some sacrifice of the resolution of the nucleoprotein species (*17*). Similarly, uncomplexed proteins would be separated from nucleoproteins by centri-fugation with metrizamide at an initial concentration of about 50% (w/v). Third, the ribonucleoprotein particles in the chromatin preparations can be separated almost completely from chromatin deoxyribonucleoprotein in metrizamide gradients (Fig. 2), so long as the chromatin has not been sheared extensively (compare Fig. 3). This separation can be enhanced by inclusion in the gradient of low concentrations of Mg^{2+}, which increases the buoyant density of the ribonucleoproteins (*22*). Thus, isopycnic band-ing of chromatin in metrizamide is a feasible method of removing most (if not all) of the ribonucleoproteins from chromatin preparations. Fourth, the buoyant density of a nucleoprotein particle is dependent on its chemical composition; in particular, Fig. 3 shows that sheared chromatin can be separated into a series of fractions which differ in the ratio of protein to DNA in the complex.

IV. Conclusions

Even in the relatively short time (about 4 years) for which it has been used as a buoyant density-gradient solute, metrizamide has proved to be an extraordinarily useful material, particularly in work involving the fractiona-tion of nucleoproteins. Metrizamide buoyant density-gradient experiments have yielded considerable amounts of information about the structure of soluble chromatin. For example, they have shown that a soluble chromatin consists of protein-poor and protein-rich regions (*17*) which are short and interspersed quite intimately throughout the chromatin (*22*). Analysis of the proteins across a gradient in which chromatin highly sheared by sonication had been banded has shown that, while the ratio of histone to

DNA was similar in both light and heavy fractions, the nonhistone proteins were concentrated in the heavy fraction (31). In addition, most (if not all) of the ribonucleoproteins in soluble chromatin prepared as described in Chart 1 are not physically associated with the deoxyribonucleoprotein, and such soluble chromatin preparations do not contain significant amounts of protein-free DNA (22). Isopycnic banding of chromatin after digestion with staphylococcal nuclease has yielded a more detailed analysis of its subunit structure (31,32). The banding pattern obtained changed significantly with increasing time of digestion, the final result after 70 minutes being the production of two distinct fractions, with protein-to-DNA ratios of 1.5 and 2.0, respectively. Again, the increased protein content of the denser fragments was due mainly (perhaps entirely) to nonhistone proteins; the lighter fragments contained little or no nonhistone proteins.

The functional significance of the different regions in chromatin revealed by these experiments is not yet clear. Although the differences between them suggested that at least a partial separation of chromatin into active and non-active fractions had been achieved, more stringent tests have disproved this notion. Reannealing of the DNA isolated from the different chromatin fractions showed that none of the DNA sequences in them were specific subsets of the sequences of the whole genome (31). Moreover, annealing of cDNA transcribed from globin mRNA or from the total polysomal mRNA of the cells from which the chromatin had been isolated failed to show that the DNA from any fraction was enriched in those sequences which were being transcribed in vivo (22,31). Despite this, it is clear that metrizamide buoyant density-gradient centrifugation does have a significant role to play in studies of the structure and function of chromatin, particularly in view of the three major advantages of metrizamide for such studies: first, its totally inert and nonionic nature means that the choice of the composition of the medium in which the density gradient is formed is entirely at the experimenter's discretion; second, chromatin is unaffected by metrizamide, and fractions recovered from metrizamide gradients are suitable for further study; third, fractionations are dependent solely on buoyant density and, hence, chemical composition, except where changes in conformation cause changes in hydration.

ACKNOWLEDGMENTS

The Beatson Institute is supported by grants from the Medical Research Council and the Cancer Research Campaign. Metrizamide used in these studies was a generous gift from Nyegaard and Co. A/S, Oslo. I am grateful to Butterworth, London, for permission to republish Chart 1 and Figs. 2 and 3, and to Nyegaard and Co. A/S, Oslo, for Fig. 1.

REFERENCES

1. Slayter, H. S., Shih, T. Y., Adler, A. J., and Fasman, G. D., *Biochemistry* **11**, 3044 (1972).
2. Doenecke, D., and McCarthy, B. J., *Biochemistry* **14**, 1366 (1975).
3. Ilyin, Y. V., Varshavsky, A. J., Mickelsaar, U. N., and Georgiev, G. P., *Eur. J. Biochem.* **22**, 235 (1971).
4. Varshavsky, A. J. and Georgiev, G. P., *Biochim. Biophys. Acta* **281**, 669 (1972).
5. Varshavsky, A. J., and Georgiev, G. P., *Mol. Biol. Rep.* **1**, 143 (1973).
6. Varshavsky, A. J., Ilyin, Y. V., and Georgiev, G. P., *Nature (London)* **250**, 602 (1974).
7. Rickwood, D., and Birnie, G. D., in "Subnuclear Components: Preparation and Fractionation" (G. D. Birnie, ed.), p. 129. Butterworth, London, 1976.
8. Clark, R. J., and Felsenfeld, G., *Nature (London) New Biol.* **229**, 101 (1971).
9. Rill, R., and van Holde, K. E., *J. Biol. Chem.* **248**, 1080 (1973).
10. Simpson, R. T., and Reeck, G. R., *Biochemistry* **12**, 3853 (1973).
11. MacGillivray, A. J., in "Subnuclear Components: Preparation and Fractionation" (G. D. Birnie, ed.), p. 209. Butterworth, London, 1976.
12. Hancock, R., *J. Mol. Biol.* **48**, 357 (1970).
13. Raynaud, A., and Ohlenbusch, H. H., *J. Mol. Biol.* **63**, 523 (1972).
14. Hossainy, E. M., Zweidler, A., and Bloch, D. P., *J. Mol. Biol.* **74**, 283 (1973).
15. Rickwood, D., and Birnie, G. D., *FEBS Letts.* **50**, 102 (1975).
16. Rickwood, D., ed., "Biological Separations in Iodinated Density Gradient Media." Information Retrieval, London, 1976.
17. Birnie, G. D., Rickwood, D., and Hell, A., *Biochim. Biophys. Acta* **331**, 283 (1973).
18. Rickwood, D., Hell, A., Birnie, G. D., and Gilhuus-Moe, C. C., *Biochim. Biophys. Acta* **342**, 367 (1974).
19. Mathias, A. P., and Wynter, C. V. A., *FEBS Letts.* **33**, 18 (1973).
20. Wunner, W. H., Buller, R. M. L., and Pringle, C., in "Biological Separations in Iodinated Density Gradient Media" (D. Rickwood, ed.), p. 159. Information Retrieval, London, 1976.
21. Freshney, R. I., in "Biological Separations in Iodinated Density Gradient Media" (D. Rickwood, ed.), p. 123. Information Retrieval, London, 1976.
22. Rickwood, D., Hell, A., Malcolm, S, Birnie, G. D., MacGillivray, A. J., and Paul, J., *Biochim. Biophys. Acta* **353**, 353 (1974).
23. Rickwood, D., Hell, A., and Birnie, G. D., *FEBS Lett.* **33**, 221 (1973).
24. Hell, A., Rickwood, D., and Birnie, G. D., in "Methodological Developments in Biochemistry" (E. Reid, ed.), Vol. IV, p. 117. Longman Group, London, 1974.
25. Birnie, G. D., and Rickwood, D., in "Biological Separations in Iodinated Density Gradient Media" (D. Rickwood, ed.), p. 193. Information Retrieval, London, 1976.
26. Young, B. D., MSE Application Information Sheet A13/6/76. MSE Scientific Instruments, Crawley, Sussex, England.
27. Schaffner, W., and Weissmann, C., *Analyt. Biochem.* **56**, 502 (1973).
28. Rickwood, D., Birnie, G. D., and MacGillivray, A. J., *Nucleic Acids Res.* **2**, 723 (1975).
29. Birnie, G. D., MacPhail, E., and Rickwood, D., *Nucleic Acids Res.* **1**, 919 (1974).
30. Buckingham, M. E., and Gros, F., *FEBS Letts.* **53**, 355 (1975).
31. Paul, J., and Malcolm, S., *Biochemistry,* **15**, 3510 (1976).
32. Malcolm S., and Paul, J., in "Biological Separations in Iodinated Density Gradient Media" (D. Rickwood, ed.), p. 41. Information Retrieval, London, 1976.

Chapter 3

Dissection of the Eukaryotic Chromosome with Deoxyribonucleases

RICHARD AXEL

*Institute of Cancer Research and Department of Pathology,
College of Physicians and Surgeons, Columbia University,
New York, New York*

I. Introduction

Over the past few years an experimentally consistent model of chromatin structure has emerged that postulates the presence of regularly repeating nucleoprotein subunits joined by short segments of DNA. Although the precise nature of the interaction between DNA and protein remains unclear, current physicochemical studies support the view that the DNA resides on the external surface of an apolar core of specific histone complexes. The core of the subunit consists of 140-base pairs of DNA plus eight histone molecules with a 40-50-base pair fiber of DNA connecting adjacent particles.

Evidence for this periodic structure initially derived from three independent experimental approaches. Electron microscopic observations (*1,2*) reveal the presence of spheroid 70–100 Å particles joined by a short filament of DNA. These particles can be observed closely opposed to one another in gently disrupted nuclei, providing evidence that these nucleoprotein particles reflect a level of structure characteristic of the chromatin fiber *in vivo*. X-ray diffraction studies reveal a series of reflections that present strong evidence for a basic structural repeat in chromatin. Recent data from low-angle (*3*) neutron scattering suggest that the 10.5-nm reflection can be attributed to the interparticle spacing.

In accord with these observations, dissection of the chromosome with a variety of deoxyribonucleases has facilitated the isolation of individual subunits. Analysis of the kinetics of nuclease digestion has permitted the elucidation of the repeating subunit structure (*4–10*) and in addition has provided information on the internal structure of the nucleosome itself

(11–13). This chapter will deal primarily with the use of various nucleases to probe the organization of chromatin.

II.   General Materials and Methods

A.   Preparation of Nuclei

Nuclei from tissue are prepared by homogenization in 5 volumes of 0.01 M Tris-HCl (pH 7.9)–0.001 M MgCl$_2$–0.25 M sucrose in a Potter–Elvehjem homogenizer. Tissue culture cells and blood cells are broken with a tight-fitting (B-type) Dounce homogenizer. The homogenate of broken cells is filtered through cheesecloth and centrifuged at 4000 rpm in a Sorvall HB-4 rotor. The resultant crude nuclear pellet is resuspended in an equal volume of the above buffer containing 0.25% Triton X-100 with the aid of a loose-fitting (A-type) Dounce homogenizer and resedimented. This washing procedure is repeated 4 times, and the nuclear pellets are then suspended in buffer without added detergent. This final nuclear pellet is then washed once and suspended in the appropriate digestion buffer (see below) at a DNA concentration of 500 μg/ml.

B.   Preparation of Chromatin

Chromatin is prepared from the final nuclear pellet by gradual swelling of the nuclei in buffers of decreasing ionic strength. Nuclei from 2 gm of tissue are suspended in 20 ml of 0.05 M Tris-HCl (pH 7.9) with a Dounce homogenizer and centrifuged at 8000 rpm in a Sorvall HB-4 rotor. This procedure is repeated consecutively with 0.01 M, 0.005 M, and 0.001 M Tris-HCl (pH 7.9). As the ionic strength is reduced, the chromatin pellet will swell to occupy perhaps 20 times the original volume of the nuclear pellet. The final chromatin pellet is suspended in 0.001 M Tris-HCl (pH 7.9) at the desired DNA concentration. Chromatin prepared in this way is stable at 0°–4°C for 2–3 weeks. Proteolytic attack can be minimized by first washing the nuclear pellet in 0.025 M ethylenediaminetetracetic acid (EDTA) (pH 6.0)–0.25 M NaCl prior to the chromatin preparation. This procedure results in the removal of a small percentage of the nonhistone proteins, some of which contain proteolytic activity.

C.   Nuclease Digestions of Nuclei and Chromatin

Digestion with staphylococcal nuclease is carried out at a DNA concentration of 500 μg/ml. Digestion of nuclei was carried out on suspensions of nuclei washed twice in 1 mM Tris-HCl (pH 7.9)–0.1 mM CaCl$_2$–0.25 M sucrose and resuspended in the same buffer at a DNA concentration of 500 μg/ml. The kinetics of DNA digestion following the addition of nuclease were assayed by measuring the amount of A_{260} absorbing material soluble in 1 M NaCl–1 M HClO$_4$. RNA contributed less than 5% of the total A_{260} of purified nuclei as determined either by alkali digestion or treatment with pancreatic ribonuclease. Nuclease reactions were terminated by the addition of NaEDTA to 5 mM.

Digestion of nuclei with pancreatic DNase I (Worthington Biochemicals, Freehold, New Jersey) is carried out in solvent containing 0.01 M Tris-HCl (pH 7.9)–0.01 M NaCl–0.003 M MgCl$_2$. Sucrose is not required under these digestion conditions since the magnesium ion concentration is sufficient to maintain the integrity of the nuclei.

Digestion with porcine DNase II is performed in 0.25 M sucrose–0.025 M NaAc–2 mM EDTA (pH 6.5) at an enzyme concentration of 20 μg/ml. It should be noted that commercial preparations of this endonuclease are contaminated with significant quantities of proteases that limit the usefulness of this enzyme without further purification.

D. Analysis of the DNA Products of Nuclease Digestion

Aliquots are removed at various times in the digestion process for analysis of the resistant DNA by polyacrylamide gel electrophoresis. Reactions were stopped by the addition of EDTA, corrected to 0.5% sodium dodecyl sulfate (SDS)–0.4 M NaCl, and incubated at 37°C for 1 hour with 100 μg/ml of Proteinase K (E. Merck, St. Louis, Missouri). This protease contains no detectable DNase activity and can be used without prior autodigestion. After incubation, the solutions were extracted 3 times with phenol-chloroform-isoamyl alcohol (14:14:1) and the DNA precipitated with 2 volumes of ethanol at −20°C. The DNA pellets are then dissolved in electrophoresis buffer at concentrations of about 1 mg/ml.

Analysis of the native DNA resulting from the digestion of chromatin or nuclei is performed in either 2.5% or 6% polyacrylamide slab gels (12 × 20 cm, E. C. Apparatus, St. Petersburg, Florida). Two and a half percent gels were supplemented with 0.5% agarose to provide greater support. Samples contained 10–15 μg of DNA in 10 μl of 0.5 X electrophoresis buffer. Two microliters of 40% sucrose–0.5% Bromophenol blue were added to each sample, and the gels were run at 200 V on a 20-cm cooled slab gel. The buffer system employed contains 0.09 M Tris-borate (pH 8.3)–0.003 M EDTA (14).

Analysis of the single-stranded products of digestion was performed on 10% acrylamide–7 M urea gels (15). Twenty μg of DNA are lyophilized and suspended in 10 μl of deionized formamide. The solution is placed in a boiling-water bath for 3 minutes, plunged in ice, and 2 μl of 0.5% Bromophenol blue are added. The buffer system in these gels is identical to that described above. Gels were stained for 30 minutes in ethidium bromide (1 μg/ml) and photographed under shortwave ultraviolet (UV) illumination using a red filter. Negatives can be cut and scanned in the gel scanning accessory of a recording spectrophotometer.

III.   Staphylococcal Nuclease Digestion of Nuclei

A.   Procedure

Digestion of chromatin from a variety of sources with the enzyme staphylococcal nuclease results in the liberation of about half of the chromatin DNA as acid-soluble oligomers (16). The associated nucleoprotein fragments contain all of the chromatin proteins but only half the original complement of DNA. The kinetics of digestion of DNA in isolated rat liver nuclei again reveal that about half of the nuclear DTA is susceptible to nuclease attack (Fig. 1). Using either chromatin isolates or whole nuclei, the reaction reaches a well-defined limit of 54% digestion with the rate of attack on chromatin only slightly greater than that observed with nuclei. Identical results are obtained with either detergent-washed nuclei or nuclei prepared

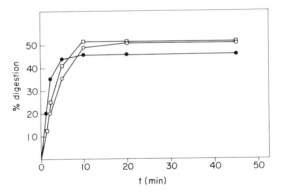

FIG. 1.   Kinetics of nuclease digestion of nuclei, chromatin, and monomer. Intact rat liver nuclei (○), isolated chromatin (□), and purified monomer (●) were incubated in 1 mM Tris-HCl (pH 7.9)–0.1 mM CaCl$_2$ at 37 °C in the presence of 5 μg/ml of staphylococcal nuclease. The initial DNA concentration was 150 μg/ml in nuclei, chromatin, and monomer. Aliquots were removed at the indicated times and the amount of DNA rendered soluble in 1 M HClO$_4$–1 M NaCl was determined.

by sedimentation through sucrose columns. Incubation of nuclei or chromatin in the absence of exogenous nuclease results in virtually no solubilization of DNA under these reaction conditions, excluding the possibility that the rat liver Ca^{2+}-, Mg^{2+}-dependent endonuclease (4) is responsible for the observed reaction.

Examination of the DNA fragments generated during the course of the digestion reveals two levels of structure within the chromatin complex: the organization of nucleosomes themselves and the arrangement of DNA within the individual nucleosomes. To this end nuclei were digested with staphylococcal nuclease. At various times the resulting nucleoprotein complexes were freed of protein and analyzed by electrophoresis in 2.5% acrylamide–0.5% agarose gels. A typical profile is shown in Fig. 2. We observe that

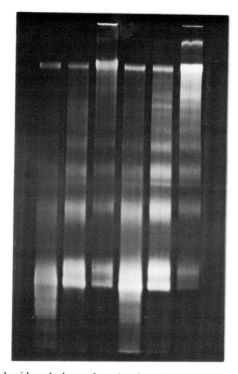

FIG. 2. Polyacrylamide gel electrophoresis of purified DNA fragments liberated upon staphylococcal nuclease digestion of metaphase chromosomes and interphase nuclei. Nuclei and chromosomes from mouse L-cells were digested with staphylococcal nuclease as described in the text. At various times, aliquots were removed and the DNA purified. Ten μg of DNA were applied to each slot of a 2.5% acrylamide–0.5% agarose slab and electrophoresed at 200 V for 2 hours. Gels were stained with ethidium bromide and photographed. Samples from right to left represent 2, 4, and 10% digestion of metaphase chromosomes and 2, 4, and 10% digestion of interphase nuclei.

early in the digestion process a series of discrete bands of DNA is present whose molecular weights correspond to integral multiples of a monomeric subunit 185-base pairs in length. Gels are calibrated with fragments of DNA of known molecular weight obtained by restriction endonuclease treatment of phage or viral DNA of defined length.

It is obvious that as the digestion proceeds, the relative proportion of the monomeric fragment increases at the expense of the multimeric components. These data would suggest that the proteins of chromatin arrange the DNA in regular conformation that repeats itself every 185-base pairs. The repeating subunit is punctuated by regularly spaced nuclease cleavage sites, presumably due to a region relatively free of protein and sensitive to nuclease attack.

It should be noted that as the digestion proceeds a series of DNA fragments of lower molecular weights is observed that provides information reflecting the internal structure of the nucleosomal subunit. A prominent fragment is observed at 140-base pairs in length. Although considerable variation exists in the size of the large monomeric subunit, this 140-base pair fragment appears in chromatin digests from virtually all eukaryotic sources and may represent the internal core of the basic repeating unit. The difference in size between the largest observable monomeric fragment and this constant 140-base pair intermediate presumably reflects the more-accessible interbead DNA.

As digestion proceeds to a limit at 50% solubilization of DNA, a second set of highly reproducible small molecular-weight bands ranging in size from 140–40-base pairs is observed (Fig. 3). Similar patterns of DNA fragments are obtained whether we use intact nuclei, nucleosomes, or isolated chromatin as the initial substrate for the enzyme. The precision with which the enzymic process produces these discrete DNA fragments suggests that there is a highly specific arrangement of the histones on the surface of the DNA in which certain well-defined portions of the polypeptide chains are in intimate contact with the DNA, or have folded the DNA in such a way that precisely determined lengths of nucleic acid are protected from digestion.

B. Comment

Treatment of nuclei from a variety of sources with the enzyme staphylococcal nuclease results in the liberation of a series of DNA fragments that are multiples of a unit length DNA. More extensive digestion of nuclei, however, reveals that this monomeric subunit is a transient intermediate and is susceptible to further specific degradation by nuclease, resulting in the generation of a unique set of DNA fragments of smaller molecular weight (160–50-base pairs). This series of fragments arising from intact

FIG. 3. Polyacrylamide gel electrophoresis (6%) of DNA liberated upon staphylococcal nuclease digestion of metaphase chromosomes and interphase nuclei. Interphase nuclei and metaphase chromosomes were digested with staphylococcal nuclease. The resistant DNA was freed of protein and 10 μg applied to each slot of a 6% polyacrylamide gel. The gel was run as described previously, stained with ethidium bromide, and photographed. Samples from right to left represent 14 and 28% digestion of metaphase chromosomes and 10 and 30% digestion of nuclei.

nuclei is virtually identical to the profile observed upon the digestion of isolated chromatin. Furthermore, digestion of the purified monomeric nucleoprotein subunit similarly liberates this highly characteristic set of lower molecular-weight DNA bands initially observed in chromatin digests.

It seems reasonable to assume that the monomeric DNA obtained early in the digestion of nuclei is contained within the regular repeating subunit of chromatin initially visualized in electron micrographs as "beads on a string" (1). One interpretation of the digestion data described here is simply that the DNA between the repeating subunits in nuclei is most accessible to nuclease attack and provides the first site of cleavage for the enzyme. Further digestion results in the accumulation of monomeric subunits at the expense of the higher molecular-weight multimeric components. Contained within the monomeric nucleoprotein complex are less accessible sites of nuclease attack that may only be cleaved after the liberation of monomer. More extensive digestion is therefore required to solubilize these internal stretches of "free" DNA. The resultant resistant DNA would reflect the points of intimate contact of histones or specific groups of histones on the DNA within the repeating subunit.

The nucleosomal array appears to be a ubiquitous means of chromatin organization in all eukaryotes examined. While variations in the size of the repeat lengths of DNA exist between species, a basic repeating unit with a 140-base pair core seems to characterize the eukaryotic genome in both yeast and man. In addition, the basic subunit structure of the chromatin fiber is retained by the mitotic chromosome. Twenty million nucleosomes are therefore present in the nucleus of a mammalian cell, and this structure is apparently maintained throughout the cell cycle.

IV.   Isolation of the Monomeric Nucleosomal Subunit

Characterization of the basic repeat unit of chromatin requires that we isolate the monomeric particle in relatively pure form. Early in the digestion of intact nuclei with the enzyme staphylococcal nuclease, a series of nucleoprotein fragments is liberated. The fragments are multimeric complexes consisting of one or more nucleosomes. These structures are readily separable from one another by sucrose gradient velocity centrifugation (5, 9, 17). Experimentally, purified nuclei are suspended and washed in 0.01 M Tris-HCl (pH 7.9)–0.1 mM CaCl$_2$–0.25 M sucrose. Nuclei are resuspended at a DNA concentration of 500 μg/ml and digested with staphylococcal nuclease at 37° to yield 2, 4, and 10% acid solubilization. The nuclei are sedimented at 4000 rpm and resuspended in 5 mM EDTA at a DNA concentration of 1 mg/ml. This procedure effectively lyses the digested nuclei. The nuclear lysate is then applied directly to a 5–20% sucrose gradient. For analytic purposes, we use a 12-ml gradient centrifuged at 27,000 rpm in the Spinco SW-41 rotor. For preparative runs, 5 mg can be loaded on a 35-ml gradient and spun in an SW-27 rotor at 26,000 rpm, 16 hours at 4°C. Larger quantities of isolated nucleosomes can be prepared by zonal centrifugation.

Typical sucrose gradient profiles of multimeric and monomeric nucleosomal subunits obtained at three levels of digestion are shown in Fig. 4. At 2% solubilization of DNA, the bulk of the nucleoprotein sediments as multimeric fragments. Digestion proceeds at the expense of the multimers, resulting in increased quantities of the monomeric subunit. At 15% digestion, we find that over 80% of the chromatin fragments sediment as the 11 S monomeric subunit. Proof that the absorbance peaks observed in Fig. 4 result from multimeric nucleoprotein fragments is obtained by simply freeing this material of protein by phenol extraction. The resultant DNA is then analyzed on 2.5% polyacrylamide–0.5% agarose gels as described above.

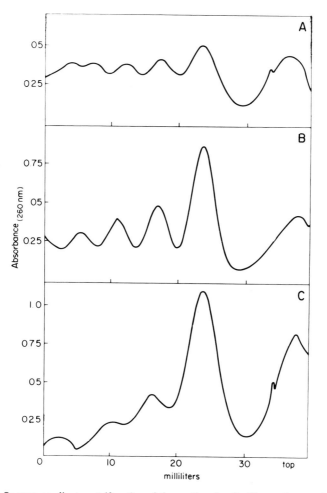

FIG. 4. Sucrose gradient centrifugation of chromatin subunits. The nucleoprotein particles generated by (A) 2, (B) 4, and (C) 10% digestion of rat liver nuclei were subjected to centrifugation in a 5–20% sucrose gradient at 110,000 *g* for 16 hours. The direction of sedimentation proceeds from right to left.

Isolation of purified monomeric nucleoprotein particles is performed by digesting nuclei with staphylococcal nuclease to 15% acid solubilization. The nuclei are lysed and the resultant nucleoprotein fractionated on sucrose velocity gradients as described. The monomeric peak is pooled, dialyzed, and concentrated by Amicon ultrafiltration using a PM 10 filter. The monomeric subunits are then reapplied to a second sucrose gradient. The monomeric particles prepared in this way are greater than 95% pure. Purity is assessed by analysis of the DNA sizes obtained on polyacrylamide gels.

An alternate procedure for the purification of monomeric subunits involves gel filtration on Biogel A-5m resins (10). A nuclear lysate is prepared following digestion as described and applied to a Bio-Rad A-5m column, 90×2.5 cm equilibrated with 0.01 M Tris-HCl (pH 7.5)–0.0007 M EDTA at $4°$C. This procedure permits the purification of large quantities of the monomeric subunit but does not permit fractionation of the multimeric forms.

For analytical purposes, nucleosomal subunits can be separated on polyacrylamide gel electrophoresis. Fifteen μg of a nuclear lysate following nuclease treatment is loaded on a 2.5% acrylamide–0.5% agarose slab gel. Electrophoresis is performed in a buffer system containing 0.04 M Tris-HCl (pH 7.9)–0.005 M sodium acetate–0.001 M EDTA. Under these conditions the nucleoprotein subunits remain intact with a linear relationship observed between migration of the multimeric subunits and the logarithm of their molecular weight. This procedure has been particularly useful in elucidating the heterogeneous forms of the monomeric particle with respect to histone Hl.

V.   Digestion with Pancreatic DNase I

Pancreatic DNase digestion of nuclei is performed in a manner analogous to that described for staphylococcal nuclease, but it reveals strikingly different kinetics of solubilization and results in the liberation of a different set of resistant DNA fragments (12,18). Nuclei are washed in DNase I buffer $[0.01$ M Tris-HCl (pH 7.9)–0.01 M NaCl–0.003 M MgCl$_2]$ and suspended at a concentration of 500 μg/ml of DNA with an enzyme concentration of 40 μg/ml. The kinetics of digestion reveal a rapid initial slope to about 10–15% digestion and a somewhat slower rate to 50% digestion. Digestion proceeds beyond 50% at an even slower rate, which is thought to reflect digestion of DNA bound to the chromatin proteins.

Analysis of the DNA resistant to digestion at various times in the reaction on native gels reveals a broad, poorly resolved distribution of DNA fragments ranging from 20 to 160-base pairs. It is striking that denaturation of these fragments prior to gel electrophoresis on urea or formamide gels reveals a characteristic series of single-stranded DNA fragments at 10-base intervals from 20 to 160 bases in length (Fig. 5).

The nature of the arrangement of DNA within the nucleosome resulting

FIG. 5. Polyacrylamide gel electrophoresis of DNA liberated upon DNase I digestion of nuclei. Nuclei were digested with DNase I for various times, and the resultant DNA was freed of protein as described in the text. Fifteen μg of DNA were dissolved in 99% formamide and placed in a boiling water bath for 3 minutes. Two μl of dye marker were added, and samples were loaded on a 10% polyacrylamide–7 M urea slab gel. Gels were run at 150 V for 3.5 hours at room temperature, stained with ethidium bromide, and photographed. Samples from right to left represent 5, 10, 20, and 26% solubilization of nuclear DNA.

in staggered endonucleolytic cleavage is unclear. One possibility put forth (5) is that DNA is wound about a histone core so that only one phosphodiester bond per strand is accessible to cleavage per helical turn. This organization, however, must be regular to generate multimer forms of the 10-base pair fragment. An alternative suggestion (19) postulates the presence of "kinks" at regular intervals that are susceptible to nuclease attack.

VI. Digestion of Nuclei with DNase II

The process of digestion of the chromatin complex with DNase II has not been as extensively studied as that described for other nucleases. Nevertheless, the potential usefulness of this enzyme warrants discussion of its properties. DNase II is an endonuclease that results in double-strand scissions yielding products ending in 3'-phosphates. The enzyme does not require divalent ions for activity, and therefore the products of digestion cannot be attributed to artifactual aggregation or rearrangement induced by divalent ions. In addition, the purified enzyme, unlike staphylococcal nuclease, is not able to degrade RNA. DNase II therefore is potentially useful for the isolation of chromatin fractions with associated RNA. Unfortunately, the commercial preparations of DNase II (Worthington) are heavily contaminated with ribonucleases and proteases that limit the usefulness of the enzyme without further purification.

Digestion of nuclei with DNase II is performed at a DNA concentration of 500 μg/ml in 0.25 M sucrose–0.025 M NaAc–0.002 M EDTA (pH 6.5), at an enzyme concentration of 20 μg/ml. Since this enzyme has an optimum activity at pH 5.0, the reaction can be essentially terminated by the addition of an equal volume of 0.1 M Tris-HCl (pH 8.5). In one report using this enzyme (20), the pattern of chromatin subunits generated with DNase II is virtually identical to that observed with staphylococcal nuclease. Interestingly, the DNA isolated from this calf monomeric subunit is 120-base pairs in length, in contrast to the 185–200-base pair fragment generated by staphylococcal nuclease. It is not clear at present whether this discrepancy results from increased susceptibility of the DNA within the nucleosomal subunit. Alternatively, this enzyme may recognize a different aspect of structure than the preceding nucleases described, perhaps generating half nucleosomes.

VII.   Cleavage of Nuclei and Chromatin with Restriction Endonucleases

The DNA restriction endonucleases have only recently been employed in the dissection of the chromatin complex. These enzymes have been used to assess the nature of the distribution of nucleosomes about specific restriction sites in the minichromosome of SV-40 (21,22). Cleavage of the SV-40 chromatin complex with specific endonucleases and examination of the resultant DNA products strongly suggest that there is a random distribution of nucleosomes about the SV-40 genome.

More recently restriction endonucleases have been used to probe the accessibility of specific sequences to endonucleolytic cleavage in intact nuclei (23). If a given restriction sequence participates in a nucleosomal structure, then the extent to which this sequence is cleaved by a specific endonuclease provides a measure of the accessibility of nucleosomal DNA to proteins that require sequence recognition for activity.

This approach has been utilized to demonstrate the accessibility of calf satellite I sequences to endonuclease Eco RI cleavage in their native conformation in intact nuclei (23). The tandemly repetitive calf satellite I has an Eco RI site at 1400-nucleotide intervals. The extent of generation of this 1400-base pair fragment from nuclei and DNA therefore provides a measure of the accessibility of nucleosomal DNA to restriction endonuclease action.

Calf thymus DNA and nuclei at a DNA concentration of 20 μg/ml were incubated for 2 hours at 37°C in 3% sucrose–0.01 M Tris-HCl (pH 7.9)–0.05 M NaCl–0.005 M MgCl$_2$ in the presence of varying concentrations of the restriction endonuclease Eco RI (Miles Laboratories, Elkhart, Indiana). Reactions were terminated by addition of Na$_2$EDTA to 0.01 M, and the resultant DNA fragments were purified by phenol extraction. The DNA was then electrophoresed through a 1.5% agarose slab gel. The gel is stained with ethidium bromide and photographed. The amount of material migrating as the 1400-base pair satellite fragment from DNA and nuclear digests is determined from scans of photographic negatives.

These studies have revealed a random distribution of nucleosomes about the Eco RI site of satellite I in intact calf nuclei. They have further indicated that about 60% of the Eco RI sites within the core of the nucleosome itself are accessible to endonuclease; this finding is consistent with a model of nucleosome structure in which DNA resides on the surface of the core, thereby permitting the DNA with the nucleosome to participate in sequence-specific interaction.

REFERENCES

1. Olins, A. L., and Olins, D. E., *Science* **183**, 330 (1974).
2. Oudet, P., Gross-Bellard, M., and Chambon, P., *Cell* **4**, 281 (1975).
3. Baldwin, J. P., Boseley, P. G., Bradbury, E. M., and Ibel, K., *Nature (London)* **253**, 245 (1975).
4. Hewish, D. R., and Burgoyne, L. A., *Biochem. Biophys. Res. Commun.* **52**, 504 (1973).
5. Noll, M., *Nature (London)* **251**, 249 (1974).
6. Kornberg, R., and Thomas, J. O., *Science* **184**, 865 (1974).
7. Sahasrabuddhe, C. G., and van Holde, K. E., *J. Biol. Chem.* **249**, 152 (1974).
8. Axel, R., *Biochemistry* **14**, 2921 (1975).
9. Sollner-Webb, B., and Felsenfeld, G., *Biochemistry* **14**, 2916 (1975).
10. Shaw, B. R., Herman, T. M., Kovacic, R. T., Beaudreau, G. S., and van Holde, K. E., *Proc. Natl. Acad. Sci. U.S.A.* **73**, 505 (1976).
11. Axel, R., Melchior, W., Sollner-Webb, B., and Felsenfeld, G., *Proc. Natl. Acad. Sci. U.S.A.* **71**, 4101 (1974).
12. Noll, M., *Nucleic Acids Res.* **1**, 1573 (1974).
13. Weintraub, H., and Van Lente, F., *Proc. Natl. Acad. Sci. U.S.A.* **71**, 4249 (1974).
14. Peacock, A. C., and Dingman, C. W., *Biochemistry* **8**, 608 (1969).
15. Maniatis, T., Jeffrey, A., and van deSande, H., *Biochemistry* **14**, 3787 (1975).
16. Clark, R. J., and Felsenfeld, G., *Nature (London) New Biol.* **229**, 101 (1971).
17. Lacy, E., and Axel, R., *Proc. Natl. Acad. Sci. U.S.A.* **72**, 3978 (1975).

18. Camerini-Otero, R. D., Sollner-Webb, B., and Felsenfeld, G., *Cell* **8**, 333 (1976).
19. Crick, F. H., and Klug, A., *Nature (London)* **255**, 530 (1975).
20. Oosterhof, D., Hozier, J., and Rill, R., *Proc. Natl. Acad. Sci. U.S.A.* **72**, 633 (1975).
21. Polisky, B., and McCarthy, B., *Proc. Natl. Acad. Sci. U.S.A.* **72**, 2895 (1975).
22. Cremisi, C., Pignatti, P. F., Croissant, O., and Yaniv, M., *J. Virol.* **17**, 204 (1976).
23. Lipschitz, L., and Axel, R., *Cell* **9**, 355 (1976).

Chapter 4

Isolation of Amplified Nucleoli from Xenopus Oocytes

HARVEY L. WAHN AND RONALD H. REEDER

Department of Embryology,
Carnegie Institution of Washington,
Baltimore, Maryland

TORU HIGASHINAKAGAWA

Mitsubishi Kasei Institute for Life Sciences,
Tokyo, Japan

I. Introduction

Previous studies on the function of chromosomal proteins have been seriously hampered by the lack of methods for isolating single genes as "native" chromatin. This chapter describes a special case in which a naturally occurring biological enrichment of a single gene makes it possible to perform such an isolation. In the frog, *Xenopus laevis*, the DNA coding for the 18 S and 28 S ribosomal RNAs (rDNA) undergoes a massive amplification at the pachytene stage of oogenesis (*1–3*). This results in a nucleus which contains about 24 pgm of rDNA plus 12 pgm of bulk chromosomal DNA (the chromosomes are tetraploid at this stage). By mid-diplotene [Dumont stages II and III (*4*)] the amplified rDNA is packaged in several thousand extra-chromosomal nucleoli which are located close to the nuclear membrane. At this stage the amplified nucleoli are actively engaged in ribosomal RNA synthesis and may be isolated completely free of other DNA contaminants.

An adult *Xenopus* ovary contains oocytes in all stages of maturation. The first part of the procedure is designed to dissociate the ovary into single cells and to isolate those oocytes which have not yet begun to accumulate pigment or massive amounts of yolk (stages II and III). Later-stage oocytes are discarded since the large amounts of yolk and pigment which they contain seriously interfere with nucleolar isolation. The immature oocytes are then homogenized, and nucleoli are isolated by banding them in buoyant-density gradients of metrizamide. Throughout the isolation, the primary bulk DNA contaminant is nuclei from the follicle cells which surround each oocyte. The majority of these follicle cells are removed during isolation of the immature oocytes, and the remainder are removed by successive metrizamide gradients.

II.   Materials

Adult female frogs are purchased from the South African Snake Farm and are maintained in the laboratory as described elsewhere (5). Metrizamide is purchased from Gallard-Schlessinger, collagenase (type I) and papain (type II) from Sigma, and Sarkosyl from Geigy.

III.   Isolation Procedure

A.   Dissociation of Ovaries

Ten frogs are anesthetized by immersing them in ice water, killed by decapitation with a heavy pair of shears, and pithed by running a needle down the spinal column. The ovaries are dissected out and washed free of blood in several rinses of 150 mM NaCl–150 mM Tris-HCl–1 mM ethylenediaminetetraacetic acid (EDTA) and finally rinsed in 100 mM sodium phosphate (pH 7.4) to remove the EDTA. The ovaries are cut into small pieces with scissors and distributed among six 200-ml siliconized screw-cap digestion tubes. To each tube is added about 150 ml of 100 mM sodium phosphate (pH 7.4) containing 100 mg of collagenase and 25 mg of papain. The tubes are capped, leaving a generous air bubble at the top, and secured to a slowly tilting platform. Digestion proceeds at room temperature for about 30 minutes or until most of the oocytes are released as single cells. As the tubes tilt back and forth the air bubble stirs the mixture and greatly improves the dissociation. However, if the ovarian tissue is not cut into small enough pieces, a single digestion step will be insufficient for complete release of the oocytes. After the initial digestion any remaining large fragments may be removed, cut into smaller pieces, and redigested for an additional 20 minutes to effect complete recovery of all oocytes. Care should be taken to avoid overdigestion since this causes the oocytes to become fragile and to break during later manipulation.

B. Isolation of Immature (Stage II and III) Oocytes

From this point onward, all operations are performed at 4° in siliconized glassware. The digestion mixture is poured into a beaker, undigested fragments are removed with forceps, and the oocytes are collected by settling. The supernatant is decanted, and the oocytes are washed 3 times by resuspension and settling in isolation buffer (IB) [90 mM NaCl–1.5 mM KCl–5 mM MgCl$_2$–10 mM Tris-HCl (pH 8)–5 mM dithiothreitol–0.1 mM EDTA]. Care is taken that all of the small, immature oocytes are fully settled each time before decantation.

The washed oocytes are resuspended in 1 liter of IB in a 1 liter separatory funnel by inverting the funnel once or twice and again allowed to settle. The more mature, pigmented oocytes sediment rapidly and are immediately drawn off by briefly opening the stopcock at the bottom of the funnel. The remaining suspended material (which contains the immature oocytes) is recovered by pouring it out through the top of the funnel, and the oocytes are collected by settling. At this point most of the pigmented oocytes have been removed but the immature oocytes are still heavily contaminated by blood cells and follicle cells. To remove most of these, the cells are resuspended in 200 ml of IB and poured into a 2-liter separatory funnel. Fresh IB is then pumped into the funnel through the bottom stopcock with sufficient flow rate to completely resuspend all the cells. When the funnel is full the stopcock is closed, the pump disconnected, and the oocytes are allowed to settle undisturbed for about 10 minutes. Most of the oocytes will settle in this time while most of the blood cells and follicle cells remain in suspension. The oocytes are collected in the smallest possible volume by drawing them off through the stopcock. With a large-bore pipette, the crude oocyte suspension is next layered over a 25-ml cushion of 1.7 M sucrose (in IB) in several 30-ml centrifuge tubes (Corex, Corning, Corning, New York). The upper one-third of the tube is gently stirred to remove any discontinuities. Centrifugation at 2000 g for 6 minutes sediments the immature oocytes into a bright yellow pellet while red blood cells, follicle cells, and mature oocytes float and are discarded. (Immature oocytes have an anomalous permeability to sucrose which they lose at the same time they begin to acquire pigment.) The pelleted oocytes are gently resuspended in a small volume of IB and spun through 1.7 M sucrose again. At this point immature oocytes are greatly enriched but bulk DNA from follicle cells is still present in 10- to 20-fold excess over rDNA.

C. Isolation of Amplified Nucleoli in Metrizamide Gradients

Oocytes are suspended in 5 ml of homogenization buffer (HB is IB with the Tris-HCl increased to 50 mM) and homogenized with five strokes each of the loose and tight pestle of a Dounce homogenizer. The homogenate is poured through four layers of cheesecloth which is then washed with 3 ml of HB. The refractive index of this homogenate is adjusted to 1.390 by addition of 65% metrizamide (w/v in HB) to achieve a final volume of about 35 ml. One-half of the adjusted homogenate (17.5 ml) is placed in the rear chamber of a linear density-gradient former, the front chamber is filled with 17.5 ml of metrizamide in HB having a refractive index of 1.402, and a linear density gradient is poured into a large (35 ml) nitrocellulose tube for the SW-27 rotor (Beckman). The other half of the homogenate is used to pour a second identical gradient. It is essential that the homogenate be distributed throughout the gradient with no sharp discontinuities. If discontinuities are present in the gradient, the nucleoli and other particles will pack together early in the centrifugation and never separate from each other. The gradients are spun at 25,000 rpm for 60 minutes at 0°. Gradients are fractionated automatically by pumping them through a Gilford recording spectrophotometer (set to scan at 650 nm since metrizamide absorbs strongly in the ultraviolet), and fractions are collected using a drop-counting fraction collector. The light-scattering (A_{650}) profile from a typical gradient is shown in Fig. 1. A drop from each fraction is examined under the phase-contrast microscope for nucleoli or other organelles, and the refractive index of every third fraction is measured. To achieve maximum purity, the fractions containing nucleoli are pooled, diluted to a refractive index of 1.388 with IB, and rebanded in a second density gradient (with metrizamide of $\eta = 1.402$ in the second chamber).

Nucleoli are recovered by diluting gradient fractions with an equal volume of IB, layering this over a cushion of 2 M sucrose in IB, and spinning for 20 minutes at 10,000 rpm (Sorvall centrifuge). The nucleolar pellet may then be used as desired.

D. Assay of DNA Purity

An aliquot of nucleoli is dissolved in 1% sodium dodecyl sarcosinate (Sarkosyl). To 90 μl of dissolved nucleoli is added 275 μl of a saturated solution of CsCl [saturated at room temperature in 1 mM EDTA–10 mM Tris-HCl (pH 8)] and 5 μl containing 0.1 μg of a suitable reference DNA as a buoyant-density marker. A suitable marker DNA is that from the bacteriophage SP01 which infects *Bacillus subtilis* and has a density of 1.742. The mixture is injected into a small sector (0.35 ml volume) cell for the Beckman Model E analytical ultracentrifuge and centrifuged to equilibrium at 44,000 rpm at 25° for about 20 hours. Bulk chromosomal DNA bands at $\rho = 1.699$, well separated from amplified rDNA at $\rho = 1.729$. After photographing the cell, the negatives are traced with a densitometer and the areas under the

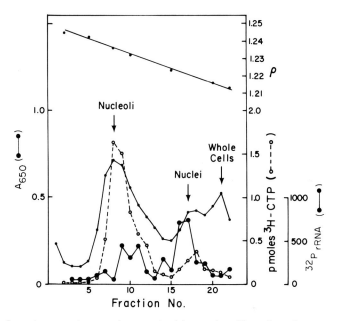

FIG. 1. Immature oocytes were homogenized in metrizamide and used to pour a linear density gradient as described in the text. Each fraction from the gradient was assayed for light scattering at A_{650} (●——●), ribosomal DNA hybridization to ³²P-labeled ribosomal RNA (●——●), and RNA polymerase activity (p moles) (○-○-○). Nucleoli, nuclei, and whole cells were identified by phase-contrast light microscopy, and buoyant densities were calculated from refractive indices.

reference, chromosomal, and rDNA peaks measured with a planimeter. Absolute amounts of chromosomal and rDNA can be determined by comparison with the known amount of reference DNA over the range 0.05–0.5 μg. Because of its high ultraviolet absorbance, all metrizamied must be removed from the nucleolar sample before performing this assay.

IV. Comments

The complete procedure as outlined requires about 8–9 hours. Starting from 10 frogs one can expect 2–3 μg of rDNA in purified nucleoli. Nuclei from follicle cells appear to be completely removed from nucleoli by the density-gradient steps. Nuclei from blood cells (*Xenopus* erythrocytes are nucleated) are more dense, however, and some may band with nucleoli. Therefore, for maximum purity it is crucial that these be eliminated during the separatory funnel settling steps.

Metrizamide is a tri-iodinated benzmido-derivative of glucose which is stable for at least 1 week in aqueous solution at neutral pH. Prolonged exposure to light causes it to turn yellow due to minor alterations in the sugar fractions with no detectable increase in free iodide (6). It has been used to produce density gradients that will band DNA, RNA, and protein within the same gradient (7). Metrizamide does not interact with DNA and only weakly with protein. The interaction appears to be completely reversible. In general it has no greater effect on enzymic activities than does sucrose. It is possible, therefore, to recover many enzymic activities associated with chromatin including RNA polymerase (8). In first-round metrizamide gradients nucleoli usually band at a refractive index (η) of 1.399 $\pm$ 0.006 with an apparent density (ρ) of 1.226 $\pm$ 0.002 gm/cm^3 derived from the relationship $\rho = 3.35\eta - 3.462$. The apparent density shifts upon rebanding in a second-round gradient to 1.221 $\pm$ 0.002 gm/cm^3. We suspect that this apparent shift is due to the large amounts of oocyte soluble protein present in the first-round gradient which contributes to the refractive index causing an anomalous estimation of density by this method. However, the relative positions of nucleoli, nuclei, and whole cells do not change upon rebanding.

Nucleoli isolated in this manner contain over 90% of their DNA as rDNA and have a normal morphology as judged by light and electron microscopy. They contain intact 40 S and 30 S precursor RNA, active RNA polymerase I, DNA relaxation enzyme, and other enzymic activities. Therefore, this method does not seem to inactivate sensitive enzymes. Although the amounts obtainable at present are small, this appears to be an excellent material for study of the organization of active chromatin and for *in vitro* studies on its transcription.

REFERENCES

1. Brown, D. D., and Dawid, I. B., *Science* **160**, 272 (1968).
2. Gall, J. G., *Proc. Natl. Acad. Sci. U.S.A.* **60**, 553 (1968).
3. Reeder, R. H., *in* "Ribosomes" (M. Nomura, P. Lengyel, and A. Tissieres, eds.), p. 489. Cold Spring Harbor Lab., Cold Spring Harbor, New York, 1974.
4. Dumont, J. N., *J. Morphol.* **136**, 153 (1972).
5. Nieuwkoop, P. D., and Farber, J., "Normal Table of *Xenopus laevis* (Daudin)," 2nd ed. North-Holland Publ., Amsterdam, 1967.
6. Rickwood, D., and Birnie, G. D., *FEBS Lett.* **50**, 102 (1975).
7. Birnie, G. D., Rickwood, D., and Hell, A., *Biochim. Biophys. Acta* **331**, 283 (1973).
8. Rickwood, D., Hell, A., Malcolm. S., Birnie, G. D. MacGillivray, A. J., and Paul, J., *Biochim. Biophys. Acta* **353**, 353 (1974).

Chapter 5

Visualization of Chromatin v-Bodies

ADA L. OLINS

*The University of Tennessee–Oak Ridge Graduate School
of Biomedical Sciences and The Biology Division,
Oak Ridge National Laboratory, Oak Ridge, Tennessee*

I. Introduction

In 1973 Woodcock (*1*) and Olins and Olins (*2*) reported that chromatin particles, *v*-bodies, could be visualized in electron micrographs. The new innovations which permitted this observation were: (1) Samples were applied to an electron microscope grid by the method developed by Miller and Beatty (*3,4*). This method used a detergent (Kodak Photo-Flo) to minimize surface tension during the drying procedure, and prevent tangling of the chromatin fibers (*2*). A negative stain was used to resolve individual particles which are very close to each other (*3*). Micrographs were made at higher magnifications so that the viewers' attention could be focused on 80-Å structures. As it turned out none of these are essential steps in visualizing *v*-bodies. Monomers produced by nuclease digestion of nuclei can be visualized without fixation, without use of Photo-Flo, by using a positive stain or shadowing and at a fairly low magnification. However, the visualization of particles in untangled chromatin fibers streaming out of an identifiable nucleus was necessary for positive identification of *v*-bodies and their correlation with the nuclease digest product.

Why didn't previous methods show the existence of the particle? Thin-section preparations suffered from (1) dehydration by organic solvents, (2) a difficulty in following fibers and knowing the direction of section with respect to individual fibers, and (3) the absence of a technique such as negative staining. Chromatin fiber spreads, however, came close to visualizing the real chromatin structure, but drying procedures involving organic solvents probably destroyed the structure of *v*-bodies. Furthermore, the use of concentrated stains with poorly extended fibers (i.e., where connecting strands are still tightly associated with the *v*-body) make the visualization of the particles impossible.

II.   Methods

A. v-Bodies in Isolated Nuclei (5, 6)

Nuclei are suspended in 0.2 M KCl to give a concentration of 10^8 nuclei/ml. A hemocytometer is used to determine the concentration. At this stage the nuclei should not be clumped or aggregated. The nuclei are then diluted 200-fold by adding 0.001 M ethylenediaminetetraacetic acid (EDTA) (pH 7.0), and the tube is gently inverted several times to allow mixing. Rough treatment at this stage could cause aggregation and should be avoided. The nuclei are allowed to swell for 10 minutes on ice, followed by the addition of fresh 10% formaldehyde, pH 7.0 (NaOH), to a final concentration of 0.9% formaldehyde. The nuclei can be kept in this state for several hours.

The microcentrifugation chamber (7) (inner diameter 4 mm and height 6 mm) is filled with 10% formaldehyde (pH 7.0). A freshly glowed carbon-coated grid is carefully inserted into the bottom of the centrifugation chamber so that no air bubbles are trapped under the grid. Two or three drops of the formaldehyde are removed from the top of the centrifugation chamber with a Pasteur pipette and replaced with one, two, or three drops of the swollen, fixed nuclei. There should be a convex meniscus standing slightly above the top level of the centrifugation chamber. This excess sample is squeezed out with a coverslip and blotted. This procedure tends to form a seal between the coverslip and the centrifugation chamber. The nuclei are then centrifuged at 800 g for 4 minutes in a swinging-bucket centrifuge. At the end of the centrifugation the coverslip is removed, 10% formaldehyde is added drop by drop to give a convex meniscus, the centrifugation chamber is turned upside down, and the grid is removed when it settles on the meniscus. Immediately thereafter the grid is washed in approximately 50 ml of water containing three drops of Kodak Photo-Flo and one drop of pH 10 borate buffer (this solution is made fresh daily) and drained by touching the edge of the grid to the edge of bibulous paper. When the grid is dry it is stained.

B. v-Bodies in Chromatin, Small Chromosomes (Viral), and Nuclease Digestion Products

The sample is diluted to an $A_{260} = 0.5$–3.0 with 0.001 M EDTA (pH 7.0). Aim for a concentration which covers the carbon film well but gives few overlapping molecules. The concentration needed varies in an unpredictable way. Apply a drop of the sample to a freshly glowed carbon film for 30 seconds, wash in dilute Photo-Flo (as with nuclei), drain with bibulous paper, and dry. Stain. Fixation is not necessary for smaller samples and may cause cross-links which make visualization of chromatin particles more difficult.

C. Stains

Many stains can be used successfully to visualize *v*-bodies. Ethanol and methanol solvents will not affect the general morphology of the *v*-bodies at this stage. Positive stains give excellent contrast but do not resolve *v*-bodies which are in contact with each other. High-concentration negative stains (1% or more) have too much contrast to visualize the small amounts of stain between close particles. Low-concentration negative stains produce less contrast but tend to be more delicate and show more detail. Our standard and most reliable staining procedure at present is to put a small drop of 0.01 *M* uranyl acetate in water, on the sample side of the grid for 30 seconds and touch the edge of the grid to the edge of bibulous paper to drain as thoroughly as possible.

Low-angle rotary shadowing has been used successfully in several laboratories. It is a high-contrast method and makes the electron microscopy much more rapid. However, the details of the internal *v*-body structure are lost, and subtle differences in morphology are unattainable.

D. Supporting Films

When low-contrast, dilute negative stains are used, it is important to use the thinnest carbon film support possible. Both 45% glycerin drained from a warm glass slide and mica have been used successfully as a smooth surface on which the carbon is evaporated. A shield 1 × 2 cm placed about 3 cm below the evaporating carbon produces a more uniform film and makes it somewhat easier to control the thickness of the film. I have kept carbon film for several months without any trouble. In order to render the carbon film hydrophilic, it is glowed at a pressure of 50 μm of mercury for 1 minute a short time before it is used.

E. Electron Microscopy

High-resolution microscopy demands the utmost from the microscope and the microscopist. It is therefore advisable to maintain the microscope at its best. A good vacuum, a cold finger, correct microscope alignment, astigmatism correction, and short exposure to the electron beam each add a small increment toward improving the quality of the data. The condensor$_1$ setting should give the smallest beam spot possible which does not give

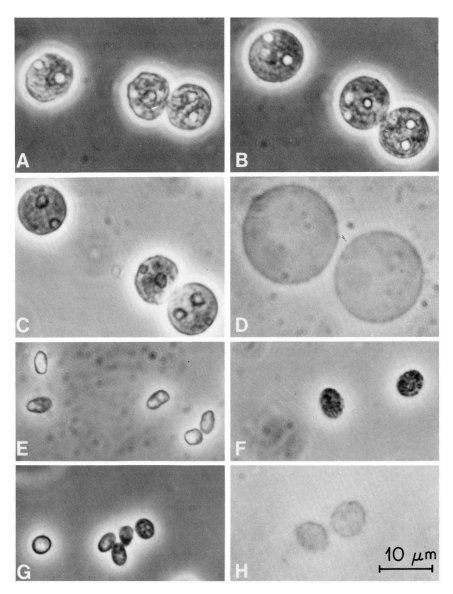

FIG. 1. (A–D) Isolated rat liver nuclei; (E–H) isolated chicken erythrocyte nuclei (8). (A) and (E), Isolated nuclei in CKM buffer; (B) and (F), nuclei from (A) and (E) swollen in water; (C) and (G), nuclei from (A) and (E) suspended in 0.2 M KCl; (D) and (H), nuclei from (C) and (G) swollen in water.

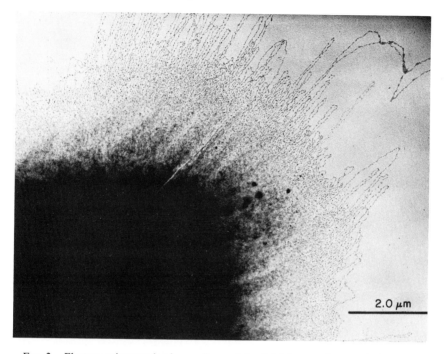

Fig. 2. Electron micrograph of a swollen and fixed chicken erythrocyte nucleus centrifuged onto a glowed carbon-coated grid, dried in Photo-Flo, and dried in 0.5% ammonium molybdate (pH 7.4). The chromatin fibers spilling out of the nucleus are well separated at the edge of the nucleus. From Olins *et al.* (*6*), with permission.

rapid contamination of the sample. Micrographs should be taken at × 40,000 magnification or higher so that astigmatism and proper focus (at or slightly under exact focus) can be checked before each micrograph is made. A pointed filament is recommended.

III. Demonstration

Isolated nuclei are extremely sensitive to variations in salt concentration. Figure 1 is a phase-light micrograph which demonstrates this point. Although the nuclei did not swell very much in 0.2 M KCl, they allow more complete swelling when placed in water than do nuclei which go directly from magnesium into water. In order to assure complete swelling we have recently used 1 mM EDTA (pH 7.0) instead of water for the final swelling; thus nuclei in CKM buffer [0.005 M MgCl$_2$–0.025 M KCl–0.05 M cacodylate (pH 7.5)]

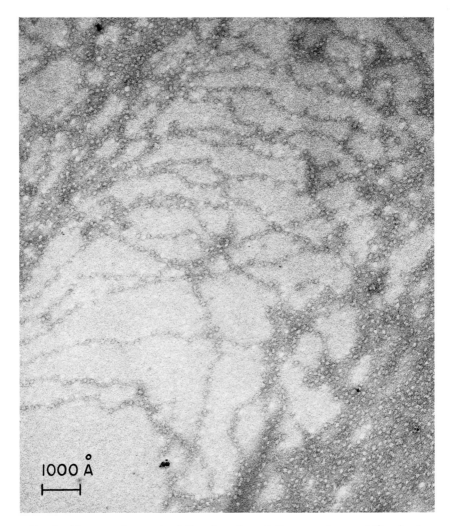

FIG. 3. Electron micrograph of the edge of a rat thymus nucleus centrifuged onto a carbon-coated grid. A few chromatin fibers are stretched out from a region of higher chromatin concentration. Negative stain with 0.5% ammonium molybdate (pH 7.4).

are spun, resuspended in 0.2 M KCl, and diluted 1:200 with 0.001 M EDTA (pH 7.0).

At low magnification it is very difficult to be certain that v-bodies are present on chromatin fibers; however, this is a useful magnification for scanning the grid and identifying the source of the chromatin fibers (Fig. 2).

Classically, a negative stain piles up around the specimen, so contrast is

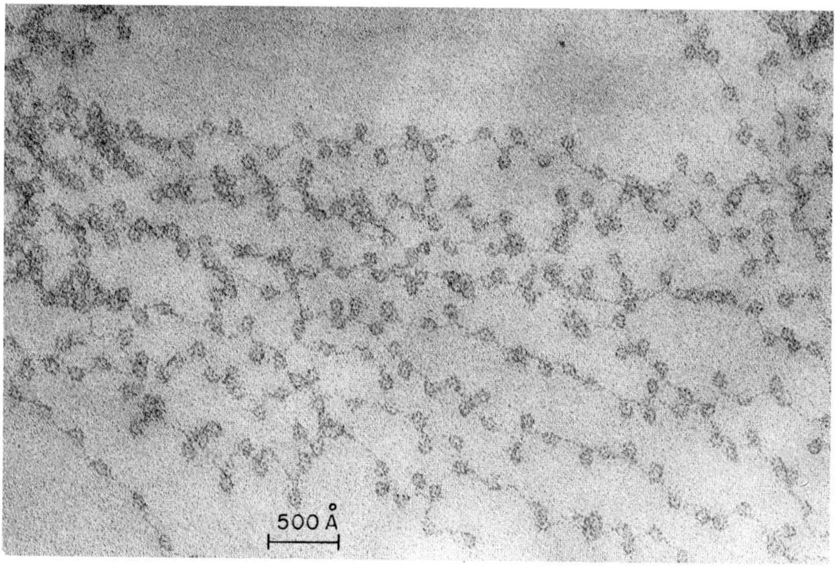

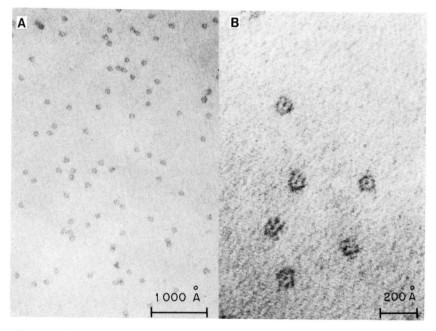

FIG. 4. High-resolution electron micrograph of a spread chicken erythrocyte nucleus. Note the ν-bodies with clear internal structure. Many chromatin fibers exhibit a zigzag configuration with the ν-bodies lying on alternate sides of the connecting strand. The specimen preparation is similar to Fig. 2, except that the sample was stained with aqueous 0.2% uranyl acetate. From Olins *et al.* (*6*), with permission.

FIG. 5. Electron micrographs of monomer ν-bodies obtained after micrococcal nuclease digestion and fractionation by sucrose gradient ultracentrifugation. From Olins *et al.* (*10*), with permission.

achieved by an outline effect (Fig. 3). A similar effect is seen by using 1–2% uranyl acetate or 1% PTA (pH 7) and drying face down on bibulous paper. If too much stain is left on the carbon film less and less structural information remains visible.

The staining method we prefer is highly reproducible and seems to give more detail than other methods (see Section IIC). Although the stain is not washed off the grid, so little stain remains on the grid that the outline formed and the direct staining of the connecting strand might not immediately identify this as a negative staining technique. It is known (9) that uranyl can bind strongly to the phosphates in DNA, and this is an alternative explanation for the staining observed in Figs. 4 and 5. We can identify more detail in the shape of ν-bodies using this method than any other method we have tried.

Isolated monomer ν-bodies, the product of micrococcal nuclease digestion, are shown in Fig. 5. Note that the internal structure of ν-bodies in chromatin fibers is still clearly visible.

ACKNOWLEDGMENT

I wish to thank O. J. Miller, Jr. for teaching me how to use the electron microscope, Howard G. Davies for helping me become a more critical microscopist, Donald E. Olins for many important discussions, and Mayphoon Hsie for excellent technical assistance.

REFERENCES

1. Woodcock, C. L. F., *J. Cell Biol.* **59**, 368a (1973).
2. Olins, A. L., and Olins, D. E., *J. Cell Biol.* **59**, 252a (1973).
3. Miller, O. L., Jr., and Beatty, B. R., *J. Cell. Physiol.* **74**, (Suppl. 1), 225 (1969).
4. Miller, O. J., Jr., and Beatty, B. R., *Science* **164**, 955 (1969).
5. Olins, A. L., and Olins, D. E., *Science* **183**, 330 (1974).
6. Olins, A. L., Breillatt, J. P., Carlson, R. D., Senior, M. B., Wright, E. B., and Olins, D. E., *in* "The Molecular Biology of the Mammalian Genetic Apparatus," Part A (Paul O. P. Tso', ed.). Elsevier/North Holland Biomed. Press, Amsterdam, 1976.
7. Miller, O. J., Jr., and Bakken, A. H., *Acta Endocrinol. Suppl.* **168**, 155 (1972).
8. Olins, D. E., and Olins, A. L., *J. Cell Biol.* **53**, 715 (1972).
9. Zobel, C. R., and Beer, M., *J. Biophys. Biochem. Cytol.* **10**, 335 (1961).
10. Olins, A. L., Senior, M. B., and Olins, D. E., *J. Cell Biol.* **68**, 787 (1976).

Chapter 6

Isolation and Characterization of Chromatin Subunits

RANDOLPH L. RILL

Department of Chemistry and Institute of Molecular Biophysics,
Florida State University,
Tallahassee, Florida

BARBARA RAMSAY SHAW

Department of Chemistry,
Duke University,
Durham, North Carolina

KENSAL E. VAN HOLDE

Department of Biochemistry and Biophysics,
Oregon State University,
Corvallis, Oregon

I.   Introduction

The past few years have witnessed a dramatic improvement in our understanding of the fine structure of chromatin. Once commonly thought to be a uniform supercoil, with histones and other nuclear proteins bound to the surface of the coiled DNA, chromatin fibrils are now known to have a particulate structure consisting of a repeated array of rather simple structural units. Specific aggregates of histone proteins appear to serve as nuclei about which the DNA is periodically folded. Thus it is possible to speak of "subunits" of chromatin, the unitary elements of this periodic structure. This chapter describes methods used in our laboratories to isolate and study such subunits.

As is often the case when new concepts develop rapidly, some confusion and debate about the nature of these subunits and their internal structure have occurred. We believe that these questions are now largely resolved, and that their resolution has provided some interesting insights into details of the structure. In this regard a brief review of recent history will be useful.

The first clear demonstration of the existence of a repetitive structure came from the work of Hewish and Burgoyne (*1*), who showed that limited digestion of rat liver chromatin within intact nuclei by endogenous nuclease produced a series of DNA fragments that were multiples of a basic size. Evidence of a different kind was developed by Rill and Van Holde (*2*) and Sahasrabuddhe and Van Holde (*3*) who isolated small nuclease-resistant nucleoprotein particles from chromatin digested with staphylococcal nuclease and found that they contained a short length of compactly folded DNA. At very nearly the same time there appeared electron microscope studies indicating the existence of a "beaded string" structure in chromatin (*4–6*). Together these different lines of research strongly indicated that the fine structure of chromatin involved repeated nucleoprotein particles or "subunits" [also termed "*v*-bodies" or "nucleosomes" by Olins and Olins (*5*) and Oudet *et al.* (*7*), respectively]. Additional studies by D'Anna and Isenberg (*8–11*), Kornberg and Thomas (*12*), and Roark *et al.* (*13*) showed that certain pairs of histones form specific dimeric (H2A, H2B) and tetrameric (H3, H4) complexes and provided further basis for the first detailed subunit models for chromatin structure proposed independently by Kornberg (*14*) and Van Holde *et al.* (*15*). Both of these models suggested that the subunit consists of a small unit length of DNA that is associated with an octameric histone complex containing two copies each of the above four histones previously found to participate in strong pairwise interactions.

There appeared, however, to be a disturbing discrepancy in reports related to the size of the unit DNA contained in each subunit. Data of Burgoyne *et al.* (*16*) and Noll (*17*) indicated that small DNA fragments obtained from nuclei slightly digested with nuclease were multiples of 200 base pairs in length, leading Kornberg (*14*) to propose that the octameric histone complex is associated with 200 base pairs of DNA to form subunits or repeating units. On the other hand, the discrete 11 S "monomer" nucleoprotein particles isolated by Sahasrabuddhe and Van Holde (*3*) and Oosterhof *et al.* (*18*) from chromatin digested with nuclease appeared to contain DNA fragments averaging only about 110–120 base pairs in length. These particles lacked histone H1. Thus Van Holde *et al.* (*15*) proposed that a smaller DNA segment (now believed to be 140–150 base pairs long) wraps about a central "core" octameric histone complex, forming a compact subunit that alternates with more accessible regions of DNA.

More recent research has helped to understand this apparent discrepancy. For example, Axel (*19*), Sollner-Webb and Felsenfeld (*20*), Van Holde *et al.* (*21*), and Shaw *et al.* (*22*) have shown that the initial scissions of chromatin in avian erythrocyte and rat liver nuclei by staphylococcal nuclease produce a broad distribution of monomer DNA fragments with a median length of about 200 base pairs and "oligomer" DNA fragments that are multiples of this unit length. However, the character of the monomer DNA changes rapidly after further digestion, yielding two to three rather discrete DNA fragments that are 160–170, 140–150, and 120–130 base pairs long; simultaneously the sizes of the oligomer fragments decrease by about 40–60 base pairs. By judicious choice of digestion time and separation method, described below, one can isolate nucleoprotein particles containing predominantly either 160–180 or 130–150 base pair lengths of DNA. Both preparations contain approximately equimolar amounts of histones H2A, H2B, H3, and H4, but the latter particles are depleted in histone H1 (and H5 in avian erythrocytes) whereas the former particles contain about the same ratio of lysine-rich histone to other histones as whole chromatin (23,24,24a,55; M. Noll, private communication).

Thus there appear to be two regions within the repeated nucleoprotein structure containing ~200 base pairs of DNA—a section of about 40–60 base pairs which is very accessible to nuclease and is probably mainly associated with lysine-rich histone that affords incomplete protection from nuclease, and a "core" portion of 140–150 base pairs which is associated strongly with, and folded by, the core octameric histone complex (21,22,25).

After longer digestion, nucleases make cuts within the core DNA. If staphylococcal nuclease is used, characteristic sets of discrete DNA fragments and nucleoprotein fragments (termed "submonomer fragments") are produced; these have been studied in detail by Felsenfeld and co-workers (20,26), Weintraub (27), and Rill et al. (28). It now seems probable that the particles first prepared by Sahasrabuddhe and Van Holde (3) and Oosterhof et al. (18) contained some such internal DNA cuts, thereby causing an underestimation of the size of the core DNA.

Because of this complexity of nuclease digestion kinetics and nucleoprotein products, discussed in more detail below, there has been some ambiguity in the terminology in this field. *In particular the term subunit has had several interpretations.* For the purposes of this contribution we will use the following definitions.

Repeating unit: The entire repetitive unit in chromatin, containing an average of about 200 base pairs of DNA, in many species, and all of the histones. Recent evidence suggests that the repeat size may be of variable length, being somewhat larger in chicken erythrocytes (29,61,62) and smaller in Chinese hamster ovary cells (29,61) and considerably smaller in lower eukaryotes such as yeast (30,62,63), neurospora (31), and aspergillus (64).

Core particle: The nuclease-resistant portion of the repeating unit, containing a core complex of two each of histones H2A, H2B, H3, and H4 and a 140–150 base pair length of DNA. The size and composition of the core particle appears to be remarkably uniform in a wide variety of eukaryotes, including such lower species as yeast (*62, 65*).

Spacer: The remaining DNA in the repeating unit, ranging from 20 to more than 60 base pairs, associated partly or entirely with lysine-rich histone (H1 and/or H5). The spacer regions are defined operationally in terms of their ready accessibility to nuclease (*22*), low melting temperature (*32*), and extensibility in low ionic strength media, as observed by electron microscopy (i.e., the string of the beads on a string structure).

It is our contention that the core particles are identical to the beadlike *v*-bodies or nucleosomes observed by electron microscopy and represent the primary structural subunits of chromatin. Most reported physical studies of chromatin subunits have, in fact, used more or less refined preparations of what we call core particles. The particles are easily prepared in quantity (because of their nuclease resistance) and can be resolved to a fairly high degree of homogeneity. On the other hand, studies of the entire repeating unit containing about 200 base pairs of DNA have been inhibited by the fact that these structures are observed as intact entities only in the initial stages of digestion, and even then exhibit a rather broad heterogeneity in DNA size. This heterogeneity presumably results from the possibility that the first cleavage within spacer regions can occur at random over a range of sites and may be compounded by heterogeneity of spacer lengths.

The techniques described herein have been largely employed for the isolation of core particles. If digestion conditions are carefully controlled they can also be used for the preparation of small quantities of repeating units (*19, 33*). In either case, use of these methods with a cell type not previously investigated will call for exploratory studies and careful characterization of the products. For this reason we include details of such characterization methods in Section IV.

II.   Nuclease Digestion of Nuclei or Chromatin

A.   General Comments

Approximately 50% of the DNA in nuclei or isolated chromatin is accessible to various endonucleases (*34*). For the most part this accessible DNA does not occur in long stretches, but instead is distributed throughout chromatin in spacer regions, or in specific small cleavage sites within core particles. Although the rate of nuclease attack, as measured in terms of the production of cold perchloric acid–soluble oligonucleotides, is dependent upon solution variables that affect enzyme activity such as pH and divalent ion concentration, the ultimate availability of these cleavage sites is rather independent of the initial degree of chromatin condensation. On the other hand, the relative rates of cleavage between and within the particles can vary greatly (*19,20,29,35*). In general it must be emphasized that the size and concentration distribution of submonomer, monomer, and oligomer fragments obtained in a given experiment depend on the state of the chromatin, the extent of digestion, and the enzyme chosen. More specific details are given below.

As in all studies of chromatin, the presence and potential influence of endogenous protease(s) must be recognized. Degradation of Hl, the most susceptible histone, does not affect the basic folded structure of the core particles or significantly alter the distinctive pattern of DNA fragments obtained after digestion with staphylococcal nuclease. On the other hand, very limited cleavage at a few specific sites on the other four histones causes a dramatic unfolding of the particles (*3,27*). Fortunately the protease activity endogenous to many commonly used tissues (e.g., rat liver and spleen, calf thymus) is minimized at or below pH 6.5 and can be inhibited by 1 mM phenylmethanesulfonyl fluoride (PMSF) or diisopropylfluorophosphate (*36,37*). Furthermore, *clean* preparations of nuclei from certain cell types (e.g., chicken erythrocytes or cultured mouse myeloma cells) appear to be nearly devoid of protease activity.

Commercial preparations of spleen DNase II and, to a lesser extent, pancreatic DNase I have been found to contain significant protease activities that can be inhibited as indicated above. Preparations of the extracellularly excreted staphylococcal (micrococcal) nuclease are generally free protease.

B.   Digestion with Staphylococcal (Micrococcal) Nuclease

Staphylococcal nuclease degrades both DNA and RNA by initial endonucleolytic, followed by exonucleolytic, attack, yielding 3′-phosphomononucleosides. The digestion process is not product-inhibited as is the case with many other nucleases. The enzyme has an absolute requirement for Ca^{2+} (at least $10^{-5}M$) and a pH optimum near pH 9, depending somewhat on the calcium concentration. Single-stranded DNA and RNA are degraded faster than native DNA; AT-rich regions of protein-free, native DNA are degraded most rapidly due to this single-stranded preference as well as a natural preference for Xp-dAp and Xp-dTp bonds. These and other properties of staphylococcal nuclease have been reviewed by Anfinsen *et al.* (*38*).

The time course of nuclease action on chromatin is most readily monitored by electrophoresis of DNA isolated from the digests on appropriate polyacrylamide gel systems (see Section IV). As is illustrated in Figs. 1 and 2, the initial DNA products of staphylococcal nuclease digestion of chicken erythrocyte and rat liver nuclei are monomer and oligomer fragments that are about 200 base pairs and $N \times (200)$ base pairs long, respectively. (Note

that these values refer to the median size of relatively broad bands.) Rather
rapidly the sizes of all of these fragments decrease by 40–60 base pairs, and
the monomer band narrows and splits into two or three distinguishable components. (These are readily observed on 6% polyacrylamide gels, but may
not be clearly resolved on lower percent gels.) The later stages of the digestion of nuclei are additionally characterized by the appearance of a series
of eight or nine discrete submonomer-size DNA fragments ranging from
approximately 50 to 130 base pairs in length, differing from each other by
about 10 base pairs (*19–22,28*). Gel patterns of this DNA electrophoresed
as single strands are very similar to those shown for double-stranded DNA,
indicating that the isolated fragments have little or no single-strand charac-

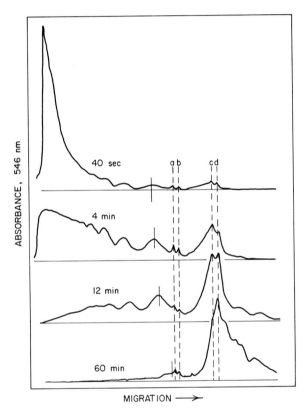

FIG. 1.   DNA electrophoresis patterns–time course of the digestion of DNA is chicken
erythrocyte nuclei by staphylococcal nuclease. Nuclei were digested at 37°C in 0.3 *M* sucrose–
0.7 m*M* CaCl$_2$–10 m*M* tris-HCl (pH 7.2) as described in Section II. DNA was electrophoresed
on 30-cm gels of 3.5% polyacrylamide, stained with toluidine blue, and scanned at 546 nm.
The small "spikes" identified by letters are Hae III fragments of PM2 DNA included as internal
markers. Sizes, in base pairs, are: a, 291; b, 265; c, 156; and d, 141. The vertical bars indicate
the progressive decrease in size of dimer DNA during digestion.

ter and virtually no single-strand nicks (22). We believe that these discrete submonomer DNA fragments result from cleavages at specific sites within the core particles that are restricted by, and reflect, domains of histone binding since we have been able to isolate several small nucleoprotein fragments that contain predominantly only one or two of these DNA species and specific subsets of histones (28,38a).

As mentioned previously, the relative accessibility of spacer and internal cleavage sites, and hence the relative rates of production of monomer and submonomer DNA fragments, appear to be governed by the degree of chro-

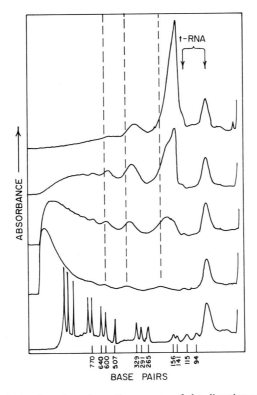

FIG. 2.   DNA electrophoresis patterns–time course of the digestion of DNA in rat liver nuclei by staphylococcal nuclease. Nuclei were digested in 1 mM Tris-HCl (pH 8.0)–0.1 mM CaCl$_2$–0.25 M sucrose at 37° for 64 minutes, 16 minutes, 4 minutes, and 15 seconds (top to bottom). DNA was isolated and electrophoresed on 16-cm gels of 6% polyacrylamide, stained with Stains-all, and scanned at 610 nm. Transfer RNA was included as a "front" marker. A minor contaminant in this preparation of tRNA migrates at about the 110 base pair position. Gels were calibrated using the pattern of simultaneously electrophoresed Hae III fragments of PM2 DNA plus tRNA shown in the lower curve. The calibration curve is expressed in terms of the mobilities of the marker fragments relative to tRNA to compensate for differences between gels.

matin condensation. Initial cleavage of chromatin *in nuclei* occurs predominantly in spacer regions, producing DNA fragments of single and multiple repeat length. The subsequent rapid reduction in size of these fragments probably occurs primarily due to the clipping of remaining "tails" of spacer regions. Significant amounts of submonomer DNA fragments do not appear in digests of nuclei maintained in near physiological ionic strength media until most of the DNA is reduced to monomer and small oligomer size, suggesting that internal cleavage sites are not accessible until nucleoprotein fragments are released from the bulk chromatin aggregate. However, digestion of chromatin gels prepared by standard methods [e.g., washing of nuclei and lysis in saline–ethylenediaminetetrocetic acid (EDTA) solutions followed by resuspension in dilute buffers containing low levels of calcium] results in the nearly simultaneous production of monomer (core) size and submonomer size DNA, indicating that cleavages occur between and within core particles at nearly equal rates in chromatin in this extended state (*19, 20, 28*). Distinctive DNA species larger than dimer or trimer are not usually observed in chromatin digests, even after very short digestion times, as would be predicted if competition between these two classes of cleavage sites is almost equal. Mechanical shearing or sonication of chromatin is not recommended as it is unnecessary for digestion. Noll *et al.* (*35*) have suggested that these procedures lead to disruption of the repeating unit.

In general, digestion of nuclei is preferable for the isolation of monomeric or oligomeric subunits due to the lower rate of intrasubunit cleavage. Digestion of chromatin is more useful if one is interested in examining the submonomer nucleoprotein fragments. The following procedures for digesting nuclei and chromatin have been used routinely in our laboratories with very consistent results.

1. ISOLATION AND DIGESTION OF NUCLEI

The method of Blobel and Potter (*39*), modified slightly as described below, has been used by Rill and co-workers to isolate nuclei from calf thymus, rat and chicken liver, chicken erythrocytes, and yeast protoplasts, and should be generally useful for isolation of nuclei from many tissues.

Tissues should be quick-frozen with Dry Ice or liquid nitrogen immediately after excision or used fresh within 30 minutes. Fresh or frozen tissues are finely minced with scissors in 2 volumes of ice-cold 0.25 M sucrose in TKMC buffer [0.05 M Tris-HCl–25 mM KCl–5 mM MgCl$_2$,–1 mM CaCl$_2$ (pH 7.5) at 20°] that is made 1 mM in phenylmethanesulfonyl fluoride (PMSF) immediately before use. (PMSF is an effective inhibitor of serine proteases and is safer to use than diisopropylfluorophosphate. Solid PMSF dissolves slowly in water but readily in isopropanol. Usually we add an aliquot of 0.1 M PMSF in isopropanol. Note that both PMSF and PMS–protein bonds are slowly hydrolyzed in water; hence protease inhibition may be lost after prolonged storage of preparations of nuclei or subunits unless fresh PMSF is added. Protease activity can be reduced further if necessary by lowering the pH of the medium to pH 6.5 with cacodylic acid.) All subsequent steps are performed in a cold room (2°–4°C) or in an ice bath. Homogenization of cells is accomplished in a Potter–Elvehjem homogenizer with a Teflon pestle driven by hand (erythrocytes) or, for tissues, by a 1/4-in. drill at 1700 rpm (controlled by a potentiometer). Ten to fifteen slow strokes are usually sufficient, with brief pauses between strokes while the homogenizer is immersed in an ice-water bath. Tissue homogenates are then filtered through eight layers of prewashed cheesecloth and two layers of rinsed Micracloth.

Separation of nuclei from cytoplasmic contaminants is accomplished by sedimentation, in one step, first through 1.62 M sucrose containing nonionic detergent and then through 2.3 M sucrose containing no detergent. Two volumes of 2.3 M sucrose–0.1 mM PMSF–0.75% Triton X-100 in TKMC buffer are added to 1 volume of homogenate in a polyallomer tube and mixed by inversion. (Volumes are measured accurately with a syringe.) This mixture is underlaid with 1 (initial homogenate) volume of 2.3 M sucrose–0.1 mM PMSF in TKMC buffer (no detergent) using a syringe with a 13-gauge needle postured on the bottom of the tube. [The final density of the upper layer, approximately 1.62 M sucrose, is chosen to barely float mitochondria and rough endoplasmic reticulum (39). Accumulation of a thick pellet at the denser sucrose interface is thereby largely avoided. The nonionic detergent removes the outer nuclear membrane and associated cytoplasmic material. Lower concentrations or other milder detergents may be used if the nuclei are particularly fragile.] Samples are centrifuged at 41,000 rpm in the International SB-283 (or Spinco SW-41) rotor for 30 minutes or at 25,000 rpm in the International SB-110 (Spinco SW-27) rotor for 2 hours. After the supernatant is carefully withdrawn with a syringe, the sides of the tubes are wiped clean with lint-free tissue paper wrapped on a spatula, and the nuclear pellet is taken up in 0.25 M sucrose–15 mM Tris–60 mM KCl–15 mM NaCl–10 mM MgCl$_2$–1 mM CaCl$_2$, 0.1 mM PMSF, adjusted to pH 6.5 with solid cacodylic acid. (The digestion with staphylococcal nuclease proceeds considerably faster at higher pHs, but protease activity is also generally higher and may become a problem.) After washing 2 times with digestion buffer, by centrifuging at 1000 g and resuspending with gentle hand homogenization using a glass homogenizer with loose-fitting pestle, the nuclei are resuspended in a small amount of buffer and diluted to give the final desired concentration. The concentration of the nuclear suspension is conveniently determined by homogenizing a small aliquot in 9 volumes of 2.2 M NaCl–5.5 M urea and measuring the absorbance at 260 nm. Suspensions containing as much as 200 $A_{260\text{ nm}}$ units/ml have been digested and resolved on sucrose gradients without difficulty.

Digestions have most commonly been done at 37°. Samples are equilibrated for 10 to 15 minutes and then inoculated with 2–10 enzyme units/A_{260nm} unit. (One A_{260nm} unit $\cong 50$ μg of DNA. Worthington preparations of staphylococcal nuclease have been used exclusively in our laboratories and by most other workers. The specific activities of these preparations range from about 15,000 to 20,000 units/mg; hence the above enzyme concentration is about 0.1 to 1% of the weight of DNA.) Digestion is terminated by addition of one-tenth volume of 0.15 M ethylenediamminetetraacetic acid (EDTA) adjusted to pH 7.5. Nuclei are lysed and the fragments are dispersed conveniently by brief dialysis (2 to 3 hours with vigorous stirring) against 10 mM Tris–1 mM EDTA, 0.1 mM PMSF (adjusted to pH 7.5 with HCl). The condition of the dialyzed nuclei can be checked with a phase-contrast microscope or with a conventional light microscope after staining. The pyronin B-methyl green stain described by Stern (40) is very useful for this purpose as well as for assessing the condition of nuclei in the original preparation. Usually lysis is spontaneous, but a few hand strokes with a homogenizer may be necessary. The resulting solutions should be nearly clear, unless the sample was digested near to the limit, and can be applied directly to a column or sucrose gradient after brief centrifugation, as described below. (The pH increase, from pH 6.5 to 7.5, increases the solubility of the subunits. Note that the *limit* digests, as one approaches 40–50% acid-soluble DNA, are insoluble!)

Nuclei dispersed in sucrose–TKMC buffer plus 25% glycerol have been stored at $-20°$ for up to 1 month without detectable adverse effects on the digestion products.

The following alternative procedure has been used by Shaw, Van Holde, and co-workers to isolate nuclei from chicken erythrocytes and does not require the use of a high-speed preparative centrifuge. Whole chicken blood is obtained from adult chickens by cardiac puncture using a large-bore syringe needle connected to Tygon tubing. Blood is drained into a flask containing roughly an equal volume of standard saline citrate [SSC: 0.15 M NaCl–0.015 M Na Citrate (pH 7.2)]. (Citrate and other chelating agents such as EDTA prevent blood clotting. The use of heparin is unnecessary.) Blood cells are collected by centrifugation at 3000 g for 10 minutes, and the plasma and upper, whitish, "buffy coat" layer of the pellet are carefully removed with a syringe. The remaining erythrocytes are further washed 2 or more times by resuspending, using a large-bore pipette, and centrifuging as above. (Thorough removal of the buffy coat containing white blood cells is important because of the high levels of digestive enzymes found in these cells.) Washed cells are frozen quickly in an equal volume of SSC and stored at $-70°C$ until needed. Cells may be stored in this manner for several months.

The following procedure for isolating nuclei starts with approximately 6 ml of *packed* frozen cells. The frozen erythrocytes in an equal volume of SSC are warmed to 37°C and lysed by repeated pipetting. The suspension is then diluted to 40 ml with cold SSC, cooled on ice, and centrifuged at 3000 g for 10 minutes to pellet nuclei. The nuclei are subsequently washed once in SSC, once in SSC containing 0.2% Nonidet P40 detergent, and twice more with SSC to remove detergent and residual hemoglobin. Washing is accomplished by resuspending the nuclear pellet, by repeated pipetting, in 40 ml of solution and centrifuging at 3000 g for 10 minutes. The sample is held on ice and stirred occasionally for 10 minutes in the presence of Nonidet P40 before centrifuging. The concentration and condition of the nuclei are determined before the final centrifugation step using a cell counter and light microscope. For counting nuclei, 50 μl of the nuclear suspension is diluted 400-fold into dilute crystal violet or trypan blue solution in SSC.

To digest with staphylococcal nuclease the nuclear pellet from the last wash step is resuspended in sufficient 0.3 M sucrose–0.75 mM CaCl$_2$–10 mM Tris-HCl (pH 7.2) to yield a final concentration of 2 $\times$ 10^8 nuclei/ml (about 0.7 to 0.8 mg DNA/ml or about 15 $A_{260\,nm}$ units/ml). About 200 ml will be required if one starts with 6 ml of packed cells. The resulting suspension is warmed to 37° and digested at this temperature with a (Worthington) staphylococcal nuclease concentration of 125 units/ml. Digestion is terminated by addition of EDTA to a final concentration of at least 2 mM. Digestion for 10 to 15 minutes reduces about 30% of the total DNA to acid-soluble oligonucleotides and produces high yields of core particles.

Nuclei digested in this manner have most commonly been treated as follows prior to fractionation of the nucleoprotein fragments by chromatography or centrifugation. After addition of the EDTA the sample is chilled on ice, then centrifuged at 12,000 g for 15 minutes to pellet nuclei. Usually the supernate is discarded since it contains only 2 to 3 $A_{260\,nm}$ units/ml, of which 20% or less are 11 S core particles. If desired this material can be dialyzed, concentrated, and fractionated. The pellet of digested, intact nuclei is resuspended in 30 ml of ice-cold 10 mM Tris–0.7 mM EDTA (pH 7.5) and disrupted by homogenization for 30 seconds to 1 minute at medium speed in a VirTis homogenizer. Insoluble nuclear debris is removed by centrifugation at 12,000 g for 15 minutes, and the clear supernatant is fractionated as described below. The supernatant should contain at least 15 to 20 $A_{260\,nm}$ units/ml if nuclei have lysed. In lieu of shearing, the digested nuclei can be dialyzed and lysed as described previously.

2. ISOLATION AND DIGESTION OF CHROMATIN GELS

We have most commonly isolated chromatin essentially by the procedure described by Axel et al. (26). Digestion of chromatin isolated by several other common methods has yielded similar results. Nuclei isolated by either of the above methods are washed by gentle hand homogenization in 0.25 M NaCl–0.025 M EDTA (pH 6.0) and centrifuged at 1000 g for 15 minutes. Nuclei are lysed and the chromatin washed by repeated gentle resuspension and centrifugation at 12,000 g for 15 minutes in decreasing concentrations of Tris–cacodyate buffer (50 mM, 30 mM, 10 mM, 5 mM, and 1 mM Tris, adjusted to pH 6.5 with solid cacodylic acid). After one or two more washes in 1 mM Tris–cacodylate containing 0.1 mM CaCl$_2$, digestion of the resulting chromatin *gel* (shearing is not necessary) is done as described above for nuclei.

Before attempting to fractionate large amounts of nuclei or chromatin digested by any of the above methods it is advisable to carry out a time sequence of small-scale trial digests. A number of methods have been used to monitor the course of the digestion process. Determination of the rate of production of cold perchloric acid–soluble DNA can be done as described in the Appendix. *Usually* yields of 11 S core particles are optimal after about 30% of the total DNA is rendered acid-soluble. Longer digestion causes more internal cleavages and precipitation of the remaining nuclease-resistant material. Considerably more information about the products can be gained if the nucleoprotein products are centrifuged on high-resolution sucrose gradients (see Section III) or if DNA is isolated and electrophoresed on polyacrylamide gels (see Section IV and the Appendix). If only the latter procedure is used it is convenient to terminate the digestion of small aliquots that are withdrawn from the digestion mixture by pipetting directly into an equal volume of 20 mM EDTA–0.3 M NaCl and 2% SDS or 4% sarcosyl. These aliquots can be treated directly with pronase or proteinase K, and phenol can be extracted to isolate DNA as described in the Appendix.

C.   Digestion with Other Nucleases (DNase I and DNase II)

Both pancreatic DNase I and spleen (acid) DNase II have been used in our laboratories to produce chromatin fragments. In general neither of these enzymes digests chromatin DNA to give as discrete and uniformly double-stranded DNA fragments as those obtained with staphylococcal nuclease. Because of its higher specificity and (usual) freedom from contaminating proteases, staphylococcal nuclease is recommended as the enzyme of choice for isolating core particles and other units. However, studies of the action of other nucleases on chromatin have been, and should continue to be, useful for providing additional insights into chromatin structure.

Digestion of nuclei by DNase II under several conditions has not, in our hands, yielded discrete monomer and oligomer DNA species such as those obtained with staphylococcal nuclease, even though the total digest can be separated into clearly distinguishable monomer and small oligomer nucleo-protein units on sucrose gradients. Gel electrophoresis of DNA isolated from DNase II digests of *chromatin* gels has yielded more distinct patterns of fragments, ranging from 40 to over 300 base pairs, than those obtained from nuclear digests, but the bands still lack the resolution of those obtained with staphylococcal nuclease. Digestion of nuclei or chromatin with DNase I, described in detail by Noll (*41*), also yields rather indistinct bands when the DNA is electrophoresed as double strands but a very distinct series of bands ranging, in 10 base intervals, from 10 bases up to over 160 bases when electrophoresed as single strands. This feature of DNase I digestion can be useful for assessing the integrity of isolated core particles since they also yield this series of DNA bands after digestion (see below). Digestion with either of these enzymes can be done under the same conditions used with staphylococcal nuclease.

Insofar as we are aware, no complete explanation of the curious differences in the cleavage products obtained with these three enzymes has been presented. Presumably the explanation lies in subtle differences in specificity, size, and cleavage mechanism. The unique ability of staphylococcal nuclease to unwind and exonucleolytically degrade DNA after initial endocleavage may be important in this regard. In addition, Von Hippel and Felsenfeld (*42*) have suggested that staphylococcal nuclease only attacks regions in which "breathing" motions can temporarily unwind DNA. If this is so, a rationale for the preference of staphylococcal nuclease for spacer regions is evident. Useful reviews of the properties of these and other nucleases are found in Volume 4 of "The Enzymes" (*43*).

III. Isolation of Core Particle Subunits and Other Nucleoprotein Fragments

From the foregoing discussion it should be evident that a definition of the subunit of chromatin and the isolation of a homogeneous population of subunits, or other species, containing identical lengths of DNA and the same protein composition, is not entirely straightforward. In general we prefer to reserve the terms "structural units" or simply subunits to refer to the core particles since these portions of chromatin have uniformly the same size, composition, and uniquely folded conformation in many species, whereas the spacer regions and total repeat lengths appear to be more variable in all of these respects.

Nuclease digests of nuclei or chromatin are readily resolved into several nucleoprotein species by chromatography on agarose columns or by sedimentation on sucrose gradients. Caution is required, however, since even very distinctly eluting or sedimenting bands may contain several DNA species due to incomplete cleavage of spacers or to cleavage within the core. (See Figs. 3 and 4.) Preliminary digestions, analyzed by electrophoresis of the isolated DNA on polyacrylamide gels, are recommended to determine the digestion time required to obtain maximal yields of the desired nucleoprotein species and minimize contamination by smaller fragments. Histone analysis of the digest by gel electrophoresis is also recommended to monitor possible proteolysis. Methods for establishing the purity and integrity of the core particle preparations are described in Section IV.

A. Isolation by Gel Filtration

Agarose gel bead columns have been used by Shaw, Van Holde, and co-workers (*22,44,45*) to isolate core particles from digests of calf thymus and chicken erythrocyte nuclei. Digests of nuclei or chromatin treated as described in Section II are made 7% in sucrose. Typically about 15 to 20 ml ($\sim$20 A_{260} units/ml) are carefully applied to a 90 × 2.5 cm column of Bio-Gel A-5m (100–200 mesh, Bio-Rad Laboratories, Richmond, California) that has been previously equilibrated with 10 mM Tris–0.7 mM EDTA (pH 7.5 at 5 °C). The sample is eluted with the same buffer at a flow rate of about 30 ml/hour. A typical elution profile obtained by this procedure is shown in Fig. 3. Selected fractions are analyzed with respect to their sedimentation coefficient, size of DNA fragments, and histone content as described in Section IV. Approximately 3 mg of core particles can be obtained from the middle third fractions of the monomer peak by the above procedure. Resolution of monomer, dimer, and trimer nucleoprotein units can be improved with longer columns, smaller sample volumes, and fine (200–400) mesh agarose. Additional purification can also be carried out by sedimentation on sucrose gradients.

B. Isolation by Sedimentation on Preformed Sucrose Gradients

Linear 5–20% sucrose gradients have been used extensively by Rill and co-workers (*18, 28*) and Axel (*19*) to resolve monomeric core particles from higher oligomers and to resolve the various submonomer nucleoprotein fragments. Nonlinear isokinetic gradients have been used by Noll (*17*) for the former purpose with similar results. Preparation of significant amounts of nearly homogeneous units of any type may require two or more centrifuge runs. Large-scale preliminary fractionation on 5–20% sucrose gradients can be accomplished by sedimentation in 25.4 × 88.9 mm (40-ml) tubes in the International SB-110 rotor (or nearly equivalent Spinco SW-27 rotor) at 25,000 rpm (4°–6 °C) for 24 hours. As shown in Fig. 4, good results have been obtained with as much as 220 $A_{260\,nm}$ units of total digest loaded on a single gradient (2 ml of a solution with $A_{260\,nm} = 110$; i.e., about 5.5 mg/ml DNA equivalents). Unlike whole chromatin, chromatin subunits appear to behave reasonably ideally, and bands remain relatively undistorted even at

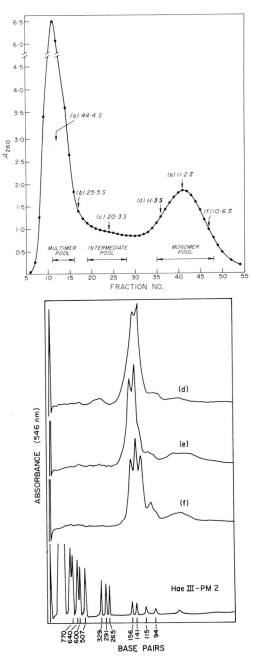

FIG. 3. (Top) Fractionation of chromatin fragments by chromatography on a 90 × 4.5 cm Bio–Gel A-5m column (100–200 mesh). Chicken erythrocyte nuclei were digested with staphylococcal nuclease for 10 minutes at 37°C in 0.3 M sucrose–0.7 mM CaCl$_2$–10 mM Tris-HCl (pH 7.2) as described in Section II. Flow rate: 30 ml/hour. (Bottom) Electrophoretic patterns of DNA isolated from column fractions (d) through (f). Lower curve gives the positions of marker Hae III fragments of PM2 DNA. Electrophoresis was done on 30-cm gels of 3.5% polyacrylamide. For the histone compositions of these fractions see Shaw *et al.* (*22*).

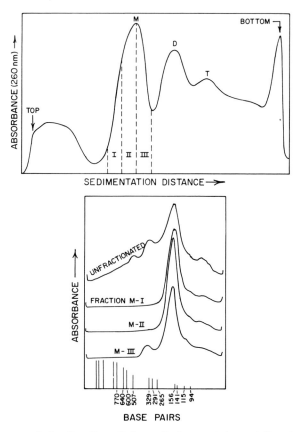

FIG. 4. Large-scale fractionation of chromatin fragments by centrifugation on sucrose gradients. (Top) Sedimentation pattern of fragments from chicken erythrocyte nuclei. Nuclei suspended in 0.25 *M* sucrose–60 m*M* KCl–15 m*M* NaCl–10 m*M* MgCl$_2$–1 m*M* CaCl$_2$–15 m*M* Tris cacodylate (pH 6.5) to a final concentration of 110 A_{260nm} units/ml were digested for 30 minutes, at 37°C, with 550 units/ml of staphylococcal nuclease. After dialysis and centrifugation to remove nuclear debris (see text) 2-ml aliquots were centrifuged on linear 5–20% (w/v) sucrose gradients at 25,000 rpm for 24 hours (6°C) in an International SB-110 rotor. Gradients were monitored using an Isco gradient fractionator and UA-5 absorbance monitor. (Bottom) Electrophoretic patterns of DNA isolated from the whole nuclear digest and "monomer" fractions M-I, II, III. Electrophoresis was done on 11-cm gels of 6% polyacrylamide. The lower bars mark the positions of marker Hae III fragments of PM2 DNA run simultaneously with the samples. Essentially equivalent resolution has been obtained with twice (~ 200 A_{260} units/ml) this amount of material on gradients, but the gradient profiles could not be monitored directly with the current flow cell.

such high loading concentrations. Because of the moderately high diffusion rate of the monomer and submonomer fragments, gradients should be handled with care and speed and use of a turbulence-free flow cell is recom-

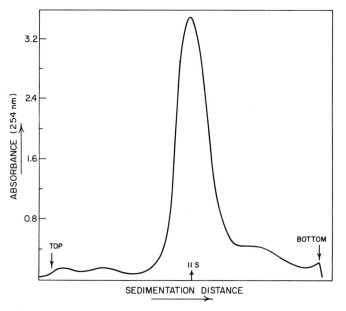

FIG. 5. High-speed resedimentation of monomer core particles on sucrose gradients. Fraction M-II from the low-speed run shown in Fig. 4 was concentrated to give $A_{260\,\text{nm}}^{1\,\text{cm}} = 40$. A sample volume of 0.5 ml was layered into an 11-ml linear gradient of 5–20% sucrose and centrifuged at 40,000 rpm for 16 hours at 16°C in an International SB-283 rotor. Gradients were monitored as described in Fig. 4. The sedimentation coefficient ($S_{20,w}^{0}$) of the peak center is approximately 11 S.

mended for monitoring. Samples can be concentrated by various conventional means. Perfusion through an agitated dialysis bag into dry Sephadex after dialysis to remove sucrose has proven useful for handling many samples simultaneously. We have also found that all of the nucleoprotein fragments can be quantitatively precipitated by addition of $MgCl_2$ to a final concentration of 3 mM followed by addition of 2 volumes of ethanol (95%). Precipitates redissolve in 10 mM Tris–0.1 mM EDTA (pH 7.5) and yield normal sedimentation patterns on sucrose gradients. Histones are not extracted by this procedure. More stringent tests of particle integrity after precipitation have not been performed.

Further resolution is best accomplished by sedimentation at 41,000 rpm in 14.5 × 96 mm tubes (14-ml) in the International SB-283 (or Spinco SW-41) rotor. Sample volumes of up to 0.5 ml and concentrations of at least $40 A_{260\,\text{nm}}$ units/ml can be used without significant loss of resolution. Choice of sedimentation time depends upon the samples desired. Optimal resolution of the 11 S monomeric core particles and slower submonomer units is obtained after 15 to 18 hours when the 11 S band is sedimented into the bottom half

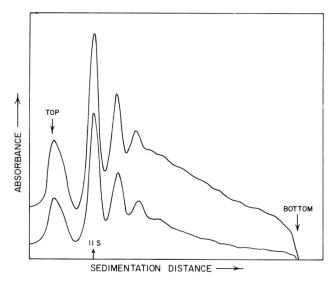

FIG. 6. High-resolution sedimentation pattern illustrating the repeating nucleoprotein structure of chromatin. Chicken erythrocyte nuclei were digested with staphlococcal nuclease as described in Fig. 4 for 16 minutes (top curve) and 64 minutes (bottom curve). Samples (0.2 ml) containing 4 A_{260} units were centrifuged on linear 5–20% sucrose gradients at 39,000 rpm for 5.5 hours (6°C) in an International SB-283 rotor.

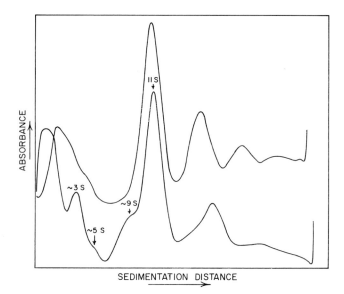

FIG. 7. Sucrose gradient sedimentation profiles. (Top) Chicken erythrocyte nuclei digested with staphylococcal nuclease for 64 minutes as described in Fig. 4 (same sample as in Fig. 6). The sample (0.5 ml, 10 A_{260nm} units) was centrifuged at 36,000 rpm for 13 hours (6° C) in an International SB-283 rotor. (Bottom) Chicken erythrocyte chromatin gel digested for 60 minutes with staphylococcal nuclease as described in Section II. The sample (0.2 ml, 4 A_{260nm} units) was centrifuged in the above rotor at 40,000 rpm for 10 hours (6°C). Note the presence of (approximately) 3 S, 5 S, and 9 S "submonomer" nucleoprotein fragments.

of the tube. Sedimentation for 10 to 12 hours is most suitable for isolation of dimer through tetramer units. Representative sedimentation profiles shown in Figs. 5–7 illustrate results obtained for digests of nuclei and chromatin centrifuged for various lengths of time.

Recently the Olins and their co-workers have developed methods using zonal rotors to prepare large quantities of nucleosomes (46a).

IV. Characterization of Core Particle Subunits

This section describes methods used routinely to characterize core particle subunits, the complexes of 140–150 base pairs of DNA and eight histone molecules, and methods employed in our laboratories as controls for preparations. Additional measurements of the physical properties and composition of core particles are summarized in Table I. Although at this stage no systematic study has been reported, there appears to be a general consensus in several laboratories that isolated subunits are relatively stable to handling if proteolytic and nucleolytic activities are absent, as judged by electron microscopy and the usual physical criteria. In our hands preparations of chicken erythrocyte core particles in 0.7 mM EDTA–10 mM Tris -HCl (pH 7.5) have shown no significant changes in properties during 1 week at room temperature.

A. Routine Tests of Core Particle Integrity

Certain experiments should be carried out with each preparation, particularly when a new chromatin source is being utilized. These include the following:

1. SEDIMENTATION VELOCITY ANALYSIS

Sedimentation velocity studies can be carried out either in the analytical ultracentrifuge or by the sucrose gradient technique. The former method is generally more exact and is preferred, especially if an ultracentrifuge with ultraviolet (UV) scanner optics is available. Low concentrations ($A_{260nm} \leq$ 1.0) should be used in such studies since the particles must be maintained at low ionic strength (they tend to aggregate at ionic strengths between 0.1 and 0.2 if any H1 or H5 are present) and perturbations from the Donnan effect are expected at high concentrations in low ionic strength buffers. We have estimated the Donnan effect under conditions used for sedimentation equilibrium on the basis of the DNA size, estimated histone charge, and a reasonable estimate of counterion binding. The effect is not likely to exceed 3%. A typical UV scan from a core particle preparation, centrifuged under

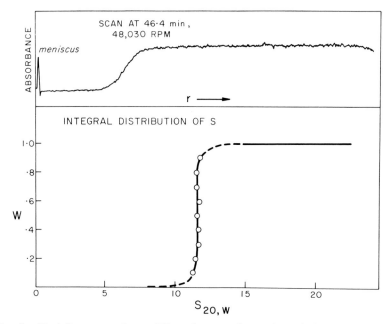

FIG. 8. (Top) Scanner tracing at 264 nm from a sedimentation velocity experiment with core particles isolated from the peak of the "monomer" band obtained by chromatography on Bio-Gel A-5m (Fig. 3, fraction e). Scan at 46.4 minutes, 48,030 rpm. The value of $S_{20,W}$ was found to be 11.0 S. (Bottom) Integral distribution of S calculated from the above scanner trace, corrected for diffusion.

conditions described in the legend, is shown in Fig. 8. Note that the boundary is symmetric and relatively sharp, and there is no evidence for either faster or slower sedimenting material. Analysis of such boundaries has shown that almost all of the spreading can be accounted for by diffusion (*45b*). The $S_{20,w}$ value of core particles from many sources is invariably about 11 S. Very nearly the same value has been found for the slightly larger particles that contain H1. For higher oligomers the $S_{20,w}$ value reportedly depends markedly upon whether or not H1 is present; higher values are found in the presence of H1 (R. Kornberg, private communication).

In summary, the presence of a single, homogeneous boundary with $S_{20,W} \cong 11$ S is a first criterion for a good preparation of core particles.

2. ANALYSIS OF DNA BY GEL ELECTROPHORESIS

The DNA in purified core particles should be examined by gel electrophoresis in double-strand form as a necessary measure of the overall homogeneity of the preparation. It is also useful to electrophorese the DNA under denaturing conditions to test for single-strand nicks. Ideally the size of the

DNA will be the same by both tests; even a small fraction of single-strand nicks will lead to pronounced differences between the electrophoresis patterns.

To accurately measure the DNA size in such experiments, "marker" DNAs of known size must be available. By far the best for such purposes are DNA fragments prepared by restriction endonuclease cleavage of a small viral DNA. Ideally, the fragments should be sequenced so that the sizes are known exactly. Unfortunately such sequenced fragments are hard to obtain in quantity, but other sets of fragments may be used if their sizes are calibrated by comparison with sequenced fragments or other fragments of known length. Several sets of restriction fragments have been employed as calibration standards. These include:

1. Sequenced Endo IV fragments of ϕX174 single-stranded DNA (17,41).
2. Synthetic oligonucleotides of known sequence (46).
3. Hin fragments of Lambda (λ), SV-40, and ϕX174 RFI DNAs (20, 46).
4. Hae III fragments of PM2 DNA (21). Recent evidence suggests that the previously reported sizes of these fragments were between 5 to 15% low (33; H. Zachau, private communication; T. R. Kovacic and K. E. Van Holde, unpublished). The values indicated in figures in the present manuscript are the most recent estimates by Kovacic and Van Holde (66).

The calibration fragments may either be electrophoresed simultaneously with the sample in a separate tube or slot or included in the sample put on the gel. Experience shows that simultaneous electrophoresis in separate tube gels is not a very accurate method unless great care is taken. Slab gels are better for this purpose and, ideally, the calibration fragments should be electrophoresed in every other channel to compensate for nonuniformity of migration across the slab. If individually isolated restriction fragments are available, the best method is to include a few of these in the sample as it is applied. Figure 1 shows such a gel pattern where certain Hae III fragments of PM2 DNA were used to provide internal calibration standards. Recently we have used a compromise procedure that has yielded very reproducible results but does not require individual isolated fragments. A small amount of commercial yeast tRNA (Calbiochem, La Jolla, California, S-RNA, A grade) is included in each sample and in two or more calibration sets that are run simultaneously on separate gels. The tRNA runs near the front, and the migration distances of sample and calibration fragment bands are converted to mobilities relative to the tRNA migration distance to compensate for the differences between gels or slots [see Fig. 2 and Rill et al. (29)]. With the latter two techniques the accuracy should be limited only by the precision with which the sizes of the calibration fragments are known and the accuracy of measurements of gel scans.

A wide variety of gel systems and gel concentrations have been used for such studies. For investigations of the size of the repeating unit, employing whole digests, we have preferred 2.5% or 3.5% polyacrylamide gels. Submonomer-size DNA fragments (about 40 to 130 base pairs) are well-resolved in double-strand form on 6–10% polyacrylamide gels. For electrophoresis in the single-strand form we have used 6% gels containing formamide. Details of the preparation of these gels are described in the Appendix.

Scanning of stained gels is much preferred over photography, particularly if quantitation is desired. Recent studies have shown that direct scanning at 265 nm, toluidine blue staining, and scanning at 546 nm, or "Stains-all" (Eastman) staining and scanning at 610 nm all give linear response to concentration over a wide range. Stains-all staining is most sensitive, but adequate precautions must be taken to prevent photobleaching of the stain. Extremely small amounts of DNA can be detected by staining with ethidium bromide and observing the fluorescence using a "blacklight," but we have not found a convenient method to reliably quantitate the fluorescence.

Electrophoretic patterns of DNA from core particle preparations are shown in Figs. 3 and 4. The DNA in such preparations is reasonably homogeneous, averaging about 140–150 base pairs in length, with some "fine structure" as described above. In the cleanest preparations only a single, sharp peak is observed.

3. ANALYSIS OF PROTEINS BY GEL ELECTROPHORESIS

Proteins in core particles or other nucleoprotein units are most easily analyzed using the sodium dodecyl sulfate (SDS)–polyacrylamide gel electrophoresis procedure of Laemmli (*47*). The proteins need not be separated from the DNA (indeed it is difficult to do so with the small DNA fragments in subunits) if the samples are treated with SDS, mercaptoethanol, and EDTA to dissociate self aggregates or mixed aggregates of proteins and DNA. The following procedure is used in our laboratories to analyze proteins of core particles and other nucleoprotein units isolated by chromatography or density gradient centrifugation. Column fraction samples in 10 mM Tris–0.7 mM EDTA (pH 7.5 at 2 °C) containing on the order of 50 μg each of DNA and protein are frozen in tubes or small vials and lyophilized. Gradient fractions are dialyzed against the above buffer to remove sucrose before lyophilization. Alternatively the particles can be precipitated with MgCl$_2$ and ethanol as described above, washed with ethanol, and dried *in vacuo*. Samples are dissolved in sufficient 1% SDS–10% glycerol (plus 5 mM EDTA if the samples were Mg^{2+} precipitated) to give a final concentration of about 1 mg protein/ml and heated immediately at 90°–100 °C for 1 minute. After cooling, 1/20 volume of β-mercaptoethanol is added, and the samples are incubated for 3 hours at 37°C. Bromphenol blue (0.001%) is usually added as a tracking dye after heating. Sample volumes of up to 25 μl (containing 25 μg or less of total protein) are applied to a 2-mm-thick slab gel prepared as described in the Appendix. (Tube gels are also satisfactory.) Concentrations of buffer salts less than 100 mM in the applied sample do not significantly affect the quality of the electrophoresis, and salt concentrations as high as 200 mM still usually yield adequate separation of the histones. If necessary the samples can be made 1% in SDS and dialyzed against "final sample buffer" containing 0.06 M Tris-HCl (pH 6.8)–5 mM EDTA–1% SDS–10% glycerol–0.001% Bromphenol blue to remove large amounts of salt or to adjust the pH close to that of the stacking gel (pH 6.8). Electrophoresis is carried out at relatively low voltages or currents to avoid band cupping and optimize gel scanning resolution. Gels are stained with Coomassie brilliant blue for maximum sensitivity and scanned at 550 nm. As little as 1 μg of a single histone can be quantitated very accurately. Details of the electrophoresis and staining procedures are given in the Appendix.

Histone subfractions can be resolved by electrophoresis on 20–25 cm, 15% polyacrylamide gels containing 2.5 M urea–0.9 N acetic acid (pH 2.7) as described by Panyim and Chalkley (48). Unfortunately, determination of the best method for quantitatively extracting histones from DNA in a form suitable for electrophoresis in this system is not straightforward and has not been explored in detail. Standard acid extraction methods using H_2SO_4 or HCl normally do *not* quantitatively extract histones from either chromatin or nucleoprotein fragments; histones H2A and H2B are usually preferentially extracted. Possibly conditions can be found that would permit total degradation of the DNA by nuclease(s).

4. DNase I Digestion as a Test of Particle Integrity

Recently Noll (41) demonstrated that digestion of chromatin in nuclei by pancreatic DNase I resulted in an unusual and characteristic cleavage of the DNA. When such digests are electrophoresed under denaturing conditions, the single-strand fragments appear as a distinctive series of discrete bands that appear to be multiples of about 10 bases. We have found that isolated core particles are digested in the same way by DNase I (22). While the exact reason for this periodicity is still unclear, this behavior can serve as a test for the integrity of chromatin subunits. A typical scan from gel electrophoresis of the DNA from DNase I-treated core particles is shown in Fig. 9 (41). The bands extend from 10 bases to over 160 bases, in multiples of 10.

5. Circular Dichroism

The isolated core particles of chromatin from a variety of sources display a circular dichroism spectrum that is readily distinguishable from that of free DNA, or even whole chromatin (2,3,44). Such spectra are shown in Fig. 10.

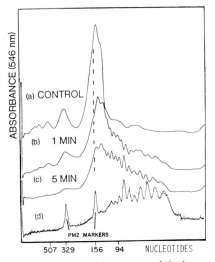

FIG. 9. DNase I digestion. Electrophoretic patterns of single-stranded DNA isolated after DNase I digestion of nucleoprotein fragments (predominantly core particles) from nuclei previously digested with staphylococcal nuclease. Samples were electrophoresed on 6% poly-acrylamide gels containing 98% formamide, stained with toluidine blue, and scanned at 546 nm.

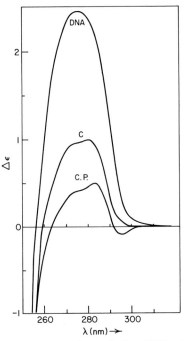

FIG. 10. Circular dichroism spectra of protein-free DNA, chromatin (C), and a core particle (C.P.) preparation.

The core particle spectrum is characterized by a number of unique features, including (1) a small negative band at about 295 nm, (2) a much lower ellipticity than free DNA at the 283 nm maximum, and (3) a "crossover point" shifted to the red from 258 nm (DNA) to 264 nm (core particle). This spectrum is very reproducible. It somewhat resembles the spectra reported for C form DNA (49,50). While this does not prove that the DNA conformation in the particles is indeed C form, the presence of this unique, reproducible spectrum is a useful analytical tool.

B. Physical Characteristics of Isolated Core Particles

The core particles (containing 140–150 base pairs of DNA and two copies each of histones H2A, H2B, H3, and H4) have been the subject of rather detailed characterization. Unfortunately, little data are available about the repeating unit that apparently contains an additional 20–60 base pairs of DNA (spacer DNA) and the lysine-rich histones as well. In Table I we tabu-

TABLE I

PHYSICAL PROPERTIES OF CHICKEN ERYTHROCYTE CORE PARTICLES

Property	Value	How measured	Reference
Sedimentation coefficient, $S_{20,w}$	11.4 S	Analytical ultra-centrifuge (scanner optics)	(60)
Partial specific volume, $\bar{v}$	0.68 cm³/gm	Kratky–Porod density meter	(60)
Extinction coefficient, $E[1 \text{ cm}/(\text{mg}/\text{ml})]$	9.3 mg⁻¹ ml⁻¹	Absorption at 258 nm, dry weight	(60)
Molecular weight	208,000 ± 10,000	Sedimentation equilibrium	(60)
	194,000	Sum of histone + DNA weight	
Composition	1.25 gm protein/gm DNA	Protein and DNA analysis	(60)
	1.24 gm protein/gm DNA	Protein and counter-ion–free DNA mol wt	
Effective diameter in solution	106 Å	Hydrodynamic studies	(21)
	107 Å	Low-angle neutron scattering	(45)

late data obtained for the core particles from chicken erythrocyte DNA. While not nearly so complete studies have been made on core particles from other sources, those that have been measured (such as sedimentation coefficients) agree very well with those reported here. Hence we assume, until more data are available, that the properties will be roughly the same for core particles from other sources.

Appendix

A.   Determination of Acid-Soluble Oligonucleotides

The time course of DNA digestion with nucleases is commonly assayed by measuring the amount of A_{260nm} absorbing material rendered acid-soluble at each time point. This method can be used on purified erythrocyte nuclei or other nuclei (or chromatin) in which RNA contributes only a small amount of the total absorbance as determined either by digestion with alkali or with pancreatic RNase (*19*).

The following procedure is used for monitoring the digestion of a nuclear suspension containing about 2×10^8 nuclei/ml (i.e., about 700–800 μg DNA/ml). At various times (including a zero time control), 200-μl to 500-μl aliquots are removed from the digestion mixture and added to an equal volume of 10 mM EDTA in 15-ml tubes sitting in an ice bath. After all samples have been taken, eight sample volumes of ice-cold 0.5 M perchloric acid–0.5 M NaCl are added to each tube. The tubes are capped and shaken vigorously at 5°C for 1 hour. Samples are then centrifuged at 12,000 g (0°–4°C) for 15 minutes and the supernate containing the small acid-soluble oligonucleotide products is carefully and completely separated from the pellet. The absorbancies at 260 nm of the supernates from digested samples are measured versus a blank consisting of the supernate from the zero time control of undigested nuclei treated in identical fashion. The perchloric acid supernates can also be heated simultaneously for 30 minutes at 70°C and their absorbancies measured in order to hydrolyze the oligonucleotides and account for any variations in hyperchromicity. Alternatively, the DNA in the hydrolyzed supernates can be determined with diphenylamine (*51*). As an additional check on the results, each pellet from the acid precipitation step can be analyzed for DNA content by resuspending the pellet in 2 to 10 ml of 2% SDS–50 mM Tris (pH 7.5) with heating at 70°C for 1 hour to solubilize the DNA. An equal volume of 20% $HClO_4$–2 M NaCl is then added, and the sample is heated at 70°C for 30 minutes. The sum of the absorbance in each heated supernate plus corresponding pellet should be the same for all samples if equal-volume aliquots were removed initially from the digestion mixture. If not, the percentage of acid-soluble material at a given time can be calculated from the quotient of the absorbance of the supernate to the total absorbance of supernate plus pellet from the aliquot taken at that time point. If the digestion mixture contains sucrose it is particularly important that all solutions be processed simultaneously because perchloric acid–sucrose mixtures react, especially at high temperatures, to give products that absorb at 260 nm.

B. Isolation of DNA for Gel Electrophoresis

DNA has been isolated from digested nuclei and chromatin and from individual nucleoprotein fragments by extraction with sodium dodecyl sulfate (SDS) or sarcosyl-chloroform, after treatment with RNase and protease. Samples are dispersed in 0.2 M NaCl–10 mM EDTA–2% SDS or sarcosyl–M Tris-HCl (pH 8.0) (final concentrations) by addition of small aliquots of appropriate concentrated stock solution(s). Pronase (Calbiochem, nuclease-free B grade) or proteinase K (E. Merck, Darmstadt, West Germany) is added to a final concentration of 3 μg/100 μg DNA, and the samples are incubated at 37°C for 4 hours. If RNA is present the sample is incubated at 37°C for 3 hours with pancreatic RNase (0.5 μg/100 μg DNA) before addition of protease and SDS. After protease digestion the sample is mixed vigorously for 10 minutes with an equal volume of phenol that has previously been saturated with the above buffer (minus SDS). One sample volume of chloroform is then added, and the mixture is mixed for an additional 5 minutes. Phase separation is facilitated by centrifuging for 5 to 10 minutes at maximum speed in a table-top centrifuge. The upper aqueous layer is carefully removed with a pipette, and the extraction with phenol–chloroform is repeated 1 or more times. DNA is precipitated by addition of 2 volumes of ice-cold ethanol, and the sample is allowed to stand overnight in a freezer at −20°C. DNA is collected by centrifugation at 10,000 g for 20 minutes (0°–5°C) and washed once or twice with ice-cold 70% ethanol to remove any excess salt. (EDTA, SDS, and other salts frequently interfere with DNA electrophoresis.) Samples are dried *in vacuo* prior to addition of electrophoresis sample buffer. Double-stranded DNA fragments at least as small as 40 base pairs are quantitatively recovered by this procedure.

C. Polyacrylamide Gel Electrophoresis of DNA

To accurately analyze both the sizes and amounts of DNA species in a sample, one must (1) choose the appropriate concentration of acrylamide for the gel, (2) employ a set of carefully sized or sequenced restriction fragments

of DNA as molecular-weight calibration standards, and (3) use an appropriate staining method if quantitation of the DNA is important. The method
of polyacrylamide gel electrophoresis used for *duplex* DNA analysis (described below) is essentially that of Loening (*52*), employing a Tris–acetate–
EDTA (TAE) buffer system. Some workers use a Tris–borate–EDTA (TBE)
buffer (*53*) or a Tris-borate-magnesium (TBM) buffer (*46*) and obtain similar
results. As a guide for choosing the proper gel concentration to determine
the sizes of *duplex* DNA digestion fragments, it should be noted that a plot of
the logarithm of molecular weight versus electrophoretic mobility is *linear*
from about 100 base pairs to 300 base pairs for 3.5% polyacrylamide TAE
gels and from about 130 base pairs to 550 base pairs for 2.5% polyacrylamide
TAE gels (*22,54*). Six percent polyacrylamide TAE gels are used for optimal
resolution of submonomer DNA fragments (40 to 140 base pairs). In comparison Maniatis *et al.* (*46*) use 12% polyacrylamide TBM gels for best
resolution of 20–100 base pairs duplex DNA, 5% polyacrylamide TBM gels
to resolve 20–200 base pairs, and 3.5% TBM gels to compare 150–400 base
pairs fragments. Polyacrylamide gels can be used for size determination in
the nonlinear range *if* appropriate-size DNA calibration standards that
bracket the region of interest are included. The 2.5% gels will resolve the oligomeric subunit pattern up to about 7 repeating units (approximately 1400
base pairs). For resolution of higher molecular weight DNA it is better to use
entirely agarose gels.

Analysis of single-stranded DNA can be carried out in polyacrylamide
gels that contain urea or 98% formamide. Although 7 *M* urea gels are preferred because they are easier to prepare and afford slightly better resolution, we use 98% formamide gels because Maniatis *et al.* (*46*) report that
complete denaturation of longer DNA molecules (>200 nt) cannot be con-
sistently maintained in polyacrylamide gels containing 7 *M* urea. Simpson
and Whitlock (*55*) have used a single polyacrylamide slab gel containing 6
M urea, Tris–borate–EDTA buffer, and 6% Cyanogum 41 to analyze simultaneously both double-stranded and denatured DNA samples from nuc-
lease-treated chromatin. The mobilities of *denatured* DNA and RNA
molecules are a linear function of the logarithm of molecular weight for
10–150 nucleotides on 12% polyacrylamide TBE gels containing 7 *M* urea.
This linear relationship is valid for DNA molecules of 10–100 nucleotides on
12% polyacrylamide–98% formamide gels. However, RNA and DNA molecules of the same molecular weight exhibit different relative mobilities on
formamide gels. They cannot be used interchangeably (*46*).

In conclusion, no one gel is optimal for resolving all sizes of DNA from
nuclease digests. If only one type of electrophoresis gel is to be run, we generally use a 3.5% polyacrylamide TEA gel for duplex DNA or a 6% polyacrylamide–98% formamide gel for denatured DNA molecules in order to follow

the progress of a digestion or look closely at the size of monomer DNA. Low percentage (<2.5%) polyacrylamide gels are best for resolving the repeating unit fragment pattern that results from nuclease treatment of chromatin; these gels, however, are nearly liquid and must be supplemented with 0.5% agarose for stability. If slab gels are used, agarose should be added to slabs of less than 3.5% acrylamide. Note that addition of agarose will alter somewhat the linear resolving region and separation properties of the gels. Gels run in Plexiglas tubes do not need to be supplemented with agarose for stability as long as the concentration of acrylamide is 2.5% or greater. However, careful handling of the 2.5–3.5% polyacrylamide tube gels must be employed (i.e., gently remove gels from tubes using a slight positive pressure into a wide 5-in. diameter staining dish; keep surfaces of the gel wet at all times, especially when manipulating them). To align gels for photography we have machined semicircular grooves about 1/4 in. diameter in a Plexiglas sheet. The gels fit nicely in the grooves. A grooved black Plexiglas sheet is used for ethidium bromide photography. Gels should be "floated" in the grooves to avoid stretching.

1. METHODS FOR ELECTROPHORESIS OF DUPLEX DNA

Polyacrylamide gel electrophoresis of double-stranded DNA has been carried out in 30 cm × 1/4 in. diameter Plexiglas tubes or on 14 cm × 14 cm × 0.2 cm slab gels containing "E" buffer (0.04 M Tris–0.02 M sodium acetate–2 mM NaEDTA adjusted to pH 7.2 with glacial acetic acid) (52). To prepare 10 3.5% tube gels, 17.5 ml of an acrylamide solution (20 gm acrylamide plus 1 gm bisacrylamide/100 ml H_2O), 33 ml of a 3 times concentrated E buffer containing 2.5 ml TEMED/liter, and 49.5 ml of water are mixed. (For slab gels the solution may have to be degassed for several minutes at this point.) The solution is polymerized by the addition of 0.5 ml of freshly prepared 10% ammonium persulfate and immediately poured into Plexiglas tubes having a double layer of parafilm wrapped around the bottom. The acrylamide gels are carefully overlayered with E buffer and allowed to polymerize at least 1 hour. Good results can be obtained with gels that are several days old as long as the gels are kept from drying out. Prior to electrophoresis the parafilm is replaced with nylon netting held on to the tube with a rubber band. The gels are placed in a homemade apparatus consisting of a top and bottom circular chamber each having an electrode suspended from the top center of its lid. Rubber electrical grommets make satisfactory seals for the top chamber from which the tubes are suspended. Polyacrylamide slab gels are prepared from the same solutions poured between two glass plates with 2-mm spacers all clamped onto a slab-gel apparatus (14 cm or 20 cm high) with a polyacrylamide dam preformed at the bottom to prevent leakage of the acrylamide. A 12–14 tooth Teflon comb is inserted into the acrylamide and the whole gel allowed to harden 30 minutes at room temperature.

Gels are pre-electrophoresed several hours at 80 V prior to sample application. DNA samples are dissolved at a concentration of 1 mg/ml DNA in 0.1 times E buffer containing 10% glycerol and 0.001% Bromphenol blue. A 25-μl sample (about 0.5 A_{260} units or ~25 μg DNA) is applied to each gel. For general calibration purposes about 15 μg of a whole Hae III–PM2 restriction digest are electrophoresed on one gel (see Fig. 2). For accurate DNA sizing, 0.2 to 0.4 μg each of several restriction fragments are coelectrophoresed with a designated sample (Fig. 1). Electrophoresis in long cylindrical gels is carried out at room temperature for 30 minutes at 70 V followed by about 6 hours at 150 V. A pumping system can be used to exchange the buffers in the two tanks to avoid pH differences. A pump is not necessary if reservoir volumes are on the order of a liter. The dye Bromphenol blue is used as an indicator and migrates with ~100 base pair fragments. In the 3.5% polyacrylamide TBE gel system, the two tracking dyes Bromphenol blue and xylene cyanol FF comigrate with 100 base pairs and 450 base pairs, respectively, of duplex DNA fragments (*46*). These tracking dyes migrate considerably faster in higher percentage gels (i.e., with 65 and 260 base pairs, respectively, in 5% gels).

2. METHODS FOR ELECTROPHORESIS OF SINGLE-STRANDED DNA

A modification of the formamide procedure of Staynov *et al.* (*56*) as discussed by Boedtker *et al.* (*57*) is described here. Reagent-grade formamide is deionized by stirring vigorously with 1/20 volume of wet Dowex AG-501-X8 (20–50 mesh, Bio-Rad) for 1 hour at room temperature, filtered, and made 0.016 M in $Na_2HPO_4 \cdot 7 H_2O$ and 0.004 M in $NaH_2PO_4 \cdot H_2O$ by dissolving the solid sodium phosphate salts in the formamide with stirring. This buffered formamide solution is used within 3–4 days and is stored in the dark. Six percent polyacrylamide–98% formamide gels are prepared from 100 ml of buffered formamide, 5.1 gm of acrylamide, and 0.9 gm of bisacrylamide; after dissolution, 1.2 ml of freshly prepared 10% ammonium persulfate in H_2O and 0.2 ml of TEMED are added. Gels are overlayered with 75% buffered formamide in water and allowed to sit overnight. Before applying the sample, gels are covered with a 1-cm layer of buffered formamide. DNA samples precipitated from ethanol are dissolved in a mixture of 90% buffered formamide–10% glycerol–0.001% Bromphenol blue at a concentration of about 2 mg/ml or greater. The samples are heated to 100° for 60 seconds to denature the DNA and immediately plunged into an ice bath prior to loading onto the gel. Aqueous 0.02 M phosphate buffer (pH 7.5) is added to the upper and lower reservoirs and is circulated between reservoirs during the run. The electrophoresis is carried out for 12 hours at 170 V in Plexiglas tubes 30 cm in length, or for about 10 hours at 80 V for 14 × 14 × 0.2 cm slab gels.

3. STAINING METHODS

For quantitative staining of DNA with toluidine blue the cylindrical gels are stained for 6 hours at room temperature with 0.05 mg/ml toluidine blue in H_2O and destained with several changes of distilled water until the background is almost clear. Gels are then removed immediately from the water and scanned. The 30-cm-long cylindrical gels that we use for accurate size and quantitative DNA analysis are stained individually in large plastic containers and diffusion destained in plastic containers (in which holes have been punched) placed in a large pan filled with distilled water and positioned on top of several magnetic stirrers.

Alternatively, gels can be stained with Stains-all (Eastman), 0.005% in 50% formamide, destained in water overnight, and photographed or scanned at 610 nm. Stains-all, which stains proteins, DNA, and RNA various shades of pink to purple, is light sensitive so appropriate precautions must be taken.

Gels prepared with electrophoretically pure acrylamide monomer can also be scanned at 260 nm for nucleotide content; however, sensitivity is reduced several-fold over that of stained gels.

Ethidium bromide staining is by far the most highly sensitive technique for detecting DNA (and RNA) in polyacrylamide gels; however, several groups have reported being unable to find conditions such that ethidium bromide staining increases linearly with DNA concentration (20,58; Shaw, unpublished). Several anomalies observed with ethidium bromide include (1) an increased uptake of dye by oligomeric-sized DNA over that of monomer and core particle-sized DNA and (2) a dependence on staining conditions; for instance, staining may or may not reveal DNA at the nodes of the DNA repeating pattern. If there is SDS in the sample applied to the gel, the SDS will migrate in the region of submonomer-size DNA and cause fluorescence of the ethidium bromide.

To get immediate feedback on the size of DNA in samples, we occasionally first stain gels with ethidium bromide, 0.5 μg/ml in running buffer, for 1 hour and photograph under "black light" UV illumination, prior to staining with toluidine blue. However, toluidine blue staining is necessary for quantitative analyses of DNA in the gels. For photography, ethidium bromide stained gels are illuminated evenly with UV lamps or blacklight lamps, one on either side of the gel. An orange glass, or Wratten filter with a 560 nm cutoff, plus a haze filter are used on a Polaroid Land camera (film type 105, 107, and 55). (A red filter with a 600 nm cutoff is better for formamide gels.) Film is exposed for 5 seconds to several minutes, depending on the ASA number of the film, the thickness of the filters, and type of lamp used. Gels stained with Stains-all, toluidine blue, or Coomassie blue can be photographed through an orange filter using Polaroid type 55 P/N, 107, or 105 film and white light illumination. If P/N film is used, the negatives can be scanned with a Joyce–Loebl densitometer. With proper exposure the measured density is linear with DNA concentration (*26*). For most accurate quantitation we prefer to scan the gels directly in a Gilford Model 2410 scanning spectrophotometer using the appropriate wavelength setting for the stain.

D. Electrophoresis of Histones on Polyacrylamide Gels Containing SDS

Histones and other small proteins are conveniently analyzed by the SDS gel method of Laemmli (*47*) using running gels of 15% polyacrylamide (acrylamide/bis ratio = 30:0.8, w/w) or 18% polyacrylamide [acrylamide/bis ratio = 18:0.09, w/w (*59*)]. The following procedure is used to prepare a 14 × 14 × 0.2 cm slab gel of 15% polyacrylamide. Twenty-five (25) ml of 30% acrylamide:bis (*N',N'*-bismethylene acrylamide) (30:0.8, w/w) in water are mixed with 12.5 ml of 1.5 *M* Tris-HCl (pH 8.8), 0.5 ml of 1% sodium dodecyl sulfate (SDS), 2.5 ml of 0.1 *M* EDTA, and 9 ml of water, and the resulting solution is thoroughly degassed. [Final running gel buffer: 0.375 *M* Tris-HCl (pH 8.8)–0.1% SDS–5 m*M* EDTA.] After adding 0.5 ml of freshly prepared 10% ammonium persulfate and 50 μl of TEMED the gel is poured and overlayered with running gel buffer. (Lower amounts of ammonium persulfate and TEMED are used for tube gels.) The stacking gel of 6% polyacrylamide is prepared about 1 hour prior to electrophoresis by mixing 2ml of 30% acrylamide–0.8% bis, 2.5 ml of 0.5 *M* Tris-HCl (pH 6.8), 0.1 ml of 10% SDS, 0.5 ml of 100 m*M* EDTA, 4.8 ml of H_2O, 10 μl of TEMED, and 100 μl of freshly prepared 10% ammonium persulfate (in that order), and slots are formed with a comb. (Final stacking gel buffer: 0.125 *M* Tris-HCl (pH 6.8)–0.1% SDS–5 m*M* EDTA.) The upper and lower chambers of the electrophoresis apparatus are filled with Tris–glycine buffer (6 gm Tris, 28.8 gm glycine, 1 gm SDS, diluted to 1 liter), and the samples, prepared as described previously, are layered on top of the slots using a microsyringe.

A voltage of 50 V is applied to the slab until samples enter the running gel, after which the voltage is raised to 90–100 V. Electrophoresis is continued for about 12 hours or until the Bromphenol blue tracking dye reaches the bottom of the slab. These low voltages allow good resolution of histones H3, H2A, and H2B.

Gels are stained overnight with 0.1% Coomassie brilliant blue in 25% isopropanol–10% acetic acid and diffusion destained in 10% acid solutions with successively decreasing concentrations of isopropanol and Coomassie blue (i.e., 10% isopropanol–10% acetic acid–0.001% Coomassie for 6 to 9 hours, 10% isopropanol–10% acetic acid–0.001% Coomassie for 6 to 9 hours, and finally 10% acetic acid).

Gels are photographed through a deep yellow filter using Polaroid Type 55 P/N film for the higher-contrast Type 107 film. Gel densitometry is done using a Gilford spectrophotometer and Model 2410 linear transport accessory, scanning through a 0.1 mM slit at 550 nm.

ACKNOWLEDGMENTS

R.L.Rill wishes to thank Daniel A. Nelson and Darlene K. Oosterhof for their helpful comments and for providing data for the figures on density gradients centrifugation.

Research related to this manuscript was supported, in part, by grant GM21126 from the USPHS (R.L.R.), by a contract between the Division of Biomedical and Environmental Research, U.S. Energy Research and Development Administration and the Florida State University (R.L.R.) and by grant B.MS73-06819 from the National Science Foundation (K.E.VH). R.L.Rill is the recipient of a National Institutes of Health Career Development Award (1975–1980).

REFERENCES

1. Hewish D. R., and Burgoyne, L. A., *Biochem. Biophys. Res. Commun.* **52**, 504 (1973).

2. Rill, R. L., and Van Holde, K. E., *J. Biol. Chem.* **248**, 1080 (1973).

3. Sahasrabuddhe, C. G., and Van Holde, K. E., *J. Biol. Chem.* **249**, 152 (1974).

4. Olins, A. L., and Olins, D. E. *J. Cell Biol.* **59**, 252a (1973).

5. Olins, D. E., and Olins, A. L., *Science* **183**, 330 (1974).

6. Woodcock, C. L. F., *J. Cell Biol.* **59**, 368a (1973).

7. Oudet, P., Gross-Bellard, M., and Chambon, P., *Cell* **4**, 281 (1975).

8. D'Anna, J. A., Jr., and Isenberg, I., *Biochemistry* **12**, 1035 (1973).

9. D'Anna, J. A., Jr., and Isenberg, I., *Biochemistry* **13**, 2098 (1973).

10. D'Anna, J. A., Jr., and Isenberg, I., *Biochemistry* **13**, 4992 (1974).

11. D'Anna, J. A., Jr., and Isenberg, I., *Biochem. Biophys. Res. Commun.* **61**, 343 (1974).

12. Kornberg, R. D., and Thomas, J. O., *Science* **184**, 865 (1974).

13. Roark, D. E., Geoghegan, T. E., and Keller, G. H., *Biochem. Biophys. Res. Commun.* **59**, 542 (1974).

14. Kornberg, R., *Science* **184**, 868 (1974).

15. Van Holde, K. E., Sahasrabuddhe, C. G., and Shaw, B. R., *Nucleic Acids Res.* **1**, 1597 (1974).

16. Burgoyne, L., Hewish, D., and Mobbs, J., *Biochem. J.* **143**, 67 (1974).

17. Noll, M., *Nature (London)* **251**, 249 (1974).

18. Oosterhof, D. K., Hozier, J. C., and Rill, R. L., *Proc. Natl. Acad. Sci. U.S.A.* **72**, 633 (1975).

19. Axel, R., *Biochemistry* **14**, 2921 (1975).

20. Sollner-Webb, B., and Felsenfeld, G., *Biochemistry* **14**, 2915 (1975).

21. Van Holde, K. E., Shaw, B. R., Lohr, D., Herman, T. M., and Kovacic, R. T., *Proc. FEBS Meet., 10th, 1975*, p. 57.

22. Shaw, B. R., Herman, T. M., Kovacic, R. T., Beaudreau, G. S., and Van Holde, K. E., *Proc. Natl. Acad. Sci. U.S.A.* **73**, 505 (1976).

23. Bakayev, V. V., Melnickov, A. A., Osicka, V. D., and Varshavsky, A. J., *Nucleic Acids Res.* **2**, 1401 (1975).

24. Varshavsky, A. J., Bakayev, V. V. and Georgiev, G. P., *Nucleic Acids Res.* **3**, 477 (1976).

24a. Olins, A. L., Carlson, R. D., Wright, E. B., and Olins, D. E., *Nucleic Acids Res.* **3**, 3271 (1976).

25. Van Holde, K. E., Sahasrabuddhe, C. G., Shaw, B. R., Van Bruggen, E. F. J., and Arnberg, A. C., *Biochem. Biophys. Res. Commun.* **60**, 1365 (1974).

26. Axel, R., Melchior, W., Sollner-Webb, B., and Felsenfeld, G., *Biochemistry* **13**, 3622 (1974).
27. Weintraub, H., *Proc. Natl. Acad. Sci. U.S.A.* **72**, 1212 (1975).
28. Rill, R. L., Oosterhof, D. K., Hozier, J. C., and Nelson, D. A., *Nucleic Acids Res.* **2**, 1525 (1975).
29. Rill, R. L., Nelson, D. A., Hozier, J. C., and Oosterhof, D. A., *Nucleic Acids Res.*, **4**, 771 (1977).
30. Lohr, D., and Van Holde, K. E., *Science* **188**, 165 (1975).
31. Noll, M., *Cell* **8**, 349–355 (1976).
32. Woodcock, C. L. F., and Frado, L.-L. Y., *Biochem. Biophys. Res. Commun.* **66**, 403 (1975).
33. Finch, J. T., Noll, M., and Kornberg, R. D., *Proc. Natl. Acad. Sci. U.S.A.* **72**, 3320 (1975).
34. Clark, R. J., and Felsenfeld, G., *Nature (London) New Biol.* **229**, 101 (1971).
35. Noll, M., Thomas, J. O., and Kornberg, R. D., *Science* **187**, 1203 (1975).
36. Ballel, N. R., Goldberg, D. A., and Busch, H., *Biochem. Biophys. Res. Commun.* **62**, 972 (1975).
37. Carter, D. B., and Chae, C.-B., *Biochemistry* **15**, 180 (1975).
38. Anfinsen, C. B., Cuatrecasas, P., and Taniuchi, H., *in* "The Enzymes" (P. D. Boyer, ed.), Vol. 4, p. 177. Academic Press, New York, 1971.
38a. Nelson, D. A., Oosterhof, D. K., and Rill, R. L., submitted.
39. Blobel, G., and Potter, V. R., *Science* **154**, 1662 (1966).
40. Stern, H., in "Methods in Enzymology," Vol. 12: Nucleic Acids (L. Grossman and K. Moldave, eds.) Part B, p. 107. Academic Press, New York, 1968.
41. Noll, M., *Nucleic Acids Res.* **1**, 1573 (1974).
42. Von Hippel, P. H., and Felsenfeld, G., *Biochemistry* **3**, 27 (1964).
43. Boyer, P. D., ed., "The Enzymes", Vol. 4. Academic Press, New York, 1971.
44. Shaw, B. R., Corden, J. L., Sahasrabuddhe, C. G., and Van Holde, K. E., *Biochem. Biophys. Res. Commun.* **61**, 1193 (1974).
45. Pardon, J. F., Worchester, D. L., Wooley, J. C., Tatchell, K., Van Holde, K. E., and Richards, B. M., *Nucleic Acids Res.* **2**, 2163 (1975).
46. Maniatis, T., Jeffrey, A., and Van de Sande, H., *Biochemistry* **14**, 3787 (1975).
46a. Olins, A. L., Breillant, J. P., Carlson, R. D., Senior, M. B., Wright, F. B. and Olins, D. E. in "The Molecular Biology of the Mammalian Genetic Apparatus" (P.O.P. Tso, ed.), Part A. Elsevier/North-Holland, Amsterdam, in press.
47. Laemmli, V. K., *Nature (London)* **227**, 1 (1970).
48. Panyim, S. and Chalkley, R., *Arch. Biochem. Biophys.* **130**, 337 (1969).
49. Tunis-Schneider, M. J. B., and Maestre, M. F., *J. Mol. Biol.* **52**, 521 (1970).
50. Johnson, R. S., Chan, A., and Hanlon, S., *Biochemistry* **11**, 4347 (1972).
51. Burton, K., *in* "Methods in Enzymology," Vol. 12: Nucleic Acids (L. Grossman and K. Moldave, eds.), Part B, p. 163. Academic Press, New York, 1968.
52. Loening, U. E., *Biochem. J.* **102**, 251 (1967).
53. Peacock, A., and Dingman, C., *Biochemistry* **6**, 1818 (1967).
54. Kovacic, T. R., Ph.D. Dissertation, Oregon State Univ., Corvallis, Oregon, 1976.
55. Simpson, R. T., and Whitlock, J. P., Jr., *Nucleic Acids Res.* **3**, 117 (1976).
56. Staynov, D. Z., Pinder, J. C., and Gratzer, W. B., *Nature (London) New Biol.* **235**, 108 (1972).
57. Boedtker, H., Cerkvenjakov, R. B., Dewey, D. F., and Lanks, K., *Biochemistry* **12**, 4356 (1973).
58. Lohr, D., Kovacic, R. T., and Van Holde, K. E., in preparation.
59. Bonner, W. M., and Pollard, H. B., *Biochem. Biophys. Res. Commun.* **64**, 282 (1975).

60. Tatchell, K., Van Holde, K. E., and Shaw, B. R., unpublished observations.
61. Compton, J. L., Bellard, M., and Chambon, P., *Proc. Natl. Acad. Sci. U.S.A.* **73**, 4382–4386 (1976).
62. Lohr, D., Corden, J., Tatchell, K., Kovacic, R. T., and Van Holde, K. E., *Proc. Natl. Acad. Sci. U.S.A.* **74**, 79–83 (1977).
63. Thomas, J. O., and Farber, V., *FEBS Lett.* **66**, 274–279 (1976).
64. Morris, N. R., *Cell* **8**, 357–365 (1976).
65. Nelson, D. A., Beltz, W. R., and Rill. R. L., *Proc. Natl. Acad. Sci. U.S.A.,* **74**, 1343 (1977).
66. Kovacic, R. T., and Van Holde, K. E., *Biochemistry* **16**, 1490–1498 (1977).

Part B.   Immunochemical Analysis of Chromosomal Properties

Chapter 7

Serological Analyses of Histones

B. DAVID STOLLAR

*Department of Biochemistry and Pharmacology,
Tufts University School of Medicine,
Boston, Massachusetts*

I.   Introduction

Serological analyses of histones have several applications in studies of the structure and functional organization of these proteins. Antibodies can provide unequivocal identification of individual histones in the absence of specific functional assays. Serological reactions can also detect slight alterations in structure, allowing comparisons of histones isolated from various sources and comparisons of the free soluble proteins with the forms in which they occur in chromatin. With immunofluorescence, antibodies can be used to study histones in chromosomes and whole nuclei, and with electron microscopy they can be used to study isolated chromatin and its subunits. While antibodies to histones should thus be very useful reagents, progress in their application requires an appreciation of the greater difficulties in the induction and assay of these antibodies than occur with most other protein antigen systems. This chapter will therefore try to evaluate some of these problems as well as to survey methods that have been successful. The technique of microcomplement fixation will then be described in detail.

II.   Sources of Sera and Procedures for Immunization

A.   Systemic Lupus Erythematosus

Antihistone antibodies that occur spontaneously in the human disease systemic lupus erythematosus (SLE) were among the first antihistone antibodies to be described (1). They were later found to show varying patterns of specificity for individual histone proteins (2). In addition, some of the sera reacted especially well with histone–DNA or histone–RNA complexes (2,3), again with varying specificity for the histone component. SLE sera are potentially valuable antihistone reagents, but they are not regularly available to research laboratories, and antibodies of a desired specificity may occur with a low frequency. They may also be accompanied by other antinuclear or unrelated antibodies, so that extra steps, such as absorption with DNA, are required for their characterization.

B.   Immunization with Free Soluble Histones

As initial reports of experimentally induced antihistone antibodies appeared, there were indications that mammalian histone preparations were only poorly immunogenic. Rümke and Sluyser (4) immunized rabbits with histones extracted from heat-treated thymus, using doses of several milligrams given with complete Freund's adjuvant. Lysine-rich histone seemed more immunogenic than others, but while whole perchloric acid (PCA)-extracted H1 was effective, chromatographically purified H1 was not (5). Mandal et al. (6) reported successful immunization of rabbits with coconut histones at 10-mg doses while calf thymus histones were not immunogenic. Similarly, Fukazawa and Shimura (7) found large doses of histones from the silkworm Bombyx mori to be immunogenic in rabbits, but found no response to calf histones. On the other hand, Pothier et al. (8) obtained antibodies to each of the mammalian histones by giving 6 weekly injections of 5 mg each of histone dissolved in water and emulsified with complete Freund's adjuvant.

The results of immunization with free histones have thus been variable. Generally, large doses have been required. It has been suggested that the low immunogenicity of most mammalian histones reflects either their great charge or their great conservation of structure which precludes their being readily recognized as foreign. Support for the latter interpretation has been given by the more regular immunogenicity of the avian H5 histone, which is not represented in the mammalian protein group (9,10).

C. Immunization with Complexed Histones

1. BACKGROUND

Since, in our experience, the free calf thymus histones were not immunogenic, either in soluble form or when adsorbed to erythrocytes, we immunized rabbits with covalently linked histone–serum albumin conjugates. This did induce antibodies with specificity for the histone (11), but the assays were difficult, since the whole conjugate was required for complement fixation reactions and the histone haptens could be measured only by inhibition techniques.

Noting the strong reactions of SLE sera with histone–nucleic acid complexes, we then immunized rabbits with such preparations (12). Complexes of whole histone with DNA elicited anti-H1 antibodies and complexes of H1 or any other histone with RNA induced antisera specific for the histone fraction used; no anti-RNA antibodies were detectable. Nearly all immunized animals have responded to histone–RNA complexes, and this approach has been used, in this laboratory and in others, for each of the major histones and for the H1 subfractions. As noted above, antibodies *have* been induced in some laboratories without the RNA, and there has been no careful comparison of dose requirements for effective immunization with one histone preparation with or without the complex formation.

2. PREPARATION OF HISTONE–RNA COMPLEXES

For preparation of the immunogens, we typically mix equal volumes of RNA (100 to 400 μg/ml in water) and histone (300–1200 μg/ml in water) in a total amount required to give about 200 μg of histone to each animal. A fine precipitate forms immediately. The complexes are emulsified with an equal volume of complete Freund's adjuvant and are injected intradermally into several sites on the back and sides of New Zealand White rabbits. The immunization is repeated with incomplete Freund's adjuvant on days 7 and 14, and an intravenous boost of complexes without adjuvant is given on day 21. Blood is drawn on day 28 and allowed to clot at room temperature. The clot is separated from the wall of the tube with a glass rod and allowed to retract in the cold for several hours to release serum, which is further clarified by centrifugation. Immunization can be continued with periodic booster injections intradermally or intravenously.

III. Serological Assays

A. Precipitation

Classical precipitin reactions with whole serum are limited by the fact that histones, at moderate concentrations, precipitate with some normal serum proteins. The serum reactants are concentrated in normal gamma globulin prepared by ammonium sulfate fractionation of the serum, but are distinct from the major antibody class, IgG, itself. The major nonimmunological precipitating protein occurs in the macroglobulin protein fraction obtained by Sephadex G-200 gel filtration of serum, though smaller proteins also give some precipitation. The most reactive protein occurs in a relatively acidic fraction obtained by O-(diethylaminoethyl) cellulose (DEAE-cellulose) chromatography of serum, but is eluted before the more acidic albumin, which is not reactive. Marked differences occur among the individual histone fractions (Fig. 1). The H4 forms the most precipitate; progressively less is formed with H3, H2A, and H2B, and very little is seen with H1. In summary of these findings (B. D. Stollar, unpublished data), it should be possible to avoid the major nonimmunological precipitation by suitable preparation of IgG from antihistone sera. True antihistone antibodies have been purified and identified as IgG molecules (*13*).

Quantitative precipitin reactions of histones have not been studied in detail, and the optimal conditions for this system are not defined. Generally,

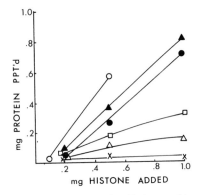

Fig. 1. Nonimmunological precipitation of calf thymus histones with normal rabbit serum. Increasing amounts of histones were added to 0.1-ml samples of serum in a total volume of 1.0 ml. The mixtures were incubated at 37 °C for 1 hour and then centrifuged at 2000 rpm for 15 minutes. Precipitates were washed twice with 0.1 M NaCl–.01 M phosphate (pH 8) and dissolved in 3 ml of 0.1 N NaOH. Protein was determined with a microbiuret assay. Histones were: H4 (○); H3 (▲); H2A (●); unfractionated histones (□); H2B (△); and H1 (×).

precipitation can be measured over a wide range of pH and ionic strength conditions and can be used to quantitate antibody or antigen and to compare antigen structures. Precipitation requires bivalent antibody and multivalent antigen, but monovalent antigen fragments can be measured by their ability to inhibit precipitation if the total number of different determinants on the intact antigen is not large. Details of the precipitin procedures as applied to other antigen–antibody systems are provided by Kabat (14) and by Maurer (15); they should be applicable to antihistone antibodies that have been freed of the major nonimmunological serum protein reactants.

B. Precipitation in Gels (Immunodiffusion)

The nonspecific reactions also may complicate immunodiffusion in gels. Early studies of antihistone antisera depended to a large extent on immunodiffusion assays (4), but it became apparent that some lines that formed were not specific immune precipitates (5). Sotirov and Johns (9), studying anti-avian H5 histone antisera, observed large nonspecific precipitates in gels; these faded after several days, leaving only the true anti-H5 precipitin line. The nonspecific serum reactants could be absorbed by mammalian histones lacking the H5. Mura et al. (10) avoided nonspecific reactions with similar sera, but only with a complex schedule for timing the addition of reagents, and in this case the specific precipitin line faded. Thus, the information gained from immunodiffusion has been limited to qualitative demonstration of precipitin lines and of some degree of specificity for H1 or H5 histones.

C. Radioimmunoassay

Partly because of nonspecific reactions and partly because the histones bind to glass and plastic, it has not yet been possible to develop a radioimmunoassay based on the use of labeled histones. A different approach has been used, however; immunospecifically purified anti-H1 antibodies were labeled with radioactive iodine, and the binding of the labeled antibodies to chromatin was measured as an approach to studying the availability of the H1 determinants (13).

D. Immunohistochemistry

Interesting beginnings have been made in the use of antihistone sera as reagents for immunofluorescence. Since a variety of substrates and fixation methods have been used, it is not yet possible to compare closely the results from different laboratories or to recommend a single procedure. Results of several studies are summarized briefly in Table I.

TABLE I

Substrates and Fixation Procedures Used in Studies of Immunofluorescence
with Antihistone Sera

Substrate	Fixation	Sera and results	References
Rat liver cell smear	Air dried, or ethanol	Anti-H1, intense diffuse nuclear stain	(4)
Hamster fibroflasts, rat embryo, HeLa cells	Cold acetone	Anti-H1, granular staining throughout nucleus	(16)
HeLa cells		Anti-H4, staining strongest at nuclear rim	
Spermatozoa	Methanol and swollen[a]	Anti-H3, -H2A, -H4, -H2B positive; anti-H1 negative	(17)
Human lymphocytes	Methanol and swollen[a]	All five sera positive	(17)
Rat liver sections	Acetone	Anti-H1, -H2B positive; anti-H2A -H3, -H4 negative	(17)
	Methanol and swollen[a]	All five sera positive	(17)
Hamster chromosomes	Methanol–acetic acid	Strongest with anti-H4, H3; weaker with anti-H2A, -H2B; negative with anti-H1	(8)
Hamster chromosomes	Methanol–acetic acid	Positive with anti-H2B; positive with anti-H1 fixation short; negative with anti-H2A, -H3, -H4	(18, 19)
	Glutaraldehyde	All five sera positive	(18, 19)
Drosophila polytene chromosomes	25% acetic acid	Anti-H3 strongly positive in bands; anti-H1 variable	(20)

[a]Methanol-fixed nuclei were swollen by their exposure for a few minutes to 0.05 M Tris (pH 9)–0.4 mM dithiothreitol–10 μg/ml trypsin.

Anti-H1 sera have most regularly stained nuclei of fixed cells, whether fixation was with ethanol or drying (4), cold acetone (16), or methanol (16, 17). Anti-H1 sera did not stain spermatozoa, even after the sperm heads were swollen and stainable by all other antihistone sera (17). Anti-H1 sera also failed to stain isolated metaphase chromosomes fixed with methanol–acetic acid in one case (8), though they did so in another study when the acidic fixation period was very short, or when chromosomes were fixed with gluteraldehyde (18, 19).

The importance of the cell fixation procedure was also apparent in experiments with other sera. For example, Tsutsui et al. (16) found that anti-H4 serum stained nuclei of acetone-fixed cells but not of methanol-fixed cells. Samuel et al. (17) similarly found anti-H4 serum negative with methanol-fixed cells; it was positive with samples that were treated so as to swell the nuclei. The swelling procedure revealed determinants for all antihistone sera, but it is not clear to what extent this involved either changes in the ability of antibodies to permeate the nuclear membrane or changes in the chromatin structure itself. A further variability with fixation occurred with chromosomes. Pothier *et al.* (8) found strong reactions with anti-H3 and anti-H4 sera even after the chromosomes were exposed to 0.2 *N* HCl for 4 hours, while Bustin *et al.* (18) found no staining of methanol–acetic acid-fixed chromosomes with these sera; all sera did stain glutaraldehyde-fixed chromosomes in the latter study. A clarification of the extent to which structure is altered by the fixation procedures will be required for the clear interpretation of the immunohistochemical experiments.

A beginning has also been made in the extension of immunohistochemical studies to the electron microscope level. Bustin *et al.* (19) have visualized the binding of anti-H2B antibodies to nucleosomes of isolated chromatin and have found this histone accessible on more than 90% of the beads examined.

E.   Complement Fixation

1.   PRINCIPLES

In quantitative complement fixation, varying amounts of antigen are added to several tubes containing a constant amount of antibody and complement. These experimental mixtures are incubated overnight at 4°C, along with control tubes that contain complement only, or complement with antibody alone or with antigen alone. Then residual complement activity is detected by the addition of sensitized erythrocytes, which serve as targets for its lytic action. If an appropriate antigen–antibody reaction has occurred during the overnight incubation, the experimental mixtures will have used up complement, leaving less available for cell lysis than is present in the control tubes. This difference, or complement fixation, is quantitated by measurement of hemoglobin released into solution by lysis. The extent of complement fixation is plotted as a function of the amount of antigen added to the experimental tubes. This describes a curve that is analogous to a quantitative precipitin curve, with zones of antibody excess, equivalence, and antigen excess. The curve can similarly be used to quantitate and compare antigens. It cannot give a value for the absolute amount of antibody in a serum, but relative serum titers can be determined from curves measured with varying serum dilutions.

As with precipitation, complement fixation requires bivalent antibody and, in nearly all cases, multivalent antigen. Again, monovalent fragments may be measured by their ability to inhibit complement fixation by the intact reactants.

Extensive analyses of antihistone antisera have been carried out with complement fixation assays. The histones are used at low concentrations that do not fix complement with normal serum proteins, and if some of the protein is lost on glass surfaces, that does not interfere with the assay as it does with a radioimmunoassay. Still, these two possible types of binding may reduce effective concentrations of the antigens; for this reason, or perhaps for unrelated reasons, relatively large amounts of histone are required for complement fixation in comparison with other antigen–antibody systems. With most systems, 2–20 ng of antigen give peak reactivity; with antihistone systems, 100 ng to 2 μg of antigen per reaction mixture may be necessary for peak reactivity.

We use the basic technique described by Wasserman and Levine (21), for which Levine (22) has described some modifications and specific applications. We have added further modifications that have markedly simplified the mechanics of performing the assays. While the technique seems complex in the first description, it requires few and simple operations for each reaction mixture.

2.   MATERIALS AND PROCEDURES

a. Buffers.   A stock buffer 1.4 M NaCl–0.1 M Tris (pH 7.4) is stored at 4°C. For working buffer, 200 ml of the stock are diluted to 2 liters with distilled water, and 6.7 ml of 0.15 M MgSO$_4$ (final 0.5 mM) and 3.0 ml of 0.1 M CaCl$_2$ (final 0.15 mM) are added, giving a buffer labeled isotris–M. Addition of 2 gm of bovine serum albumin (final 0.1%), which stabilizes complement, gives a buffer labeled isotris-BSA. Gelatin (0.1%) can be used in place of BSA.

b. Complement.   Complement is purchased as lyophilized guinea pig serum (Colorado Serum Co., Denver, Colorado), reconstituted with the diluent provided, and stored at −20°C in portions of about 1 ml. With careful thawing, the complement can be frozen and thawed several times without significant loss of activity.

c. Hemolysin.   Rabbit anti-sheep cell hemolysin is purchased as a 50% serum solution in glycerol (Baltimore Biological Labs, Cockeysville, Maryland). This is diluted 25-fold with isotris–M to give a 1/50 stock solution.

d. Erythrocytes.   As a source of erythrocytes, sheep blood in Alsever's solution is used beginning a week after it is drawn; this can then be used for several weeks, as long as little or no spontaneous hemolysis occurs.

e. Histones.   These are diluted from stock solutions into isotris–M to give a concentration of 10–20 μg/ml, the highest amounts used in the assay. Serial 2-fold dilutions are then made with isotris–M.

f. Antisera.   Antisera are heated at 60°C for 15 minutes. Samples are diluted into isotris–BSA on the day of use. New sera are usually diluted 1/100 for the initial test of complement fixation reaction; this stock will be diluted 6-fold more in the reaction mixtures, as described below.

g. Reaction mixtures. In the first stage, the antigen–antibody–complement mixtures and suitable controls are prepared and incubated at 4 °C overnight (16–18 hours). Complement-containing reagents and the mixtures are prepared in the cold, with tubes kept in an ice-water bath. It is convenient to use 16 × 100 mm or 13 × 100 mm glass tubes. Either acid-washed or clean disposable tubes may be used; they should be of uniform size and shape and should not be scratched on the inner surface.

Table II presents a sample protocol for testing one histone antigen with one serum sample. A total volume of 1.2 ml is added to each tube at this stage. Tubes 1 through 6 contain experimental mixtures of antigen, antibody, and complement. Tube 7, with serum and complement only, tests for anticomplementary activity of the serum. Tubes 8, 9, and 10 similarly test for anticomplementary activity of the antigen alone. These four control tubes (7 through 10) should retain the same lytic activity as the one that receives only buffer and the standard amount of complement (tube 12). There are two additional controls: tube 11 receives a double dose of complement and will show early and complete lysis; tube 13 receives only buffer and will control for spontaneous cell lysis without complement. The last three control tubes (11–13) are prepared once for each full experiment, while the experimental and the antigen control tubes are repeated for each antigen tested.

The procedure is markedly simplified by the preparation of premixed solutions containing either buffer and complement (4:1, v/v) or buffer,

TABLE II

PROTOCOL FOR QUANTITATIVE MICROCOMPLEMENT FIXATION ASSAY OF ANTIHISTONE SERA

Reagent Tube No.:	1	2	3	4	5	6	7	8	9	10	11	12	13
1. Isotris–BSA buffer													1.2 ml
2. Premixed													
Isotris–BSA buffer 4 ⎫							1.0 ml[a] —————→						
Diluted complement 1 ⎭													
3. Premixed													
Isotris–BSA buffer 3 ⎫													
Diluted serum 1 ⎬ 1.0 ml[a] —————→													
Diluted complement 1 ⎭													
4. Isotris–M							0.2					0.2	
5. Diluted complement											0.2		
6. Histone antigen[b]	0.2 ml of: —————→							0.2 ml of:					
solution of (μg/ml):	10	5	2.5	1.25	0.62	0.31		10	5	2.5			

[a] Added with a Cornwall repetitive syringe.

[b] Added with a plastic-tipped dispenser, proceeding from lower to higher antigen concentrations.

serum, and complement (3:1:1, v/v) and the use of a 2.0-ml Cornwall repetitive syringe to dispense 1.0-ml samples to corresponding tubes. The last 0.2 ml is conveniently added to each tube with a plastic-tipped dispenser. The net effect is the same as the separate additions of 0.2-ml portions of diluted serum, diluted complement, and antigen, with enough buffer to make up the constant volume of 1.2 ml. The vertical numbers at the left of the protocol in Table II also indicate a convenient order of addition of reagents.

   h. Sensitized erythrocytes.   These are prepared as described in the procedure of Wasserman and Levine (*21*). Sheep blood cells are washed with isotris–BSA. About 0.5 ml of packed cells are then added to 9.5 ml of buffer to give a 1/20 suspension. A 0.5-ml sample of this suspension is lysed by its addition to 7.0 ml of 0.1% Na_2CO_3. The absorbance of this lysate at 541 nm should be between 0.660 and 0.700. If it is not, more packed cells or more buffer are added to the 1/20 cell suspension so that lysis of the sample in this way does produce such an absorbance; this standardizes the cell concentration.

   A volume (such as 2 ml) of the standardized cell suspension is measured into a small flask, and an equal volume of diluted hemolysin is added to it while the suspension is swirled. The hemolysin is usually diluted 1/15 from the stock, to give a 1/750 dilution before its addition to the cells (see Section III,E,2,*l*, below). The erythrocyte-hemolysin suspension is incubated at 37°C for 10 minutes. The sensitized cells are then diluted 10-fold with isotris–BSA and are ready for use.

   i. Cell lysis.   Sensitized cells, in 0.2-ml portions, are added to each of the complement fixation reaction mixtures that have been incubated overnight at 4°C. The mixtures are kept cold during this addition, which is conveniently made with a 1.0-ml Cornwall repetitive syringe. They are then mixed well and incubated in a 37°C water bath. Lysis occurs in the excess complement control tube in 12–14 minutes and in other control tubes in 30–40 minutes. When lysis reaches about 80% in the control tubes, as judged visually, both the experimental and control mixtures are removed to an ice-water bath, chilled for 5 minutes, and then centrifuged in the same tubes in a refrigerated centrifuge. The supernatants are either poured off directly into standard 1-ml cuvettes for reading of hemoglobin concentration at 412 nm or drawn into a micro flow-through cuvette of the Gilford 300N spectrophotometer; the latter is a much more rapid procedure. The tube with complete lysis should read about 0.680 at 412 nm, while the spontaneous cell lysis control supernatant should read less than 0.02.

   j. Calculations.   The antibody control (tube 7, Table II) serves as the basis for calculation. Readings of the absorbance at 412 nm for the experimental tubes are subtracted from that of the antibody control mixture. The difference in each case is then expressed as a percentage of the value of the antibody control (tube 7) minus the spontaneous cell lysis control (tube 13). This is expressed as percentage complement fixation. It is not precisely the percentage of available complement actually fixed, since the degree of lysis at a given time is not a simple linear function of the amount of complement present.

k. Choosing the dilution of complement. The sensitivity and ease of analysis of the micro assay depends on the use of a small amount of complement—only slightly more than enough to lyse the dose of sensitized cells used. To standardize this control level of complement activity for a new lot of complement, a series of complement control mixtures (Table II, tube 12), containing 1.0 ml of isotris–BSA and 0.2 ml of varying dilutions of complement (ranging from 1/200 to 1/400) are incubated overnight at 4 °C. Sensitized cells are then added (0.2 ml/tube), and the suspensions are incubated at 37 °C. The complement dilution that causes about 80% lysis in 40 minutes is chosen as the amount to be used in the assays. This dilution may be adjusted from day to day if the activity of a given lot changes, as seen in daily controls. With careful storage and handling during thawing, 7 ml of stock of undiluted complement will suffice for about 50 experiments of up to 100 tubes each.

l. Choosing the dilution of hemolysin. To determine the amount of a new lot of hemolysin that is required for optimal sensitization of the sheep cells, a standardized erythrocyte suspension is prepared as described above (Section III,E,2,*h*), and 0.2-ml portions are distributed into each of several tubes; then 0.2 ml of hemolysin of varying dilution (1/6, 1/9, 1/12, 1/15, 1/18, and 1/21 from the stock 1/50 dilution) is added to each. These suspensions are incubated for 10 minutes at 37 °C and then diluted with 3.6 ml of isotris–BSA. Then 0.2-ml samples of each of these sensitized cell suspensions are added to tubes containing 1.0 ml of isotris–BSA and 0.2 ml of diluted complement. The complement is used here at a dilution that gives only partial hemolysis in 40 minutes with the cells sensitized by the 1/18 hemolysin solution. A 1/350 to 1/450 dilution of complement is usually suitable, as can be determined in a preliminary incubation. All tubes are then incubated at 37 °C. More rapid lysis will be noted in samples prepared with more hemolysin, but a plateau is reached where the increase in the rate and extent of lysis is slight. A hemolysin concentration just higher than that required for this turning point is chosen. If this approach to a plateau does not occur, the hemolysin is not satisfactory and will introduce additional variability. The hemolysin is stable, and a standardized stock prepared from 10 ml as purchased will last for more than 1000 experiments of up to 100 tubes each; this standardization is therefore done infrequently.

m. Anticomplementary sera and antigens. If a serum is anticomplementary, the cells in the serum control tube (tube 7, Table II) lyse more slowly than those in the complement control (tube 12). All cell lysis mixtures containing that serum sample may be left at 37°C for a longer incubation, after the antigen and complement control tubes have been removed to the cold. As long as the antigen is not anticomplementary also, the serum control can still serve for all the tubes containing that serum sample; they are all removed to the cold together. When the anticomplementary activity is marked, it may be necessary to repeat the assay with more complement (to ensure that a measurable amount is free in the serum control tube) or with more dilute serum. Alternatively, the antibodies may be partially purified by DEAE-cellulose chromatography or Sephadex G-200 gel filtration (23).

If antigens show anticomplementary activity by themselves, it is usually not possible to correct for it in the incubation or calculations, but only in the purification of the antigen. As added to the reaction mixtures, for example, the antigen solutions should be at the same pH and ionic strength as the isotris–M and should not contain chelating agents.

IV.   Applications of Complement Fixation Assays with Antihistone Antisera

A.   Comparisons of Histones

Differences in the heights of the complement fixation curves indicate qualitative differences in the antigen structures. In order to be sure of detecting such differences, it is necessary to use a serum dilution that gives only partial (optimally about 60%) complement fixation with a standard homologous antigen. If structural variations among antigens are slight, a single serum dilution may then demonstrate the full range of their relative reactivities (Fig. 2) (12,24,25). The suitable dilutions, as used in Fig. 2, were determined by preliminary tests of the full curve for standard calf thymus histones with each of several serum dilutions. The microcomplement fixation test is a very sensitive measure of structural differences (22); the close similarities of the curves of H2A, H2B, and H4 histones of species as diverse as human and lobster with their corresponding antisera reflect their marked conservation of structure (26).

If structures vary more significantly, as among H1 subclasses, it may not be possible to measure them all with a single amount of serum, and each antigen may have to be tested with several dilutions (Fig. 3) (27). One can then determine the serum dilutions required for a given level of reaction with two antigens and express the ratio of these dilutions as an "index of dissimilarity" (28). Usually one determines the dilutions required to give a curve with a peak of 50% complement fixation, as read by interpolation (Fig. 4)

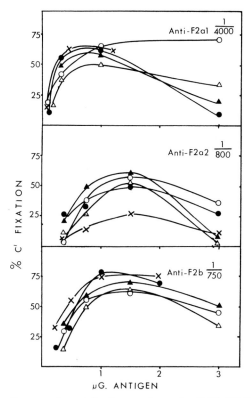

FIG. 2. Comparative complement fixation reactions of anti-H4 (F2al), anti-H2A (F2a2), and anti-H2B (F2b) sera with the corresponding histone fractions from human spleen (O), calf thymus (●), chicken erythrocyte (△), frog liver (▲), and lobster hepatopancreas (×). Serum dilutions refer to solutions before their addition to the reaction mixtures; final dilutions were 6-fold higher. Reprinted from Stollar and Ward (12) with permission of the publisher.

(27). The relationship of this index to the extent of amino acid sequence change and its use in determining evolutionary relationships has been described by Prager and Wilson (29). It was applied to comparisons of H1 histones by Bustin and Stollar (27).

This same reading of a 50% maximal complement fixation titer can be applied to evaluating the relative strengths of different sera with a given antigen.

B.   Characterization of Antiserum Specificity

If antisera are to be used in techniques such as immunofluorescence, it is important to determine whether they are truly specific for a given histone. While sera may appear to be specific at high serum dilution, tests with a wide

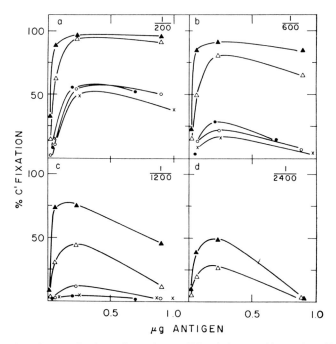

Fig. 3. Complement fixation of rat thymus H1 subclasses with varying dilutions of antiserum to subclass V. Antigens were subclasses V (▲), IV (△), III (×), II (●), and I (○), as separated by 1RC-50 column chromatography. Serum dilutions refer to solutions before their addition to reaction mixtures. Reprinted from Bustin and Stollar (27) with permission of the publisher.

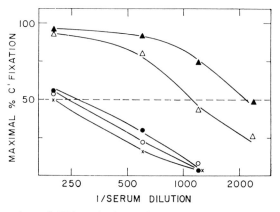

Fig. 4. Comparison of 50% maximal complement fixation titers for varying rat thymus H1 subclasses with antiserum to subclass V. Peak complement fixation levels were obtained from the data in Fig. 3; symbols refer to the same H1 subclasses as in Fig. 3. An index of dissimilarity can be calculated as the ratio of the titer for the homologous subclass V to the titer for a different subclass. Reprinted from Bustin and Stollar (27) with permission of the publisher.

TABLE III

CROSS-REACTIONS OF HISTONE FRACTIONS WITH ANTISERA TO CALF THYMUS
H2A, H2B, AND H4

		Serum dilution[a] required for 50% maximal complement fixation with		
Test antigen		Anti-calf H2A	Anti-calf H2B	Anti-calf H4
Calf	H1	<50	65	200
	H2A	800	65	<200
	H2B	50	900	250
	H3	170	50	<200
	H4	80	55	5500
Lobster	H1A	650		
	H2B		925	
	H4			5500

[a]Serum dilutions refer to solutions before their addition to reaction mixtures; final dilutions were 6-fold greater. Data from Stollar and Ward (12).

range of serum concentrations may be necessary to clarify the degree of specificity; cross reactions may occur with the high concentrations (Table III) (8,12).

C. Measuring Chemical or Physical Alterations

Complement fixation was used to measure changes caused by maleylation of histone amino groups (30), relating serological changes to decreased histone–DNA interaction. It also measured changes caused by enzymic cleavage or nitration of H1 (31,32), giving some insight into the location of antigenic sites.

D. Measuring the Accessibility of Histones in Chromatin

Chromatin reacts weakly or not detectably in complement fixation tests with antibodies to individual histones (3,12,33). This indicates that some of the protein determinants are altered or masked in the complex structure. It has been possible, however, to adapt the technique to measure the accessibility of histones in chromatin by using chromatin as a solid-phase immunoadsorbant (33,34). Complement fixation activity of a standard amount of antibody and a given histone was measured before and after incubation of the serum with insoluble chromatin. The chromatin readily removed antibodies to H1 and H2B; it was much less effective in removing anti-H3 or anti-H4 antibodies, indicating the lesser accessibility of the latter.

E.   Identification and Quantitation of Histones

Complement fixation assays with specific antisera can selectively identify and quantitate homologous isolated proteins, as in analysis of fractions eluted from gel slices after electrophoresis (Fig. 5) (12). For quantitation, one measures the level of complement fixation for a given dilution of the test sample and reads the absolute amount of a purified histone that gives the same level of reactivity in a standard curve. At least two dilutions of the unknown must be tested so that it is clear whether the measured level of complement fixation is on the antibody excess or the antigen excess side of the curve.

Identification and quantitation in mixtures, however, is not as straightforward, since interactions among the histones can reduce their reactivity with antibodies to an individual fraction. The reactivities of H1 and H3 are least sensitive to this interference; to measure the other histones in mixtures it may be necessary to use high serum concentrations or pooled sera. Interestingly, while the anti-H4 antiserum reacts very weakly with total acid-extracted histones, it reacts well with the H3–H4 complexes isolated by the salt–dissociation procedure of Van der Westhuysen and Von Holt (35,36).

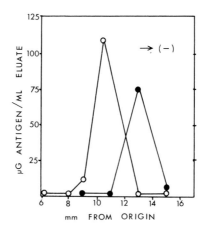

F𝖎g. 5.   Immunochemical localization of H4 and H2A in fractions eluted from sections of of polyacrylamide gel (12 × 45 mm), after electrophoresis of a 1-mg sample of the mixed proteins. Cross-sections of 1 mm length were cut and eluted with 0.15 M NaCl. Each fraction was assayed with anti-H4 antiserum (●) and anti-H2A antiserum (○). The amount of antigen in each fraction was calculated by comparison with the amounts of standard H4 and H2A preparations required for similar complement fixation reactivity as a given dilution of the fraction. Reprinted from Ward and Stollar (12) with permission of the publisher.

V. Conclusions

The application of immunochemical techniques to the study of histones and other chromatin components is still at an early stage. As the specificities of antisera become more narrowly focused, antibodies can serve as sharp selective probes for corresponding structures. It should be possible to prepare subpopulations of antihistone antibodies, for example, that will measure only a selected portion of a given histone molecule; such antibodies could determine whether the structure of that segment is maintained during interaction with other histones, with DNA, or with nonhistone proteins. As antibody monovalency is approached in this sense, the major assays will have to depend on measurements of simple binding rather than on complex secondary reactions such as precipitation or complement fixation; the latter assay remains very useful, however, for studies of multivalent systems because of its requirement for relatively small amounts of material and its great sensitivity to structural change in the antigens.

ACKNOWLEDGMENT

Research performed in the author's laboratory has been supported by grants from the National Science Foundation.

REFERENCES

1. Holman, H. R., Deicher, H., and Kunkel, H. G., *Bull. N.Y. Acad. Med.* **35**, 409 (1959).
2. Stollar, B. D., *Arthritis Rheum.* **14**, 485 (1971).
3. Stollar, B. D., *J. Immunol.* **99**, 959 (1969).
4. Rümke, P., and Sluyser, M., *Biochem. J.* **101**, 1c (1966).
5. Sluyser, M., Rümke, P., and Hekman, A., *Immunochemistry* **6**, 494 (1969).
6. Mandal, R. K., Mondal, H., Bom, S., and Biswas, B. B., *Indian J. Biochem. Biophys.* **8**, 50 (1971).
7. Fukazawa, C., and Shimura, K., *Biochim. Biophys. Acta* **154**, 619 (1968).
8. Pothier, L., Gallagher, J. F., Wright, C. E., and Libby, P. R., *Nature (London)* **255**, 350 (1975).
9. Sotirov, M., and Johns, E. W., *J. Immunol.* **109**, 686 (1972).
10. Mura, C., Huang, P. C., and Levy, D. A., *J. Immunol.* **113**, 750 (1974).
11. Sandberg, A. L., Liss, M., and Stollar, B. D., *J. Immunol.* **98**, 1182 (1967).
12. Stollar, B. D., and Ward, M., *J. Biol. Chem.* **245**, 1261 (1969).
13. Bustin, M., and Kupfer, H., *Biochem. Biophys. Res. Commun.* **68**, 718 (1976).
14. Kabat, E. A., "Experimental Immunochemistry," Thomas, Springfield, Illinois.
15. Maurer, P., *in* "Methods in Immunology and Immunochemistry (C. A. Williams and M. W. Chase, eds.), Vol. III, p. 1. Academic Press, New York, 1971.
16. Tsutsui, Y., Suzuki, I., and Hayashi, H., *Exp. Cell Res.* **94**, 63 (1975).
17. Samuel, T., Kolk, A. H. J., Rümke, P., Aarden, L. A., and Bustin, M., *Clin. Exp. Immunol.* **24**, 63 (1976).
18. Bustin, M., Yamasaki, H., Goldblatt, D., Shani, M., Huberman, E., and Sachs, L., **97**, 440 (1976).

19. Bustin, M., Goldblatt, D., and Sperling, R., *Cell* 7, 297 (1976).

20. Aesai, L. S., Pothier, L., Foley, G. E., and Adams, R. A., *Exp. Cell Res.* **70**, 468 (1972).

21. Wasserman, E., and Levine, L., *J. Immunol.* **87**, 290 (1961).

22. Levine, L., *in* "Handbook of Experimental Immunology" (D. M. Weir, ed.), 2nd ed., p. 22.1. Blackwell, Oxford.

23. Fahey, J. L., and Terry, E. W., *in* "Handbook of Experimental Immunology" (D. M. Weir, ed.), 2nd ed., p. 7.1. Blackwell, Oxford.

24. Bustin, M., and Stollar, B. D., *J. Biol. Chem.* **247**, 5716 (1972).

25. Sluyser, M., and Bustin, M., *J. Biol. Chem.* **249**, 2507 (1974).

26. Elgin, S. C. R., and Weintraub, H., *Annu. Rev. Biochem.* **44**, 726 (1975).

27. Bustin, M., and Stollar, B. D., *J. Biol. Chem.* **248**, 3506 (1973).

28. Prager, E. M., and Wilson, A. C., *J. Biol. Chem.* **246**, 5978 (1971).

29. Prager, E. M., and Wilson, A. C., *J. Biol. Chem.* **246**, 7010 (1971).

30. Burnotte, J., Stollar, B. D., and Fasman, G. D., *Arch. Biochem. Biophys.* **155**, 428 (1973).

31. Bustin, M., and Stollar, B. D., *Biochemistry* **12**, 1124 (1973).

32. Hekman, A., and Sluyser, M., *Biochim. Biophys. Acta* **295**, 613 (1973).

33. Bustin, M., *Nature (London) New Biol.* **245**, 207 (1973).

34. Goldblatt, D., and Bustin, M., *Biochemistry* **14**, 1689 (1975).

35. Van der Westhuysen, D. R., and Von Holt, C., *FEBS Lett.* **14**, 333 (1971).

36. Feldman, L., and Stollar, B. D., *Biochemistry* **16**, 2767 (1977).

Chapter 8

Immunochemical Analysis of Nonhistone Proteins

F. CHYTIL

Departments of Biochemistry and Medicine,
Vanderbilt University School of Medicine,
Nashville, Tennessee

I. Introduction

This chapter introduces the methodology of immunochemistry of chromosomal nonhistone proteins. Attempts also will be made to review some applications of the immunochemical analysis in chracterization not only of the nonhistone per se, but also in characterization of chromatin. The material discussed here concerns mostly antibody raised against heterogeneous mixtures of these proteins. The immunochemical analysis will certainly become more meaningful when an antibody against a single homogenous nonhistone protein preparation is obtained.

II. Immunogenicity of Nonhistone Proteins

In contrast to histones the heterogeneous mixtures of nonhistone proteins are very good immunogens (1–3). To produce antibody, nonhistone proteins can be injected into the experimental animal complexed with DNA (1,2). Administration of whole chromatin also has been shown to give antibody against nonhistone proteins as well as against histones (3–5). Finally, free (5) as well as sodium dodecyl sulfate (SDS)-treated nonhistone protein preparations (6) are good immunogens. No difficulties have been reported in producing antibody against nonhistone proteins from various eukaryotes.

III.   Choice of Animal for Production of Antihistone Antibody

Rabbits as well as chickens are suitable for inducing antibodies. Guinea pigs were not reported to produce antibody against nonhistone proteins, but these animals were used for production of antinuclear antibody without major complications (7). It is therefore no *a priori* reason to eliminate this animal as a potential antinonhistone antibody producer. Both rabbits and chickens produce and then carry these antibodies without apparent pathological signs. In the past 5 years we have not experienced deaths of immunized animals which could be attributed directly to the immunization by nonhistone proteins.

A.   Rabbits

Rabbits seem to be the choice of most workers because the handling and housing of these animals does not need special skills. Moreover the quantity of blood which can be obtained is usually sufficient for many experiments. The nonhistone protein antibodies have been produced in most instances against heterogeneous mixtures of these proteins. Because the antibody response to such a heterogeneous immunogen may vary from one animal to another, it is not advisable to pool the antisera after immunization of several animals. Also, sera from different bleedings of the same animal should not be pooled since the antibody specificity might undergo variations. But the serum should be tested before immunization for the presence of natural antibody against nuclear material because recently it was found that some nonimmunized rabbit sera contain complement-fixing antibodies directed against nuclear components (8).

The most common rabbit strain, New Zealand White, is used. The rabbits usually weigh at least 3 kg; male sex is preferable as it eliminates complications with accidental pregnancies occurring in the animal facilities.

B.   Chickens

Chickens have been shown to be good producers of nonhistone antibodies when injected with whole chromatin (3,4). This animal might be preferred when precipitation antibody is to be produced. Chicken antibody optimally forms precipitin complexes in the presence of 1.5 M NaCl. This concentration of sodium chloride facilitates the solubility of chromosomal proteins (9).

IV. Immunization Procedure

A. Preparation of the Immunogen

Nonhistone proteins are prepared as immunogens similar to the other proteins. Usually these proteins—regardless of whether injected free, SDS-treated, or as the whole chromatin or in the form of complexes with DNA—are emulsified with complete Freund adjuvant. An equal volume of Freund adjuvant is stirred with the nonhistone protein preparation in a small beaker using a magnetic stirrer. The resulting white emulsion is administered to the animal. The mixing of the immunogen with the adjuvant also can be accomplished by vigorously shaking the mixture in a syringe.

B. Administration Dose

Microgram quantities of the nonhistone protein mixtures (see Table I), are sufficient to elicit antibody response. Some workers have been using milligram quantities. The use of smaller quantities has been adopted in this laboratory for two main reasons. First, the quantity of the nonhistone protein mixture available often represents a limiting factor; second, by using larger quantities of the immunogen there is a higher probability of producing antibody against some undesired impurities.

C. Route of Administration

Various laboratories administer the immunogens differently, as is evident from Table I. A subcutaneous, intramuscular, or intradermal injection is usually the starting procedure. Sometimes the immunization is completed by intravenous injection of the immunogen. In our laboratory we use the following procedure.

Fifty to 100 μg of total nonhistone protein mixed with complete Freund adjuvant (1:1 by volume) are injected under the skin of the toe pads of one of the hind legs of the rabbit. Usually 0.25 to 0.5 ml of the mixture is injected into the pads, and the rest is administered in small doses intramuscularly over the whole rabbit body. After a week, a second injection is given in the toe pads of the other foot. Then, in weekly intervals (usually two more doses are required), the immunogen is administered intramuscularly in small doses over the whole body. When antibody is detected in the serum taken from these animals an intravenous injection (booster) of the immunogen (usually one-fourth of the original quantity) diluted with saline is given intravenously (major marginal vein of the ear) without adjuvant. Seven days later the animal is bled from the marginal ear vein. Thirty milliliters of the blood can be obtained from a good "bleeder" rabbit in one bleeding.

TABLE I

PRODUCTION OF ANTINONHISTONE PROTEEIN (NHP) ANTIBODY

Problem studied	Immunogen	Animal	Mode of injection[a]	Dose per injection (mg)	Intervals (days)	Number of injections	Booster dose	Reference
Tissue specificity	NHP–DNA complex	Rabbit	SC and IM	0.1	7	3–6	25 μm	(1,2,10–12)
Eu and hetero chromatin	NHP–DNA complex	Rabbit	NI	Approximately 2 mg	7	3	NI	(13)
	Chromatin	Chicken	ID and SC	0.4	3–4	5	400 μg	(9)
		Rabbit	NI	Approximately 8 mg	7	3		(13)
Structure of chromatin	NHP–DNA complex	Rabbit	SC and IM	0.1	7	3–6	25 μg	(14–16)
	Chromatin	Chicken	ID and SC	0.4	3	5	400 μg	(17)
Liver development	NHP–DNA complex	Rabbit	SC and IM	0.1	7	3–6	25 μg	(18)
Estrogen-mediated differentiation	NHP–DNA complex	Rabbit	SC and IM	0.1	7	3–6	25 μg	(15)
Malignant growth	NHP–DNA complex	Rabbit	SC and IM	0.1	7	3–6	25 μg	(19,20)
	Nucleoli	Rabbit	ID and IM	10, 20, 50 total amount	7	3	30 + 30 mg	(29)
Tissue culture	Chromatin	Rabbit	ID and SC	0.4	7	5	400 μg	(3)
		Chicken	ID and SC	0.4	3	5	400 μg	(3)
Drosophila	NHP and SDS	Rabbit	SC and IP	9–10	NI	NI	NI	(6)
	NHP and SDS	Rabbit	SC	2	28	2	6 mg	(21)
Newt oocytes	NHP	Rabbit	SC	NI	NI	NI	NI	(22)

[a] IV = Intravenously; IP = intraperitoneally; IM = intramuscularly; SC = subcutaneously; ID = intradermally; NI = not indicated.

D.   Treatment of the Antisera

The blood collected into a centrifuge tube is left to coagulate for about 1 hour at room temperature and then stored in a refrigerator overnight. After centrifugation of the blood at about 2000 g for 10 minutes the serum is siphoned off and kept frozen at $-20°C$. To avoid infection, the serum can be sterilized by using Millipore filtration. It is our experience that hemolyzed sera also can be used.

Usually the whole serum can be used. When necessary, the immuno-globulin fraction G (IgG) can be purified from the sera by ammonium sulfate precipitation (23) or by O-(diethylaminoethyl) cellulose (DEAE-cellulose) column chromatography (24).

V.   Testing Antigenicity

One of the most popular methods for testing for the presence and charac-teristics of the nonhistone protein antibodies is quantitative microcomple-ment fixation as described originally by Wasserman and Levine (25). The method used in my laboratory is described below.

A.   Quantitative Microcomplement Fixation

Principle: The microcomplement fixation method is an adaptation of complement (C') fixation which is based on the capacity of the complement system in fresh guinea pig serum to combine irreversibly with antigen–

TABLE II

QUANTITATIVE COMPLEMENT-FIXATION ASSAY[a]

Stage 1: Fixation of C' by $AbAg$ complex		Stage 2: Lysis of EA by C'	
$Ab + C'$	$Ab + C'$	$+ EA$	Lysis
$Ag + C'$	$Ag + C'$	$+ EA$	Lysis
$Ab + Ag + C'$	$[Ab - Ag - C']$	$+ EA$	No lysis

Stage 3: Quantitative measurement of lysis by determining hemoglobin concentration (optical density 413 nm). Results are expressed in percentage of C' fixed.

[a]C', complement (guinea pig); Ab, antibody; Ag = corresponding antigen; E = erythrocytes (sheep); A = antibodies (rabbit) to E (sheep); EA = sensitized erythrocytes (E complexed with A).

antibody complexes. If the antigen is associated with the sheep erythrocyte cell surface, such a combination may result in lysis of the erythrocyte. The lysis of hemolysin-sensitized sheep erythrocyte determined quantitatively by measuring hemoglobin spectrometrically is an excellent indicator system for C' activity (the principle of this method is outlined in Table II).

If C' is allowed to incubate with the antigen–antibody system under study, then its combination can be estimated by the residual hemolytic activity it possesses when a known quantity of antibody-coated (hemolysin-sensitized) erythrocyte is added at a later time. The method is not only very sensitive for detecting the presence of antibody, but also by using the same antibody against two or more antigens, one can determine the extent of similarity or conformational differences of the protein molecules by comparing the amount of each protein which elicits maximum fixation (26).

1. Materials

a. Diluent. Isotris stock buffer: To 81.6 gm of NaCl add 12.1 gm of Tris (hydroxymethyl) aminoethane base and dissolve in about 800 ml of deionized H_2O. Add 6.6 ml of concentrated HCl, mix, and add 33 ml of 0.15 M $MgSO_4 \cdot 7\ H_2O$ (3.7 gm/100 ml) and 15 ml of 0.1 M $CaCl_2$ (1.1 gm/100 ml). Dilute to 1 liter after adjusting the pH to 7.4.

b. Isotris diluent—working solutions. To 100 ml of the above stock buffer add 1 gm of bovine serum albumin and mix with a magnetic stirrer until dissolved. Make up the volume to 1000 ml with water.

c. 0.1% Na_2CO_3.

d. Sheep cells. These can be purchased commercially (50% whole blood, 50% Alsevers solution). They are kept refrigerated and can be stored up to 1 month.

e. Anti-sheep hemolysin. This is purchased commercially, frozen at $-20\,°C$, and can be stored over 4 years. The commercial hemolysin is in 50% glycerol solution. Two milliliters of this preparation are diluted to 50 ml with isotris diluent and stored frozen in 0.5 ml aliquots. The 1/1000 dilution is prepared fresh daily: 0.2 ml of 1/50 stock dilution plus 3.8 ml of isotris diluent.

f. Guinea pig complement. Complement (C') should be kept cold all the time (not exceeding 4 °C). The commercial product (lyophilized guinea pig serum) is reconstituted with 5 ml of the supplied diluent. The serum is kept frozen in 0.25-ml aliquots. Once an aliquot is unfrozen for use, any complement remaining is discarded. Complement will retain its original titer approximately $1\frac{1}{2}$ months if stored at $-20\,°C$ ($-40\,°C$ is preferable). When commercial frozen preparation of complement is used, the sample is first thawed and kept frozen in aliquots of 0.25 ml or more.

2. PROCEDURES

a. Complement titer determination. With any new batch of complement it is necessary to determine the titer as follows: (1) dilute the frozen complement aliquots 1/100 (0.1 ml is diluted to 10 ml with isotris diluent); (2) prepare different dilutions of C' according to Table III. Remember that all dilutions must be made in an ice bath. Use disposable glass culture tubes (13 × 100 mm). Pipette different dilutions of C' according to Table IV. Again, all operations have to be performed in an ice bath. After mixing the above components, store the tubes in a cold room on ice overnight. Next day add 0.2 ml of sensitized sheep cells to each tube (see Section V,A, 2,*b*).

After the addition of sensitized cells the tubes are placed in a constant-temperature water bath and incubated 40–60 minutes at 37°C. The last tube is the sheep cell blank (lysis in the absence of C'). The tubes are then transferred to an ice bath for 10 minutes to stop hemolysis and centrifuged 10 minutes at 2000 rpm in a clinical centrifuge.

The optical density (OD) of the supernatants is measured at 413 nm against isotris. The highest dilution of complement that gives 90% lysis is used for the fixation procedure (see Fig. 1). We find that the titer of the complement fluctuates between 1/125 and 1/225, depending on the batch and the supplier.

b. Standardization of sheep cells (prepare daily).

1. Three milliliters of cell suspension are centrifuged at 1500 rpm for 5–10 minutes in a clinical centrifuge (refrigerated).

TABLE III
DILUTION OF COMPLEMENT C'

C' dilution	1/100	1/125	1/150	1/175	1/200	1/225	1/250	1/275	1/300
C' (1/100) (ml)	1	1	1	1	1	1	1	1	1
Isotris (ml)	—	0.25	0.50	0.75	1	1.25	1.50	1.75	2

TABLE IV
COMPLEMENT TITER

	1/100	1/125	1/150	1/175	1/200	1/225	1/250	1/275	1/300	Cell blank
C' dilution ml	0.2	0.2	0.2	0.2	0.2	0.2	0.2	0.2	0.2	0.0
Isotris (ml)	1	1	1	1	1	1	1	1	1	1.2

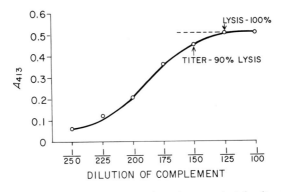

FIG. 1.   Determination of the complement titer. A_{413} = optical density at 413 nm.

2. The supernatant is aspirated off, and the cells are resuspended in 4–5 ml of isotris diluent. Repeat twice more or until the supernatant is clear.

3. Next 0.5 ml of washed, packed cells is suspended in 9.5 ml of isotris diluent.

4. Then 1 ml of this 5% cell suspension is added to 14 ml of 0.1% Na_2CO_3; 5 minutes are allowed for complete hemolysis.

5. At 541 nm the desired optical density is 0.680. The range of acceptability is 0.665–0.695.

6. The concentration of sheep cells is adjusted to the proper value by isotris diluent. The formula for adjusting is

$$\frac{OD\ sample}{0.680} \times V_1 = V_2.$$

V_1 = initial volume of the preparation after removing 1 ml for estimating optical density; V_2 = adjusted final volume.

   c. *Sensitization of sheep cells (prepare fresh daily).*

1. In a 50-ml Erlenmeyer flask, to 1 ml of standardized sheep cells add with a swirling motion 1 ml of hemolysin (1/1000) [0.2 ml of the stock solution (1/50) of hemolysin is diluted with 3.8 ml of isotris diluent].

2. Incubate for 15 minutes at 37°C to allow maximum sensitization. Then 18 ml of isotris diluent is added.

3. The sensitized cells may be kept ½ day if they are refrigerated.

   d. *C′ fixation procedure.*   The nonhistone protein preparation of known protein content to be tested for antigenicity should be kept frozen in small aliquots.

Before use in complement fixation the antibody serum is incubated at

60°C for 20 minutes to inactivate the endogenous C'. The dilution of antibody is to be determined by complete complement fixation run on each dilution (1/200, 1/400, 1/800, 1/1600, etc.). The antibody is stored in the freezer at not less than 1/200 dilution.

1. To a series of disposable glass culture tubes (13 × 100 mm) add in the following order:

0.6 ml of diluent;

0.2 ml of diluted serum;

0.2 ml of antigen solution serially diluted as shown in Table V;

0.2 ml of diluted C' (see C' titer determination).

Table V shows an example of this determination.

2. Appropriate dilutions of antigen and C', antibody and C', diluent and C', and diluent in a total volume of 1.2 ml are included in every experiment (Table V, tubes 7–17).

3. After incubation at 2°–4°C for 16–18 hours, 0.2 ml of sensitized cells is added and hemolysis allowed to proceed, with occasional swirling

TABLE V

COMPLEMENT FIXATION

Tube no.	Isotris (ml)	Antisera (ml)	Antigen (ml) and content of nonhistone protein	C' (ml)	Cells (ml)
1	0.6	0.2	0.2 (10 μg)	0.2	0.2
2	0.6	0.2	0.2 (5 μg)	0.2	0.2
2	0.6	0.2	0.2 (5 μg)	0.2	0.2
3	0.6	0.2	0.2 (2.5 μg)	0.2	0.2
4	0.6	0.2	0.2 (1.25 μg)	0.2	0.2
5	0.6	0.2	0.2 (0.62 μg)	0.2	0.2
6	0.6	0.2	0.2 (0.31 μg)	0.2	0.2
7	0.8	—	0.2 (10 μg)	0.2	0.2
8	0.8	—	0.2 (5 μg)	0.2	0.2
9	0.8	—	0.2 (2.5 μg)	0.2	0.2
10	0.8	—	0.2 (1.25 μg)	0.2	0.2
11	0.8	—	0.2 (0.62 μg)	0.2	0.2
12	0.8	—	0.2 (0.31 μg)	0.2	0.2
13	0.8	0.2	—	0.2	0.2
14[a]	1.0	—	—	0.2	0.2
15	0.8	—	—	0.4	0.2
16	1.2	—	—	—	0.2
17	1.2 Na$_2$CO$_3$	—	—	—	0.2

[a] OD obtained in tube 14 is used as 100% of lysis.

in a water bath at 35°–37°C until controls are estimated to be 80–90% hemolysed. This requires about 40–60 minutes.

4. After immersion in an ice bath to stop hemolysis reaction (10 minutes), the tubes are centrifuged 10 minutes at 2000 rpm in a clinical centrifuge.

5. The optical density of the supernatant is determined at 413 nm.

e. Calculation.

$$\frac{\text{OD sample}}{\text{OD 100\% lysis}} \times 100 = \text{percent lysis}$$

The OD obtained in tube 14 is used as 100% of lysis. Percent C' fixed = 100 − percent lysis.

3. COMMENTS

The quantitative complement fixation assay has been the most widely used method for determination of antigenicity of nonhistone proteins. It is very reproducible, sensitive, and allows the detection of conformational changes of antigens (*26*). Some problems may be encountered, however. As shown in Table II, one of the prerequisites for quantitative evaluation of the interaction of the antigens with antibody is that the antisera, at the dilution used, should not bind complement in the absence of the antigen. However, some antisera exert this type of binding. This becomes a complicating factor mainly when low dilutions of antisera have to be employed. To minimize this problem it is advisable to screen the sera before immunization of the animals by determining the amount of complement fixed by various dilutions of the serum starting at a 1/200 dilution followed by 1/400, 1/800, etc. Usually those animals whose sera shows binding of complement (anticomplementarity) at 1/400 dilution are not used for immunization. Occasionally, this anticomplementarity develops during the immunization of the animal. There are several ways to remedy this complication. Purification of the IgG fraction might help. Sometimes centrifugation of the serum diluted with 9 volumes of 0.14 M sodium chloride at 105,000 g at 4°C for 1 hour removes the anticomplementarity. Finally, the amount of C' used in the assay can be raised by an increment which eliminates the binding by antiserum (*27*).

A more severe problem is the anticomplementarity of the antigens, which might be concentration dependent. Whole chromatin preparations usually do not show anticomplementarity. But nonhistone proteins as well as nonhistone protein–DNA complexes may show limited binding C'. Then the amount of C' fixed in the fixation assay in the absence of antiserum (see Table V) can be subtracted from that observed in the presence of antigen-antibody complexes. We have noted that repeated thawing and freezing of chromatin preparations sometimes induces anticomplementarity. Storing antigens in small aliquots is therefore advisable. Chromatin preparations treated with acid to remove histones usually show high anti-complementarity. Removing histones with a high concentration of urea and salt seems to have little effect.

A typical complement fixation curve when antinonhistone protein–DNA antisera are reacted with the preparation used for immunization and with the corresponding whole chromatin preparation is shown in Fig. 2. The fact that the nonhistone protein antisera also react with the whole chromatin lets one use such an antibody in the characterization of chromatin preparations. An example of this use is plotted in Fig. 3. The amount of antigen can be expressed in the amount of DNA or nonhistone protein in the assay.

B.   Other Methods of Testing Antigenicity

Recently a double diffusion technique was employed (13,28). Sodium dodecyl sulfate–polyacrylamide gel electrophoresis also has been used (6).

C.   Applications

By the use of the quantitative complement fixation method tissue specificity of antigenic properties of nonhistone protein–DNA complexes from chick oviduct nuclei has been described (1). Wakabayashi and Hnilica (2) have shown that an antiserum against nonhistone protein–DNA complexes isolated from Novikoff hepatoma will distinguish between this tumor and similar complexes isolated from rat liver and calf thymus. Furthermore, Zardi et al. (4) found that whole chromatin, when used as an immunogen, can elicit complement-fixing antibodies which recognize differences in chromatin and nonhistone proteins not only between WI-38 human diploid fibroblast and 3T6 mouse fibroblasts grown in culture, but also between chromatin of WI-38 cells and their SV-40 transformed counterparts, the 2-RA cells (3).

Changes in antigenicity of nonhistone protein–DNA complexes during perinatal development of rat liver were noted (18) as well as during estrogen-induced growth and differentiation of chick oviduct (14,15). Regeneration of rat liver apparently induces changes in antigenic properties of these complexes (19). Bush et al. (29) have described production of antibody showing specificity of nucleolar proteins.

Considerable success has been had in analyzing antigenic properties of

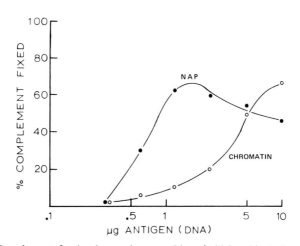

FIG. 2.   Complement fixation by varying quantities of chick oviduct chromatin (○) and nonhistone protein–DNA complex (●) in the presence of antiserum against chick oviduct nonhistone protein–DNA complex. From Chytil and Spelsberg (1).

nonhistone proteins during malignant growth. Wakabayshi and Hnilica (2) have found that anti-Novikoff-hepatoma nonhistone protein–DNA anti-serum also reacted with chromatins prepared from Walker carcinoma or AS-30D hepatomas. Furthermore, it was reported (20) that chromatins from fast-growing and poorly differentiated Morris hepatoma 7777 and 3924-A

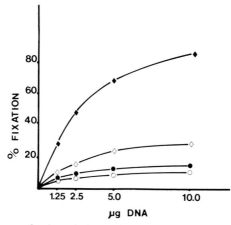

FIG. 3.   Complement fixation of chromatin preparations from Novikoff hepatoma (◇ or ◆) and normal rat liver (○ or ●) in the presence of antiserum against Novikoff hepatoma nonhistone protein–DNA complex. The antiserum was absorbed with normal rat liver chromatin (◆ or ●). Assays (◇ or ○) were performed in the presence of anti-serum against nonhistone protein–DNA complex from Novikoff hepatoma absorbed with Novkioff hepatoma chromatin. From Chiu et al. (12), with permission.

are more immunoreactive than better differentiated and slow-growing hepatomas 7800 and 7787 when anti-Novikoff-nonhistone protein–DNA antibody was employed. Normal liver chromatin differs antigenically from tumor chromatins; for instance, from transmissible sarcoma and Ehrlich ascitic tumor (20). Immunochemically specific sites inherent to nonhistone proteins can be induced by administration of a carcinogen-N-N-dimethyl-p-(m-tolylazo)aniline (20).

The character of the antigenic sites belonging to nonhistone protein is still obscure. McClure and Mayer (31) have found a nonuniform distribution of the antigenic sites within chromatin isolated from oviduct of estrogen-treated chicks. A tissue-specific immunochemically reactive fraction of non-histone proteins which has very high affinity to the homologous DNA (rat) showing two major and one minor band (12,000–15,000 MW) has been described (11). Low molecular fractions of nonhistone proteins with a tissue-specific antigenic characteristics binding very tightly to the homologous DNA has been recently found in rat liver and Novikoff hepatoma (30).

Experiments using reconstituted chromatins have shown that tissue-specific antigenic sites belonging to nonhistone proteins can be transferred from one DNA to another. Using antisera specific for the nonhistone–DNA complexes from a differentiated oviduct, it was found that the conformation of nonhistone proteins after transfer from chick oviduct to DNA of other organs is similar to that found in native chromatin (14). Wakabayashi and Hnilica (2) have used complement fixation to show that transfer of antigenic nonhistone proteins from Novikoff hepatoma chromatin requires presence of homologous (rat) DNA. Reconstitution of nonhistone proteins to other anions such as sea urchin DNA, yeast RNA, dextran sulfate, polyethylene sulfonate, or polyglutamic acid did not produce complexes reactive immunospecifically with anti-Novikoff-hepatoma nonhistone protein–DNA antibody; this suggests that the antinonhistone antibody can see only the conformation of nonhistone proteins influenced by the presence of the homologous DNA (20). These results were in contrast to Zardi et al. (3,4) who found that free nonhistone protein can react with antichromatin antibody. No presence of DNA was therefore necessary to the antigen–antibody reaction. Recent results obtained by Zardi (9) suggest that immunization with chromatin might induce a spectrum of nonhistone protein antibodies, some of which have higher affinity to the nonhistone protein–DNA complex than to free nonhistones. It is quite possible that individual differences among animals of the same species might produce various types of antibody.

The complement fixation method has been also used in assessing conformational changes which occur in chromatin after removing histone with acids (16).

VI.　Immunohistochemical Localization of Nonhistone Proteins

A.　Horse Radish Peroxidase Method

Principle: Tissue slices are incubated with a nonhistone protein antiserum produced, for instance, in rabbit. Then an immunoglobulin fraction (IgG) isolated from sheep (goat) serum immunized with rabbit serum IgG fraction conjugated with horse radish peroxidase (HRP) is added. The attachment of antinonhistone antibody to the cell components is localized by light microscopy after adding to the slide diaminobenzidine and hydrogen peroxide. Horse radish peroxidase attached to the second antibody catalyzes oxidation of diaminobenzidine by hydrogen peroxide. The reaction product exhibits a gold-brown color.

1.　PREPARATION OF PEROXIDASE-LABELED ANTIRABBIT IgG

The methods described here were introduced by Nakane and Pierce (*32*) and reviewed recently by Nakane (*33*). The method of labeling antirabbit IgG by HRP was worked out by Nakane and Kawaoi (*34*).

Principle: The antirabbit IgG fraction is conjugated with HRP in two steps. First the carbohydrate portion of HRP is oxidized to aldehyde. Sodium periodate is employed to react with HRP previously treated with fluorodinitrobenzene to block amino groups. The resulting HRP–aldehyde is then coupled with IgG.

a. Materials

1. Horse radish peroxidase (Sigma Type VI).
2. Freshly prepared sodium bicarbonate solution (0.3 M).
3. Fluorodinitrobenzene (1% in absolute alcohol).
4. Sodium periodate (0.06 M in water).
5. Ethylene glycol solution (0.14 M–dilute 0.9 ml of ethylene glycol, spe gr 1.113, to 100 ml with water).
6. 0.01 M sodium bicarbonate (pH 9.5).
7. 0.01 M phosphate-buffered saline (PBS): To 42 gm of NaCl add 6.89 gm of $NaH_2PO_4 \cdot H_2O$; dissolve in water, adjust pH to 7.1–7.2, and dilute to 5 liters.

b. Procedure

Preparation of horse radish peroxidase–aldehyde:

1. Ten milligrams of HRP are dissolved in a 30-ml centrifuge tube in 2 ml of freshly prepared sodium bicarbonate solution (0.3 M). A brownish solution is formed.
2. Then 0.2 ml of the fluorodinitrobenzene solution is added. If a precipitate is formed, it should be removed by centrifugation at 10,000 g for 5 minutes.

3. The clear solution is shaken for 1 hour at room temperature and 2 ml of sodium periodate solution are then added.

4. The solution is again shaken for 1 hour at room temperature.

5. Then 2 ml of the ethylene glycol solution are added and shaking is continued for another 1 hour.

6. The solution is dialyzed against four changes of 100–200 volumes of the sodium bicarbonate buffer (0.01 M), pH 9.5.

Conjugation of IgG with HRP–aldehyde:

1. To 5 mg of the IgG fraction 3 ml of the above HRP–aldehyde are added.

2. The mixture is then shaken for 2–3 hours and dialyzed against four changes of 100–200 volumes of 0.01 M phosphate-buffered saline (PBS).

3. The conjugate is then stored in aliquots at −20°C.

2. Localization Procedure

a. Materials

1. 0.01 M phosphate-buffered saline (PBS), pH 7.2.

2. 0.05 M Tris-HCl buffer (pH 7.6).

3. Diaminobenzidine staining solution [90 mg of diaminobenzidine hydrochloride are dissolved in 300 ml of 0.05 M Tris buffer (pH 7.6) and 0.05 ml of 30% hydrogen peroxide is added]. The solution is used immediately.

4. Other reagents: Ames O.C.T. Compound (Ames Company, Elkhart, Indiana), acetone, denatured ethyl alcohol, ethyl alcohol (50, 70, 95, and 100%), xylene, permount, Dry Ice, aluminium foil.

b. Procedure

Tissue preparation:

1. A piece of tissue is placed in a small cylinder prepared from the aluminum foil.

2. The tissue is covered by slowly pouring the O.C.T. compound into the cylinder.

3. The sample is then frozen by immersing the cylinder in a mixture of denatured alcohol and Dry Ice.

4. Sections 6 μm thick are prepared with a cryostat, placed on microscopic slides, and left for 30 minutes at room temperature to dry.

Staining procedure:

The following steps are performed on ice.

1. The slides with the tissue sections are immersed in ice-cold acetone for 10 minutes and removed.

2. Then they are washed 3 times in fresh PBS buffer, each time for 3 minutes.

3. The liquid around the tissue sections is wiped off with a piece of cotton.

4. The slides are then transferred to a wet chamber which is held moist by putting on the bottom a thick filter paper soaked with **PBS**.

5. Then a drop of sheep serum is applied on each section and left for 20 minutes. This treatment lowers the nonspecific staining.

6. The slides are again washed 3 times in **PBS** (as under 2).

7. A drop of the nonhistone antibody is then applied (usually diluted 1:10 or 1:20 with **PBS**) for 15–20 minutes.

8. The slides are again washed 3 times, as under 2.

9. Now a drop of HRP-conjugated IgG is placed on each slide for 15–30 minutes.

10. Washing is repeated as under 2.

The next steps are performed at room temperature.

11. The slides are placed in freshly prepared diaminobenzidine solution for 10 seconds–10 minutes.

12. The slides are washed 3 times in fresh **PBS** for 5 minutes and in deionized water for 5 minutes once.

13. Then the slides are placed for 5 minutes each time sequentially into 50, 70, 95% and absolute alcohol.

14. The slides are immersed in xylene for clearing and mounted with permount.

3. COMMENTS

The analysis of the attachment of antinonhistone antibody to the cell components by the above method requires careful experimental assessment of the incubation times employed. This depends on the antibody, sometimes on the batch of diaminobenzine, or on the actual concentration of hydrogen peroxide. The absence of naturally occurring antimembrane antibodies in some nonimmunized animals (8) always should be determined prior to the immunization of the animal. The optimum dilution of nonhistone protein antibody is a prerequisite for successful staining. As can be seen from Fig. 4, this method can show good interaction of the antisera with the nucleus with no signs of appreciable cytoplasmic staining. This method is also suitable for electron microscopy of the interaction of the antinonhistone protein antibody with nuclear components. The staining of the nucleus will be refined by using antibody against a single nonhistone protein species. The preparation of the HRP-IgG conjugate does not require special skills. Though Nakane and Kawaoi (34) have recommended the use of conjugate separated by gel filtration from the free **HRP**, we have found that this separation is not necessary to obtain good staining.

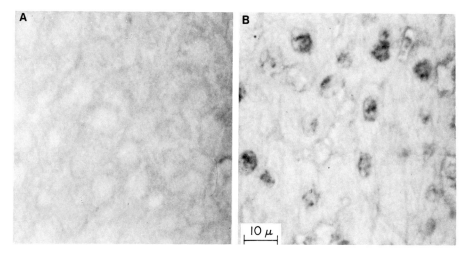

FIG. 4. Histochemical localization of anti-rat liver nonhistone protein–DNA antisera by the horse radish peroxidase technique. (A) Liver slice incubated with serum before immunization (B) Liver slice incubated with antiserum against liver nonhistone protein–DNA complex. From Chytil (5).

B. Other Methods of Localization

Coleman and Pratt (35) have compared the use of the method of Steinberger *et al.* (36) which employs antiperoxidase–peroxidase complex to localize nonhistone protein antibody. A fluorescein-labeled antibody technique has been used by Okita and Zardi (37), Scott and Sommerville (22), Silver and Elgin (21), and Elgin *et al.* (38). This method is also described by Silver *et al.*, this volume, Chapter 10.

C. Applications

Unlike the quantitative complement fixation technique, the localization methods have not yet gained popularity. The horse radish peroxidase method has been used to assess possible cytoplasmic contaminations of nonhistone protein preparations. Okita and Zardi (37) have used the fluorescein-labeled antibody method to localize antibodies against chromatin from 3T6 mouse fibroblasts and WI-38 human diploid fibroblasts. Species specificity of these antibodies was demonstrated. The nucleus was diffusely stained, and a diffuse fluorescence in the cytoplasm was observed which the authors interpreted as a suggestion of the presence of a cytoplasmic pool of chromosomal proteins. Because localization of antibody raised against nonhistone protein–DNA complex failed to show any cytoplasmic localization (5), the above results also could be interpreted to mean that the chromatin preparations, unlike the nonhistone protein–DNA complexes, did contain a cytoplasmic contamination. The indirect immunofluorescence method has been used very elegantly for *in situ* localization of *Drosophila* nonhistone proteins (*21,38*). Localization of nuclear proteins on chromosomes of newt oocytes has been demonstrated (*22*). The fluorescence technique facilitated the localization of rabbit antibodies in nucleoli of normal rat liver and Novikoff hepatoma (*29*).

ACKNOWLEDGMENTS

This work was supported by U.S. Public Health Service grants HD-05384, HD-09105, and HL-15341. I wish to thank L. Chytil for collaboration and discussions.

REFERENCES

1. Chytil, F., and Spelsberg, T. C., *Nature (London) New Biol.* **233**, 215 (1971).
2. Wakabayashi, K., and Hnilica, L. S., *Nature (London) New Biol.* **242**, 153 (1973).
3. Zardi, L., Lin, J. C., and Baserga, R., *Nature (London) New Biol.* **245**, 211 (1973).
4. Zardi, L., Lin, J. E., Peterson, R. O., and Baserga, R., *Cold Spring Conf. Cell Proliferation*, p. 729, 1974.
5. Chytil, F. *in* "Methods in Enzymology," Vol. 40: Hormone Action, Part E, Nuclear Structure and Function (B. W. O'Malley and J. G. Hardman, eds.), p. 191. Academic Press, New York, 1975.
6. Stumpf, W. E., Elgin, S. C. R., and Hood, L., *J. Immunol.* **113**, 1752 (1974).
7. Miescher, P., Cooper, N. S., and Benacerraf, B., *J. Immunol.* **85**, 27 (1960).
8. Wilson, E. M., and Chytil, F., *Biochim. Biophys. Acta* **426**, 88 (1976).
9. Zardi, L., *Eur. J. Biochem.* **55**, 231 (1975).
10. Wakabayashi, K., Wang, S., Hord, G., and Hnilica, L. S., *FEBS Lett.* **32**, 46 (1973).
11. Wakabayashi, K., Wang, S., and Hnilica, L. S., *Biochemistry* **13**, 1027 (1974).
12. Chiu, J. F., Hunt, M., and Hnilica, L. S., *Cancer Res.* **35**, 913 (1975).
13. Ashmarin, I. P., Gavriljuk, I. P., Konarev, V. G., Resnik, S. E., Sidorova, V. V., and Fedorova, N. A., *Cytologia* **16**, 1488 (1974).
14. Spelsberg, T. C., Steggles, A. W., Chytil, F., and O'Malley, B. W., *J. Biol. Chem.* **247**, 1368 (1972).
15. Spelsberg, T. C., Mitchell, W. M., Chytil, F., Wilson, E. M., and O'Malley, B. W., *Biochem. Biophys. Acta* **312**, 765 (1973).
16. Spelsberg, T. C., Mitchell, W. M., and Chytil, F., *Mol. Cell. Biochem.* **1**, 243 (1973).
17. Nicolini, C., *Physiol. Chem. Phys.* **7**, 571 (1975).
18. Chytil, F., Glasser, S. R., and Spelsberg, T. C., *Dev. Biol.* **37**, 295 (1973).
19. Chiu, J. F., Wakabayashi, K., Craddock, C., Morris, J. P., and Hnilica, L. S., *in* "Cell Cycle Controls" (G. M. Padilla, I. L. Cameron, and A. Zimmerman, eds.). Academic Press, New York, 1974.
20. Chiu, J. F., Craddock, C., Morris, H. P., and Hnilica, L. S., *FEBS Lett.* **42**, 94 (1974).
21. Silver, L. M., and Elgin, S. C. R., *Proc. Natl. Acad. Sci. U.S.A.* **73**, 423 (1976).
22. Scott, S. E. M., and Sommerville, J., *Nature (London)* **250**, 680 (1974).
23. Fudenberg, H. H., *in* "Methods in Immunology and Immunochemistry" (C. A. Williams, and M. W. Chase, eds.), Vol. 1, p. 307. Academic Press, New York, 1967.
24. Fahey, J. L., *in* "Methods in Immunology and Immunochemistry" (C. A. Williams, and M. W. Chase, eds.), Vol. 1, p. 321. Academic Press, New York, 1967.

25. Wasserman, E., and Levine, L., *J. Immunol.* **87**, 290 (1961).
26. Levine, L., *in* "Handbook of Experimental Immunology" (D. M. Klir, ed.), p. 701. Davis, Philadelphia, Pennsylvania, 1967.
27. Cikes, M., *J. Immunol. Methods* **8**, 89 (1975).
28. Roberts, D. B., and Andrews, P. W., *Nucleic Acids Res.* **2**, 1291 (1975).
29. Bush, R. K., Daskal, I., Spohn, W. H., Kellermayer, M., and Bush, H., *Cancer Res.* **34**, 2362 (1974).
30. Chiu, J. F., Wang, S., Fujitani, H., Hnilica, L. S., *Biochemistry* **14**, 4552 (1975).
31. McClure, M. E., and Mayer, J. L., *J. Cell. Biol.* **63**, 215 (1974).
32. Nakane, P. K., and Pierce, G. B. Jr., *J. Cell. Biol.* **33**, 307 (1967).
33. Nakane, P. K., *in* "Methods in Enzymology," Vol. 37: Hormone Action, Part B, Peptide Hormones (B. W. O'Malley and J. G. Hardman, eds.), p. 133. Academic Press, New York, 1975.
34. Nakane, P. K., and Kawaoi, A., *J. Histochem. Cytochem.* **22**, 1084 (1974).
35. Coleman, R. A., and Pratt, L. H., *J. Histochem. Cytochem.* **22**, 1039 (1974).
36. Steinberger, L. A., Hardy, P. H., Jr., Cucullis, J. J., and Meyer, H. G., *J. Histochem. Cytochem.* **28**, 315 (1970).
37. Okita, K., and Zardi, L., *Exp. Cell Res.* **86**, 59 (1974).
38. Elgin, S. C. R., Silver, L. M., and Wu, C. E. C., *in* "The Molecular and Biological Mammalian Genetic Apparatus" (P. O. P. Ts'o, ed.), Vol. 1, p. 127. North-Holland, Amsterdam, 1977.

Chapter 9

Radioimmunoassay of Nonhistone Proteins

MICHAEL E. COHEN AND LEWIS J. KLEINSMITH

Division of Biological Sciences,
The University of Michigan,
Ann Arbor, Michigan

A. REES MIDGLEY

Reproductive Endocrinology Program,
The University of Michigan,
Ann Arbor, Michigan

I. Introduction

The use of antibodies directed against nonhistone chromosomal proteins has provided a powerful new approach for characterizing and analyzing these proteins (1–6). Previous investigations in this field have employed the complement fixation assay to quantitate the reaction between these antibodies and the nonhistone protein antigens. Although many useful results have been obtained using complement fixation, development of a radioimmunoassay would theoretically offer several advantages. First, radioimmunoassay has a greater potential sensitivity, for it is capable of detecting proteins in nanogram or even picogram levels. In addition, radioimmunoassay is inherently more specific, for it is not subject to interference by nonspecific factors as is complement fixation. For these reasons we have undertaken the development of a radioimmunoassay procedure for the quantitation of specific nonhistone proteins. Extensive preliminary studies using iodinated nonhistone proteins as antigens were largely unsuccessful, due to the varying stabilities of nonhistone proteins after iodination. However, we have developed an indirect procedure based on an initial reaction of unlabeled nonhistone proteins with rabbit antinonhistone protein antibodies, followed by detection of the antigen–antibody complexes with radioiodinated anti-rabbit gamma globulin.

143

II. Methods

A. Preparation of Nonhistone Proteins

Nuclei were prepared from various rat tissues by homogenization in 3 volumes of 0.32 M sucrose–3 mM MgCl$_2$ with a motor-driven Teflon–glass homogenizer. The homogenate was filtered through two layers of cheese-cloth and centrifuged at 700 g for 10 minutes. The crude nuclear pellet was suspended in 0.25 M sucrose–1 mM MgCl$_2$–1% Triton X-100, collected by a second centrifugation under the same conditions, and then resuspended in 2.4 M sucrose–1 mM MgCl$_2$. Centrifugation for 1 hour at 70,000 g yielded a purified nuclear pellet which was resuspended in 0.25 M sucrose–1 mM MgCl$_2$ and collected by centrifugation at 700 g for 10 minutes.

After a second identical wash with 0.25 M sucrose–1 mM MgCl$_2$, the nuclei were suspended in 0.14 M NaCl and again collected by centrifugation at 700 g. The pellets were resuspended in 0.14 M NaCl (8 ml/ml packed nuclei), and an equal volume of 2.0 M NaCl–0.3 M Tris-HCl (pH 7.5) was then added. The suspension was homogenized with a Polytron PT-10 for 20 seconds at 60 V to dissolve the chromatin. The salt concentration was then lowered to 0.4 M by adding 1.5 volumes of 20 mM Tris-HCl (pH 7.5), and the DNA, histones, nuclear membranes, and nucleoli were removed by centrifugation at 95,000 g for 2 hours. The supernatant containing nonhistone proteins was stored at $-70\,°C$.

B. Immunization Procedures

Antisera against rat liver nonhistone proteins were prepared in New Zealand White virgin female rabbits by weekly intraperitoneal injections of 600 μg of protein emulsified in complete Freund's adjuvant (2 volumes adjuvant to 1 volume of protein solution). Six weeks after the beginning of immunization, the rabbits were bled from the marginal ear vein.

Antisera against rabbit gamma globulin (anti-RGG) was prepared according to the method of Niswender *et al.* (7). Adult male sheep were injected subcutaneously every 3 weeks with 50–100 mg of rabbit gamma globulin (fraction II, Pentex) emulsified in complete Freund's adjuvant (1 volume adjuvant to 1 volume protein solution). After a satisfactory titer of antibody was obtained, sheep were bled from the jugular vein at weekly intervals.

C. Purification of Anti-RGG

Antibodies directed against rabbit gamma globulin were purified from the serum of immunized sheep by affinity chromotography on columns of

Sepharose containing covalently bound rabbit gamma globulin. All procedures were carried out at 0°–4°C unless otherwise specified. Rabbit gamma globulin was prepared by adding an equal volume of ammonium sulfate, saturated at 0°C, to rabbit serum. The solution was allowed to stand for 20 minutes, and the precipitate was then collected by centrifugation at 10,000 g for 10 minutes. The pellet was redissolved in one-half volume of 0.14 M NaCl–0.01 M sodium phosphate, pH 7 (PBS). An equal volume of saturated ammonium sulfate was added, and the protein precipitate was again collected by centrifugation. Ammonium sulfate precipitation was repeated three more times, and the final pellet was then dissolved in PBS. Residual ammonium sulfate was removed by dialysis against 0.02 M sodium phosphate (pH 8.0), and any undissolved material was removed by centrifugation at 10,000 g for 10 minutes.

This rabbit gamma globulin preparation was further purified by chromatography on a 2.2 × 30 cm column of O-(diethylaminoethyl) cellulose (DEAE-cellulose) that had been equilibrated with 0.02 M sodium phosphate (pH 8.0). The dialyzed sample (15–20 ml) was applied to the column and eluted with the equilibration buffer at a flow rate of 60–70 ml/hour. Immunoglobulin fraction G (IgG) is eluted from the column in the flow-through peak.

The purified IgG was linked to cyanogen bromide-activated Sepharose that had been swollen in 1 mM HCl and washed with 200 ml of 1 mM HCl/gm of dry gel on a sintered glass funnel. IgG (12 ml, 1.8 mg/ml) which had been dialyzed against 0.5 M NaCl–0.1 M NaHCO$_3$ (pH 8.3) was mixed with the swollen gel (6 ml packed volume) and incubated with shaking at room temperature for 2 hours. The gel was collected on a sintered glass funnel and washed with 6 ml of coupling buffer. The gel was then incubated for 2 hours at room temperature with 6 ml of 1.5 M ethanolamine (pH 9.0) to block any active groups remaining on the Sepharose. The gel was then filtered again and noncovalently bound protein removed by alternatively washing with 0.5 M NaCl–0.1 M borate (pH 8) and 0.5 M NaCl–0.1 M acetate (pH 4) (three washes each, 10 ml of wash/ml of packed gel). The IgG–Sepharose complex was then washed 3 times with 10 volumes of PBS. The final preparation, which contained about 1 mg of IgG/packed ml of Sepharose, is stable at 0°–4°C for at least several months.

Sheep anti-RGG was purified by binding to the above rabbit IgG–Sepharose complex. After preliminary purification of sheep anti-rabbit serum by ammonium sulfate fractionation as described above for rabbit serum, the sheep globulin fraction (1 ml, 38 mg/ml in PBS) was incubated with the rabbit IgG–Sepharose complex (0.2 ml packed volume) with shaking at room temperature for 3–4 hours. The anti-RGG–IgG–Sepharose complex was then washed by suspension and centrifugation in 4-ml aliquots of PBS (once), 1 M NaCl–0.01 M glycine (pH 10) (twice), and 0.5 M sodium phosphate (pH 7). The anti-RGG was iodinated while still bound to the rabbit IgG–Sepharose complex.

D.   Iodination of Anti-RGG

The iodination reaction mixture contained 0.2 ml packed volume of anti-RGG–IgG–Sepharose, 4 mCi of Na^{125}I (carrier-free) in 5–20 μl, and 240 μg of chloramine T in 0.12 ml of 0.5 M sodium phosphate (pH 7). The mixture was incubated at room temperature with shaking for 3 minutes. The entire mixture was then transferred to a Pasteur pipette containing 0.5 ml of Bio-Gel P-10 (serving as a support). The column was washed with 5 ml of 1.0 M KI–0.01 M acetate (pH 7) to remove unreacted iodide, damaged anti-RGG, and nonspecifically adsorbed protein. The ^{125}I-labeled anti-RGG was then eluted with 1.0 M KI–0.01 M acetate (pH 3). Ten-drop fractions were collected into 1-ml aliquots of 0.5 M sodium phosphate (pH 7.0) containing 0.1% gelatin. Tubes containing the iodinated anti-RGG were pooled, dialyzed against PBS, and stored at 0°–4°C. The material remained immunoreactive for at least a month.

E.   Radioimmunoassay of Nonhistone Proteins

Nonhistone proteins (NHP) were linked to Sepharose via the same procedure described for rabbit IgG. Fifty microliters of packed volume of NHP–Sepharose in 0.2 ml of PBS were mixed with 0.2 ml of rabbit antiserum against NHP and 0.1 ml of PBS containing 5% Triton X-100. After 2 hours' incubation at room temperature with shaking, 4 ml of 1 M NaCl–0.01 M glycine (pH 10) was added and the gel collected by centrifugation at 1000 g for 5 minutes. After removing the supernatant by aspiration, the gel was washed by suspension and centrifugation in 4 ml of 1.0 M NaCl–0.01 M glycine (pH 10) (once) and 4-ml aliquots of PBS containing 1% Triton X-100 (twice). The washed gel was then resuspended in 0.5 ml of PBS containing 5% normal sheep serum and ^{125}I-labeled anti-RGG (30,000–60,000 cpm). After 2 hours' incubation at room temperature with shaking, the gel was collected by centrifugation at 1000 g for 5 minutes and washed twice by suspension and centrifugation in 1.0 M NaCl–0.01 M glycine (pH 10). Labeled antibody was then eluted from the NHP–Sepharose complex by suspending the gel in 1 ml of 3 M ammonium thiocyanate. The suspension was transferred to a scintillation vial, mixed with 10 ml of Triton-toluene (1:2) scintillation fluid, and counted in a liquid scintillation spectrometer.

In competition experiments, antisera were absorbed by incubating them

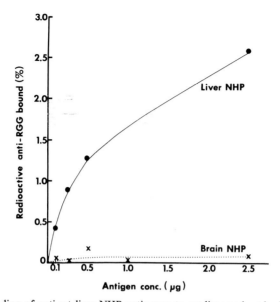

FIG. 1. Binding of anti-rat liver NHP antiserum to rat liver and rat brain NHP as measured by the indirect radioimmunoassay described in text. Antiserum was preabsorbed by incubating with brain NHP–Sepharose (5 μg) prior to testing in this radioimmunoassay.

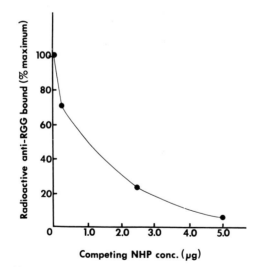

FIG. 2. Competition curve for binding of anti-rat liver NHP antiserum to rat liver NHP. Competition was accomplished by prior absorption of antisera with varying amounts of rat liver NHP–Sepharose, removing reacted antibody by centrifugation. The absorbed antisera were then tested for their ability to bind to fresh rat liver NHP–Sepharose (2.5 μg protein), using the indirect radioimmunoassay described in the text.

with appropriate NHP–Sepharose preparations and removing the reacted antibodies by centrifugation prior to using them in the routine binding assay.

III. Typical Results

The binding of anti-rat liver NHP antibodies to rat liver and brain NHP preparations as detected by the indirect radioimmunoassay is illustrated in Fig. 1. Not only does this figure reveal that as little as 100 ng of rat liver NHP can be easily detected, but the specificity of the reaction is also evident by the great difference in binding between liver and brain NHP. This degree of specificity requires that the anti-liver NHP antibodies be first absorbed with brain extract prior to running of the assay.

Figure 2 illustrates the type of competition curve obtained when anti-liver NHP antibodies are reacted with competing liver NHP prior to assay for binding to liver NHP. The ability to demonstrate competition in this way permits this system to be used to compare the similarities between different NHP preparations. As is shown in Table I, thymus and kidney NHP compete only about half as well as does liver NHP under the same conditions. Interestingly, mouse and liver NHP barely compete at all, suggesting the existence of prominent species specificity in the antigenic composition of nonhistone proteins.

TABLE I

SPECIFICITY OF ANTIGEN–ANTIBODY REACTION AS DETERMINED BY COMPETITION

Competing protein	Radioactive anti-RGG bound (% of maximum)
Blank gel	100
Rat liver nonhistone protein [a]	6
Rat thymus nonhistone protein	59
Rat kidney nonhistone protein	45
Bovine liver nonhistone protein	83
Mouse liver nonhistone protein	90

[a] Antiserum to rat liver NHP was absorbed with 5 μg of competing protein prior to radioimmunoassay against rat liver NHP

ACKNOWLEDGMENTS

These studies were supported by National Institutes of Health grant HD-08333 from the United States Public Health Service.

REFERENCES

1. Chytil, F., and Spelsberg, T. C., *Nature (London) New Biol.* **233**, 215 (1971).
2. Wakabayashi, K., and Hnilica, L. S., *Nature (London) New Biol.* **242**, 153 (1973).
3. Zardi, L., Lin, J.-C., and Baserga, R., *Natuure (London) New Biol.* **245**, 211 (1973).
4. Wakabayashi, K., Wang, S., Hord, G., and Hnilica, L. S., *FEBS Lett.* **32**, 46 (1973).
5. Wakabayashi, K. Wang, S., and Hnilica, L. S., *Biochemistry* **13**, 1027 (1974).
6. Zardi, L., *Eur. J. Biochem.* **55**, 231 (1975).
7. Niswender, G. D., Reichert, L. E., Midgley, A. R., and Nalbandov, A. V., *Endocrinology* **84**, 1166 (1964).

Chapter 10

Immunofluorescent Techniques in the Analysis of Chromosomal Proteins

L. M. SILVER

The Committee on Higher Degrees in Biophysics,
Harvard University,
Cambridge, Massachusetts

C. E. C. WU

Department of Biology,
Harvard University,
Cambridge, Massachusetts

S. C. R. ELGIN

Department of Biochemistry and Molecular Biology,
Harvard University,
Cambridge, Massachusetts

I. Introduction

Progress in the analysis of the nonhistone chromosomal proteins (NHC proteins) has been significant, but still disappointingly slow, during the last decade. Difficulties include the complexity of this class of proteins and the fact that many of the major NHC proteins apparently play structural roles for which no direct assay is available. The use of specific antibody techniques appears likely to assist in resolving these problems. Specific antibodies against chromosomal proteins can be used to determine the distribution of these proteins *in situ* using indirect "staining" techniques. The *in situ* distribution patterns obtained can suggest as well as test hypotheses concerning structural and active roles of these proteins. Specific antibodies, once obtained, can also be used for rapid purification of components from complex mixtures and as probes in enzyme assays. This chapter will primarily discuss the production of antibodies against chromosomal proteins of *Drosophila* and their use in an indirect immunofluorescent assay to determine the *in situ* distribution of these proteins.

II. Preparation of Antigens and Antisera

Drosophila melanogaster is the obvious organism of choice for these studies sine the polytene chromosomes of the third instar larva provide the best substrate for tests of the organization of macromolecules in chromosomes. The 1000-fold amplification of the euchromatic material in polytene chromosomes provides an increase in the sensitivity of the test, while the reproducible spatial localization inherent in the banding pattern allows identification and resolution at the level of the chromomere, the apparent unit of genetic organization. Consequently, all the following work was carried out using *Drosophila* chromatin and chromosomal proteins. We see no inherent reason, however, that would prevent many analogous studies of interest from being carried out using other organisms and other test substrates, including interphase nuclei, metaphase chromosomes, and lampbrush chromosomes [see Scott and Sommerville (*1*) for an example of the latter].

Drosophila melanogaster (Oregon R) flies are mass cultured as described by Elgin and Miller (*2*). Purified embryo chromatin is prepared essentially as described by Elgin and Hood (*3*) with minor modifications (*4*). Chromosomal proteins can be prepared from chromatin by any of several techniques, including fractionation of salt–urea dissociated chromatin on hydroxyapatite (*4,5*), or acid extraction of histone followed by SDS solubilization of the NHC proteins and removal of the DNA (*6*). In the former case the protein fractions are dialyzed extensively against 10 mM acetic acid, lyophilized, and stored at $-20°$ for subsequent use. *Drosophila* NHC proteins isolated by this technique, but not the second, appear to be completely free of histone contamination as indicated by analysis using gel electrophoresis.

Antiserum against total NHC proteins can be prepared as follows. Two milligrams of NHC protein dissolved in phosphate-buffered saline [10 mM sodium phosphate (pH 7.2)–0.15 M NaCl; PBS]–0.1% SDS are mixed in a 1:1 emulsion with Freund's complete adjuvant (Difco Laboratories, Detroit, Michigan). The preparation is injected subcutaneously into the neck region of an adult female rabbit. Four weeks later the injection is repeated using 6 mg of NHCP in PBS–0.1% sodium dodecyl sulfate (SDS). Rabbits are bled from the major ear arteries between 6 and 14 days after the booster injection. Antiserum is collected and stored at $-20°$ with 0.02% NaN$_3$. When desired, immunoglobulin fraction G (IgG fraction) is purified from the sera by precipitation with 1.75 M ammonium sulfate, dialysis to 17.5 mM phosphate buffer, and chromatography on O-(diethylaminoethyl) cellulose (DEAE-cellulose) (7).

Antisera against specific NHC protein molecular-weight subfractions are produced from SDS–polyacrylamide slab-gel bands essentially as described by Tjian *et al.* (8). Gel electrophoresis in SDS is carried out according to the procedure of Laemmli (9) (see below). About 20 mg of total NHC protein in 2 ml of Laemmli sample buffer (concentrated if necessary by pressure filtration using an Amicon P10 filter) are loaded onto the flat surface of a 5-mm thick by 15-cm wide slab gel. The sample is electrophoresed at 125 V until the running front has nearly reached the gel bottom. Major protein bands are observed following staining with Coomassie blue for 5 minutes at 37° and destaining at ambient temperature with distilled water for 30–60 minutes. These bands are sliced out, lyophilized, and ground to a fine powder with a mortar and pestle. Then 0.5 ml of powder is added with 0.4 ml of Freunds adjuvant (Difco) and 0.8 ml of PBS–0.1% SDS to a 5-ml syringe. This syringe is linked with a short piece of rubber tubing to a second syringe, and the sample is shot back and forth to obtain an even suspension. The suspension is injected subcutaneously into the neck of a rabbit with a 16 gauge needle. This primary injection is repeated every 2 days, until a total of 0.6–1.0 mg of protein has been received (usually 2 to 3 injections). (The amount of protein in the gel may be estimated by reelectrophoresis of a weighed portion and comparison to known standards by densitometry.) A single booster injection, containing up to 1 ml of powder, without Freund's adjuvant, is given 4 weeks after the first primary injection; the rabbits are bled 7–14 days later. Antibodies to histone 1 of high titer have also been produced by this technique (T. Ashley, personal communication).

III. Characterization of Antiserum by Indirect Immunofluorescent Staining of SDS–Polyacrylamide Gels

Stumph *et al.* (*10*) have demonstrated that SDS–polyacrylamide gels can be used in an indirect immunofluorescent staining technique to assess the specificity of an antiserum for a component of a complex mixture. We have slightly modified this protocol in an attempt to increase resolution and economize on time and serum quantities used by scaling down to gels of smaller diameter and by fixing the proteins by precipitation in the gel prior to indirect immunofluorescent staining.

A. Gel Electrophoresis

Polyacrylamide gel electrophoresis was performed according to the discontinous SDS–Tris–Glycine system of Laemmli (9). NHC proteins and protein standards are dissolved in sample buffer [62.5 mM Tris-HCl (pH 6.8)–2% SDS–10% glycerol–5% β-mercaptoethanol] and heated to 100 °C for 1 minute. The gel solution [final concentrations 0.375 M Tris-HCl (pH 8.8)–0.1% SDS–10% acrylamide–0.025% TEMED–0.025% ammonium persulfate] is introduced by Pasteur pipette into siliconized (Siliclad, Clay Adams, Parsippany, New Jersey) glass tubes of length 7 cm and inner diameter 2.5 mm, sealed well at one end with Parafilm and taped vertically to a stand. Tubes are filled to 6.2-cm height and overlayered with water-saturated isobutanol. Prior to electrophoresis, the isobutanol is removed by flicking the tubes, and both gel surfaces are rinsed extensively with electrode buffer [25 mM Tris base–0.192 M glycine–0.05% SDS (pH 8.3)].

After sample application, gels are electrophoresed at 25 V (constant voltage) for 5½ hours. Gels are extruded by carefully rimming with a narrow-gauge needle (27 guage or higher) using a syringe filled with electrode buffer and subsequently gently applying compressed air pressure from a large rubber bulb. Gels for immunofluorescent staining are immediately fixed in 40% methanol for 2 hours and then washed in PBS for 3 hours with hourly changes (Robert Jackson, personal communication). Gels for comparison purposes are immediately stained with Coomassie blue (0.025% Coomassie blue–25% isopropyl alcohol–10% acetic acid, overnight; 0.0025% Coomassie blue–10% isopropyl alcohol–10% acetic acid, 6–9 hours; 0.0025% Coomassie blue–10% acetic acid, overnight) and destained in 10% acetic acid (11).

B. Indirect Immunofluorescent Staining

Antisera against native proteins or SDS-solubilized proteins can be used directly for gel staining. Antisera produced against polyacrylamide–protein emulsions as described above must be adsorbed with polyacrylamide prior to use for gel staining. Typically, equal volumes of extensively PBS-washed pulverized polyacrylamide and antiserum are incubated at 4° for 12 hours. The polyacrylamide particles are then removed by centrifugation and the adsorbed serum decanted. Nonadsorbed serum causes intense immunofluorescent staining over the whole gel.

The antibody staining procedure is as follows. Each gel is placed in a siliconized glass tube, 6.5 cm long with a 4-mm internal diameter, sealed at one end with Parafilm. (Note: Keep fingerprints off the gels!) Most of the PBS that enters the tube with the gel is drained onto tissue paper. The antiserum is then pipetted into the tube almost to the brim. The top is sealed with Parafilm and the tube placed horizontally on a shaker and left to equilibrate at room temperature for 24 hours with agitation. The volume of serum used is about 0.7 ml/gel. Gels are then individually washed at room temperature with agitation in PBS (40 ml/gel) for 48 hours; the PBS is changed every 12 hours. Gels are stained with a 1:10 dilution of fluorescein isothiocyanate (FITC)-conjugated goat anti-rabbit IgG (Miles Labs, Elkhart, Indiana) in PBS as above. This and all subsequent steps involving FITC – IgG are performed in the dark or under red light. The gels are washed as above in PBS for 48 hours.

For photography the gels are placed on a grooved Plexiglass plate (painted dull black) to prevent curling during long exposure periods. Ten-minute exposures are taken under ultraviolet (UV) light (Mineralight, UV Products, San Gabriel, California) with Polaroid Type 105 film using a Kodak Wratten #3 gelatin filter. The negatives are scanned at 550 nm using a spectrophotometer with linear carriage. After photography, gels can be stained in Coomassie blue for direct comparison.

C.   Results

Figure 1A and B illustrates the validity of the method. In Fig. 1A, total NHC proteins have been electrophoresed on each gel. About 50 μg of protein in 18 μl of sample buffer are normally used. Gel b demonstrates that anti-bovine serum albumin (BSA) does not bind to any NHC protein bands specifically, although there is background fluorescence. The two fluorescent bands at the running front in gels a, b, and c are an artifact of FITC–goat IgG binding; gel a, which was not pretreated with rabbit antiserum before secondary staining with FITC–goat IgG, also shows this staining. Gel c indicates the heterogeneity of the antibodies produced against total NHC proteins purified by hydroxyapatite chromatography. Gel d is gel c stained with Coomasie blue.

In Fig. 1B, the inverse control, it is clear that anti-NHC protein sera does not cause staining of standard proteins, nor does anti-BSA sera cause the staining of the three other standard proteins, ovalbumin, chymotrypsinogen, and cytochrome c.

An analysis of the sera elicited using molecular-weight subfractions of the NHC proteins in SDS–polyacrylamide gels is presented in Fig. 1C. It is clear that the majority of the components of the antisera elicited using these molecular-weight subfractions specifically stain those subfractions of the NHC proteins, although there is background staining and the anti-ρ serum has components that stain the π polypeptides. Background staining might be reduced by the inclusion of nonspecific carrier bovine IgG with the rabbit antiserum during the primary interaction.

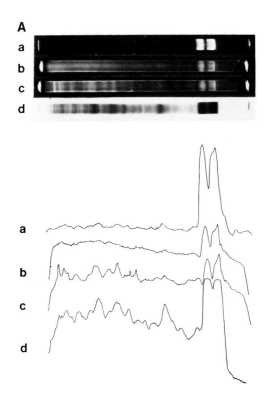

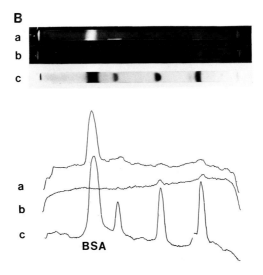

BSA

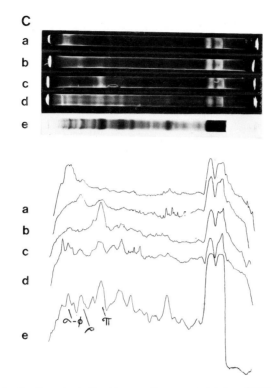

FIG. 1. Analysis of antisera by indirect immunofluorescent staining of SDS–polyacrylamide gel. All gels run left to right. (A) Analysis of antibodies produced using total NHC proteins. See text for details. (B) Control analysis of antibody specificity. A mixture of four standard proteins (BSA, ovalbumin, chymotrypsinogen, and cytochrome c, in order of decreasing molecular weight) were electrophoresed on each gel. Gel a was treated with anti-BSA serum; gel b was treated with anti-NHCP serum. Both gels were subsequently treated with goat fluorescein-conjugated anti-rabbit IgG. Gel c is gel a stained with Coomassie blue. (C) Analysis of specific antisera. Total NHC proteins were electrophoresed on all gels. Gel a was treated with anti-σ-ϕ sera, gel b with anti-ρ sera, gel c with anti-π sera, and gel d with anti-NHCP sera; all gels were subsequently treated with goat fluorescein-conjugated anti-rabbit IgG. Gel e was stained with Coomassie blue.

IV.   Chromosome Fixation and Squashing

A.   Traditional Techniques of Chromosome Fixation

The traditional techniques for preparing polytene chromosome spreads of good morphology use a squashing solution containing 45% acetic acid without other fixation. Unfortunately, about 50% of the total histone content of chromatin is extracted by 45% acetic acid within 30 seconds (*12*). Formaldehyde has often been used as an effective agent for cross-linking chromatin (*13*). However, polytene chromosome spreads are not readily prepared from salivary glands which have been extensively fixed with formaldehyde. A 45% acetic acid–3.3% formaldehyde fixation solution allows one to prepare chromosome spreads of good morphology, but such a solution also causes the extraction of most histones and 10% of the NHC proteins from purified *Drosophila* chromatin (*14*). Such a result is explained by the observations that acetic acid penetrates rapidly into tissue, while formaldehyde reacts at a much slower rate (*15*).

We have developed a new technique for the fixation and spreading of polytene chromosomes for subsequent staining with antisera directed against chromosomal proteins. The fixation appears to preserve the arrangement of macromolecules in the chromosomes as it exists *in vivo*, but at the same time allows the spreading out of chromosome arms, necessary for cytological visualization of specific chromosomal regions. The morphological integrity of the chromosomes is not adversely affected by the process.

B. Formaldehyde Fixation Procedure

All buffers and fixatives are made fresh from stock solutions on the day of the experiment. All stock solutions are stored containing 0.01% NaN_3. Coverslips and slides are washed for 10 minutes in 5% Chem-Solv (Mallinkcrodt, St Louis, Missouri). Slides are then rinsed in distilled water and baked dry. Coverslips are rinsed in distilled water and either baked dry immediately or placed in a 1% Siliclad solution for 10 seconds before rinsing in distilled water again and baking dry.

Larvae are grown in plastic half-pint bottles containing approximately 50 ml of corn flour media (*2*). Each bottle is inoculated with approximately 50 eggs and placed in a humid incubator at 17°C. After 14 to 15 days, third instar larvae begin to crawl up the sides of the bottle. Each larva is removed from the bottle with a needle and placed into a well on a three-spot depression plate containing a drop of Cohen and Gotchel (*16*) Medium G (25 m*M* disodium glycerophosphate–10 m*M* KH_2PO_4–30 m*M* KCl–10 m*M* $MgCl_2$– 3 m*M* $CaCl_2$–162 m*M* sucrose).

The larva is washed in this drop and then transferred to a second drop of Medium G containing 0.5% Nonidet P40 (Shell Chemicals) (*p-tert*-octylphenyl polyoxyethylene with an average of 9 ethylene units). Using No. 5 jeweller's tweezers, the salivary glands are dissected out of the larva. At this point, the procedure should be continued only if the nuclei within the glands appear to be fully replicated. Large nuclei are essential for good results. Care must be taken not to damage the glands during this initial incubation. Intact glands remain transparent; injured glands become cloudy.

After a 10-minute incubation in Medium G–Nonident P40, the glands are transferred on a small (2-mm square) piece of nylon material (Nitex) to the fixative solution [100 mM NaCl–2 mM KCl–10 mM MgCl$_2$–0.5% Nonidet P40–2% formaldehyde–19 mM sodium phosphate (pH 7.2)]. The glands are fixed for 30 minutes at 22°–24° and then transferred to the squashing solution (45% acetic acid–10 mM MgCl$_2$). Extraneous material associated with the glands should be dissected away at this time. After a period of at least 2 minutes, glands are transferred to a 10-μl drop of squashing solution on a dust-free siliconized coverslip. If desired, the two gland lobes can be separated and placed on individual coverslips for subsequent comparative analysis using different sera. Glands are incubated for a total time of from 10 to 25 minutes in squashing solution before squashing as described below.

C.   Preparation of Polytene Chromosome Spreads

A clean microscope slide is touched to the drop on the coverslip, and the slide is placed, coverslip up, on top of a piece of black filter paper to make the glands visible. A rubber-tipped needle is used to rapidly move the coverslip back and forth horizontally on the slide, applying minimal vertical pressure. This action breaks open the glands, disperses the nuclei, and is sometimes sufficient to shear open the nuclear envelope. If the nuclei have not broken, gentle tapping with a metal needle, allowing some horizontal movement, is helpful. Finally, extensive, but not intensive, tapping with the eraser end of a pencil, not allowing coverslip movement, is used to spread apart and flatten the chromosome arms. Flat spreads are essential for good phase-contrast morphology, but oversquashing causes a loss of contrast between chromosomal bands and interbands. Thumb pressure is never applied to the slides.

When the chromosome preparations are spread sufficiently, the slide is plunged into liquid nitrogen for 15 seconds. The coverslip is immediately pried off with a razor blade, and the slide is quickly placed horizontally into a petri dish of postfixative solution, PBS containing 3.7% formaldehyde. After 5 minutes, the slide is placed vertically into a rack in TBS [Tris-buffered saline, 0.01 M Tris, 0.15 M NaCl prepared from stock, 1 M Tris buffer (140.4 gm Trizma HCl plus 13.4 gm Trizma base per liter), pH 7.71 at 4° and 7.15 at 25°]. When laboratory humidity is low, it is more difficult to obtain well-spread chromosomes.

Slides can be used immediately for staining or stored for future use. In the latter case, the rack of slides is removed from the TBS solution and washed in TBS two times, five minutes each, with agitation. The rack is then placed into a holder containing a 67% glycerol–33% PBS solution, and this is stored at $-20°$. Slides have been successfully stored in this condition for up to 24 days.

D. Principles of Fixation

Cohen and Gotchel Medium G has a total osmolarity (0.356) and pH (6.78) which is close to that of *Drosophila* hemolymph (*16*). Glands retain excellent nuclear morphology in this medium for extended periods of time, as indicated by a maintenance of chromosomal banding structure within the nucleus. When glands are incubated for 10 minutes in Medium G containing 0.5% Nonidet P40, cytoplasmic membrane structures disappear, while nuclear morphology remains excellent as observed by phase-contrast microscopy.

Nonionic detergents, in the presence of divalent cations, have long been used to isolate nuclei (*17*). Electron microscopic observations and biochemical analyses indicate that 0.5% Triton X-100 or 0.5% Nonidet P40 [two detergents that are almost identical (*18*)] solubilize cytoplasmic membrane structures and at least the lipid components of the double membrane of the nucleus (*17, 19, 20*). However, the peripheral lamina (dense lamella) appears to act as a nuclear skeleton which maintains chromosomal integrity, even when both outer and inner nuclear membranes are partially, perhaps completely, dissolved (*20–22*). In addition, nonionic detergents do not appear to extract or greatly rearrange chromosomal proteins (*3, 19, 23*).

Following incubation in Medium G containing Nonidet P40, glands are placed into the formalydehyde fixation buffer for a period of 30 minutes. Only nuclear components can become extensively cross-linked, since most other cellular components are now solubilized. The ionic strength of the fixation buffer is sufficiently high (0.18) to insure irreversible fixation of the chromosomal proteins (*24*). Time course studies on the extent of histone fixation in calf thymus chromatin indicate that the action of formaldehyde is nearly complete after 20 minutes at ambient temperature (*15*).

To verify the fixation of chromosomal proteins by this procedure, purified *Drosophila* embryo chromatin in Medium G containing Nonidet P40 was mixed with an equal volume of 2 times concentrated fixation buffer. After the appropriate fixation period, the chromatin was recovered by centrifugation (17,000 g, 10 minutes) and sequentially extracted with the squashing solution (45% acetic acid–10 mM MgCl$_2$) and with 50 mM Tris (pH 8)–1% SDS. Proteins extracted by the fixative were collected by precipitation with 5% perchloroacetic acid; the pellet then was washed with 95% ethanol and lyophilized. The squashing solution extracts were dialyzed and lyophilized. The Tris–SDS extracts were dialyzed directly to SDS gel sample buffer. All extracts were analyzed by SDS gel electrophoresis (*9*).

As determined by disc gel electrophoresis, no detectable protein (less than 1%) was extracted by either the fixation buffer or the acetic acid squashing solution using this protocol (*14*). Most of the histones and approximately 10% of the NHC proteins were extracted from unfixed *Drosophila* chromatin by the squashing solution even if it contained 3.3% formaldehyde. About 50% of the NHC proteins could be subsequently extracted from unfixed chromatin with 1% SDS, while less than 10% of the NHC proteins from formaldehyde fixed chromatin could be extracted with 1% SDS.

A fixation procedure utilizing 0.6% glutaraldehyde in place of 2% formaldehyde, with a fixation time of only 15 minutes, allows no detectable extraction of chromosomal proteins from purified chromatin by either squashing solution or 1% SDS. However, it is extremely difficult to obtain flat, well-spread polytene chromosomes, and therefore good phase-contrast morphology, from glutaraldehyde-fixed salivary glands.

V.   Indirect Immunofluorescent Staining of Polytene Chromosomes

A.   Procedure

The following operations are carried out at 22°–24°, with the exception of the wash steps which use TBS at 4°. The rack of slides is removed from the postfixative solution, or the glycerin–PBS storage solution, and washed in TBS 3 times for 5 or 7 minutes each. It is critical that the squash region of the slide remain wet during the entire washing and incubation procedure. Slides are removed, one at a time, from the final TBS wash; the bottom side of the slide and the top half of the slide furthest from the squash are quickly blotted dry on a paper towel, and the slide is placed onto a polystyrene frame within a humidity tank containing water-soaked sponges. One hundred microliters of an appropriate dilution of antiserum is quickly released onto the squash. Slides are incubated for 30 minutes and are then washed 3 times for 5 minutes each in TBS. Antiserum was used at dilutions from 1:2 to 1:40 (depending on titer) in TBS containing 1–2 mg/ml of bovine gamma globulin (Sigma) as carrier to reduce nonspecific staining.

The second antibody incubation is performed as described above, using a 1:50 dilution of the FITC-conjugated IgG fraction of goat anti-(rabbit gamma globulin) serum (Miles Labs) in TBS. This and subsequent operations can be carried out in a darkroom with red light. The slides are again washed 3 times in TBS. A drop of a 9:1 solution of glycerin: 1 *M* Tris-HCl (pH 8.1) is placed directly onto the squash area of each wet slide, and a dust-free non-siliconized coverslip is mounted using intensive thumb pressure over a tissue blotter. Slides are viewed with incident UV illumination and with phase-contrast optics using a ×40 oil-immersion lens (Leitz). Photographs are taken using Tri-X film (Kodak). Slides stored in the dark at −20 °C for future viewing remain in good condition for many months.

B.   Results

We have demonstrated that our fixation procedure will prevent extraction of chromosomal proteins from purified *Drosophila* chromatin. As an *in vivo* test for fixation we have stained polytene chromosomes using antiserum against histone 1. Polytene chromosomes from salivary glands "fixed" with 45% acetic acid–3.3% formaldehyde are not stained by antihistone 1, while polytene chromosomes spreads prepared using formaldehyde fixation stain brightly using antihistone 1 serum (L. M. Silver, T. H. Ashley, and S. C. R. Elgin, unpublished results).

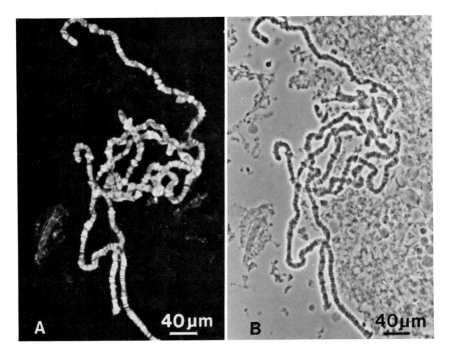

FIG. 2.   Polytene chromosome spread prepared by formaldehyde fixation and stained as described using antiserum against total *Drosophila* NHC proteins, viewed using (A) UV dark field and (B) phase-contrast optics.

Squashes stained using antiserum directed against total NHC proteins show specific fluorescence of the polytene chromosomes; occasionally fragments of what appears to be the ghost of the nuclear envelope (22) are also stained (see Fig. 2). Each chromomere of the polytene chromosomes is stained (or not stained) uniformly along the width of the chromosome, as anticipated; the polytene chromosomes are the consequence of replication without chromatid separation, so that each chromomere consists of a lateral amplification of the chromatin by a factor of up to 1024 (25). The complete fluorescent pattern of each polytene chromosome spread is reproducible from one set of polytene chromosomes to the next in a given salivary gland squash.

The staining pattern obtained with third instar larvae from different puffing stages are very similar. The differences we have observed are found primarily at bands which are puffed in one stage, but not the other. Most puffs stain brightly over the level of general chromosome staining. This observation is in concurrence with experiments demonstrating an accumulation of "acidic protein" at puff sites (26).

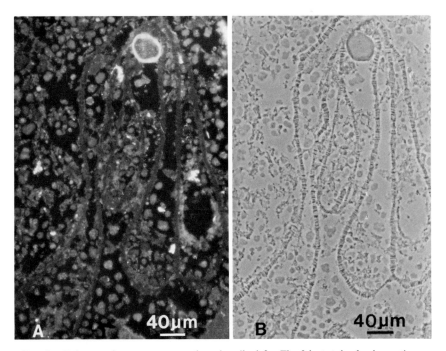

Fig. 3. Polytene chromosome spread as described for Fig. 2 but stained using antiserum against BSA.

That the staining pattern observed is a consequence of specific antibody–antigen interactions is substantiated by several control experiments. No staining is observed if the chromosome preparation is initially incubated with PBS, preimmunization rabbit serum, or rabbit antiserum against bovine serum albumin (Fig. 3) before incubation with the fluorescein-conjugated goat antibody against rabbit IgG. Adsorption of the anti-NHC protein serum with chromatin, or a suspension of total NHC protein (0.5 mg/ml) in 10 mM Tris (pH 8) prior to use eliminates staining. However, preincubation of such an antiserum with calf thymus DNA (0.8 mg/ml) or yeast RNA (0.5 mg/ml) has no observable effect on the staining pattern.

A comparison of the fluorescent patterns obtained with chromosomes treated by each of the fixation techniques used indicates that the formaldehyde and gluteraldehyde techniques yield the same qualitative fluorescent patterns, at the present level of resolution, with two different anti-NHC protein antisera (Fig. 4). The fluorescent pattern obtained with traditionally fixed chromosomes is of lower contrast and resolution.

The fluorescent pattern obtained by staining with antiserum against NHC protein molecular-weight subfraction ρ is most interesting (Fig. 5). A very low percentage of the genome is prominently stained—primarily puffs and a few other bands. This result suggests that at least some of the antigenic determinants of the nonhistone chromosomal proteins are distributed in a limited and specific fashion. Scott and Somerville (*1*) have observed similar results in one case of staining newt lampbrush chromosomes using antibodies against a particular class of proteins from ribonucleoprotein particles. Results using antibodies produced by several immunized animals should be checked before conclusions are drawn correlating a given distribution pattern with a given protein; different rabbits can respond differently to different antigenic determinants. It should be noted that unstained regions could be a consequence of limited accessibility of some antigenic determinants following fixation, as well as a consequence of differential distribution. It has been observed that prominent, dense chromomeres that are heavily stained using anti-NHC protein sera are not well stained using antihistone sera. There is no reason to think that these chromosome regions do not include histones; rather it seems likely that a high concentration of NHC proteins can obscure the histones in the fixed chromosomes. Alfagame *et al.* (*27*) have also observed extensive staining using antihistone sera following a milder fixation technique. Current models of chromatin structure suggest that the DNA may be wound around the outside of the histone v-body core, with the nonhistone chromosomal proteins associated with the other DNA surface, forming the outer layer of the fiber structure [see Elgin and Weintraub (*28*) for review]. If correct, this model would imply that the NHC proteins should be much more accessible than the histones to an antibody probe.

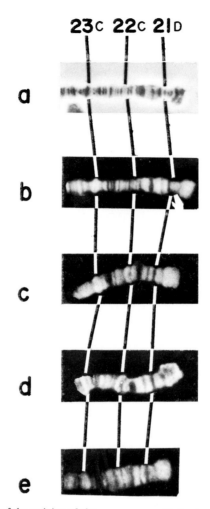

FIG. 4. Comparison of the staining of chromosome arm 2L (regions 21–23) using two different anti-NHCP sera (b,c) following preparation by different techniques: phase-contrast view of formaldehyde-fixed chromosome stained using anti-NHCP-2 (a); UV dark field view of the same chromosome (b); formaldehyde-fixed chromosome stained using anti-NHCP-1 (c); glutaraldehyde-fixed chromosome stained using anti-NHCP-2 (d); traditionally fixed (45% acetic acid) chromosome stained using anti-NHCP-2 (e). NHCP-1 was prepared from acid-extracted chromatin; NHCP-2 was prepared by hydroxylapatite chromatography.

Given the above caveats, all available evidence indicates that the antibody staining technique described displays the distribution of the nonhistone chromosomal proteins as they occur *in vivo*. It has not yet been possible to assess accurately the ultimate sensitivity of the technique; however, work

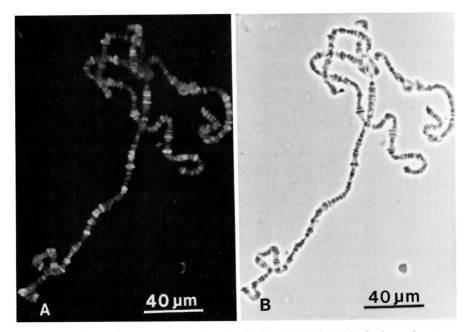

Fig. 5. Polytene chromosome spread as described for Fig. 2 but stained using antiserum against NHCP subfraction ρ.

by others (29) as well as theoretical calculations, indicates that with specific antibodies, one could detect a group of 1024 protein molecules, which in the polytene chromosomes results from the binding of a protein to one specific site of the haploid genome.

In addition to determining the distributions of purified major structural NHC proteins and relevant enzymes, this technique should also serve as an assay for the success of attempts to obtain *in vitro* reconstitution of defined segments of DNA with chromosomal proteins; the constituents of the complex at a minimum should show overlapping distributions in the native polytene chromosomes. Many extensions of such experiments could be suggested, analogous to the ongoing study of the arrangement of DNA sequences using *in situ* hybridization. The results obtained indicating a nonuniform distribution of some chromosomal proteins suggest that the correlation of cytological, biochemical, and genetic techniques possible using the *Drosophila* polytene chromosomes will lead to increased understanding of the molecular structure/function relationships in the eukaryotic genome.

ACKNOWLEDGMENTS

This work has been supported by grants to S. C. R. Elgin from the National Institutes of Health (GM 20779) and the American Cancer Society (NP-184); by a National Institutes of

Health Training Grant for L. M. Silver (STO1 GMOO782–17); and by a scholarship from the Harvard University Faculty of Arts and Sciences to C. E. C. Wu. For a more general description of results obtained using these techniques please see "The *in situ* Distribution of *Drosophila* Nonhistone Proteins" in *The Molecular Biology of the Mammalian Genetic Apparatus*, edited by P. O. P. Ts'o (North Holland Publishing Co., Amsterdam, 1977).

References

1. Scott, S. E. M., and Sommerville, J., *Nature (London)* **250**, 680 (1974).
2. Elgin, S. C. R., and Miller, D. W., *in* "The Genetics and Biology of Drosophila" (M. Ashbruner and Novitski, eds.), Vol. 2, Academic Press, London, 1978, in press.
3. Elgin, S. C. R., and Hood, L. E., *Biochemistry* **12**, 4984 (1973).
4. Elgin, S. C. R., and Miller, D. W., *in* "The Genetics and Biology of Drosophila" (M. Ashbruner and T. R. F. Wright, eds.), Vol. 2, Academic Press, New York, 1977. (In press).
5. MacGillivray, A. J., Cameron, A., Krauze, R. J., Rickwood, D., and Paul, J., *Biochem. Biophys. Acta* **277**, 384 (1972).
6. Elgin, S. C. R., *in* "Methods in Enzymology," Vol. 40: Hormore Action, Part E, Nuclear Structure and Function (B. W. O'Malley and J. G. Hardman, eds.), p. 144. Academic Press, New York, 1975.
7. Williams, C. A., and Chase, N. W., *in* "Methods in Immunology and Immunochemistry," Vol. I: Preparation of Antigens and Antibodies (C. A. Williams and M. W. Chase, eds.), p. 318. Academic Press, New York, 1967.
8. Tjian, R., Stinchcomb, D., and Losick, R., *J. Biol. Chem.* **250**, 8824 (1975).
9. Laemmli, U. K., *Nature (London)* **227**, 680 (1970).
10. Stumph, W. E., Elgin, S. C. R., and Hood, L., *J. Immunology* **113**, 1752 (1974).
11. Fairbanks, G., Steck, L. L., and Wallach, D. F. H., *Biochemistry* **10**, 2606 (1971).
12. Dick, C., and Johns, E. W. *Exp. Cell Res.* **51**, 626 (1968).
13. Brutlag, D., Schlehuber, C., and Bonner, J., *Biochemistry* **8**, 3214 (1969).
14. Silver, L. M., and Elgin, S. C. R., *Proc. Natl. Acad. Sci. U.S.A.* **73**, 423 (1976).
15. Chalkley, R., and Hunter, C., *Proc. Natl. Acad. Sci. U.S.A.* **72**, 1304 (1975).
16. Cohen, L. H., and Gotchel, B. V. *J. Biol. Chem.* **246**, 1841 (1971).
17. Hymer, W. C., and Kuff, E. L., *J. Histochem. Cytochem.* **12**, 359 (1964).
18. Helenius, A., and Simons, K., *Biochem. Biophys. Acta* **415**, 29 (1975).
19. Hancock, R. *J. Mol. Biol.* **86**, 649 (1974).
20. Faiferman, I., and Pogo, A. O., *Biochemistry* **14**, 3808 (1975).
21. Aaronson, R. P., and Blobel, G., *Proc. Natl. Acad. Sci. U.S.A.* **72**, 1007 (1975).
22. Riley D. E., Keller, J. M., and Byers, B., *Biochemistry* **14**, 3005 (1975).
23. Tata, J. R., Hamilton, M. J., and Cole, R. D., *J. Mol. Biol.* **67**, 321 (1972).
24. Jackson, V., and Chalkley, R., *Biochemistry* **13**, 3952 (1974).
25. Rudkin, G. T., *in* "Developmental Studies on Giant Chromosomes" (W. Beerman, ed.), p. 58. Springer-Verlag, Berlin and New York, 1972.
26. Berendes, H. D., *Chromosoma* **24**, 418 (1968).
27. Alfagame, C. R., Rudkin, G. T., and Cohen, L. H., *J. Cell Biol.* **67**, 6a (1975).
28. Elgin, S. C. R., and Weintraub, H., *Annu. Rev. Biochem.* **44**, 725 (1975).
29. Goldman, R. D., Lazarides, E., Pollack, R., and Weber, K., *Exp. Cell Res.* **90**, 33 (1975).

Part C.   Sequencing of Histones

Chapter 11

Peptide Mapping and Amino Acid Sequencing of Histones

ROBERT J. DELANGE

Department of Biological Chemistry,
UCLA School of Medicine,
Los Angeles, California

I.   Introduction

There are generally five major histone fractions (H1, H2A, H2B, H3, and H4) in most eukaryotes (*1–3*), but a few species appear to have only three or four (*3*). Some tissues or species have additional histone fractions, such as H5 (histone V or F2C) in nucleated erythrocytes (*4*) and H6 (histone T) in trout tissues (*5,6*). The major fractions often can be separated into subfractions which differ from other subfractions of the same major fraction in (1) amino acid sequences at specific sites and/or (2) the numbers and types of side-chain modifications (*1–3*).

Four (H4, H3, H2B, and H2A) of the five major histones in calf thymus have been completely sequenced, and H1 has been partially sequenced (*1–3*). Complete sequences for histones H4, H3, and H2A and partial sequences for histones H2B, H1, H5, and H6 also have been reported for other species (*1–3*). The comparative studies have demonstrated a remarkable conservation of histone sequences throughout evolution, particularly for the arginine-rich histones H4 and H3. The order of sequence conservation appears to be H4 > H3 ≫ H2A > H2B > H1, with the H4 and H3 sequences being the most conserved protein sequences known to date (*2*).

Because of the considerable sequence conservation of histones (see above), it seems likely that few complete sequence determinations will be attempted after those currently in progress are finished. However, sequence studies will continue to be of general application and importance in the following areas: (a) identification of sites of amino acid substitution in histones from new species; (b) determination of sites of sequence heterogeneity in a histone fraction from one species; (c) identification of sites of naturally occurring side-chain modifications in histones; (d) establishment of the residues which are chemically modified during *in vitro* studies; (e) identification of sites cross-linked to other histones or nonhistone chromosomal proteins, during studies on chromatin structure; and (f) etc. This article is written primarily to aid those involved in such studies. Note that the sequence studies on histones have played a major role in formulating our current models of chromatin structure. In fact, based on sequence studies of histone 4, a prediction (g) of the types of interactions which the different regions of histones would have in chromatin has been substantiated by the more recent chromatin model experiments (*3*).

II.   Preparation and Characterization of Histones for Sequence Work

A.   Extraction and Fractionation

Histones to be used for sequence studies can be extracted from chromatin by any of the usual methods (acid, salt, denaturant, or combinations of these) used in other types of studies (*1,8*). In some of these methods certain histones are selectively extracted, but often they need to be purified further for sequence studies. Note that some histone modifications (e.g., *N*-phosphohistidine) are not stable under certain extracting conditions (e.g., acid), and other conditions may be required, if these modifications are to be studied.

Many methods for fractionating histones have been described (see the references above), but we prefer to use gel filtration on Bio-Gel P-60 in 0.02 *N* HCl–0.1 *M* NaCl to obtain homogeneous fractions of histones 1, 2B, and 4; the fraction containing histones 2A and 3 is then subjected to gel filtration on Sephadex G-100 in 0.05 *M* sodium acetate–0.05 *M* NaHSO$_3$ (pH 5.4) as described by Böhm *et al.* (*9*). Since the histone 1 fraction generally consists of several subfractions with limited sequence variation, it is often necessary when doing sequence studies to separate these fractions by a procedure such as that used by Bustin and Cole (*10,11*). In some sequence studies it also may be advantageous to separate histone subfractions which differ only in the extent of side-chain modifications. Procedures such as those used by Wangh *et al.* (*12*) for histone 4 can be used for this purpose. The homogeneity of a histone fraction to be studied by sequence methods should be verified by several criteria including its behavior during column chromatography (gel filtration and/or ion exchange), gel electrophoresis (13), analysis (see Section II,B), and end groups (see Sections VI,B and C).

B.   Amino Acid Analysis

Amino acid analysis of a histone fraction can be used as a criterion of purity and also to compare the fraction to one that has been analyzed previously. We have found that since histones do not contain tryptophan, hydrolysis of duplicate samples (1–2 mg each) *in vacuo* for at least two different times (e.g., 24, 48, and 72 hours) at 110°C in 5.7 N glass-distilled HCl (1–2 ml), with 1–2 drops of 5% phenol added, will generally give satisfactory values for all residues except cysteine (and cystine) and sometimes methionine (due to partial oxidation). Good values for cysteine are usually obtained if a cysteine derivative (see Section II,C) has been prepared prior to analysis. In general we recommend that a sample of the histone be subjected to performic acid oxidation (Section II,C) and then hydrolyzed and analyzed to verify the methionine values (as methionine sulfone) and to determine the number of cysteines (or half cystines) present, if any.

Since Val-Val, Val-Ile, Ile-Val, and Ile-Ile bonds are only cleaved to the extent of 50–70% in 24 hours of hydrolysis under the conditions specified, the longer times of hydrolysis (48–72 hours) will also give more accurate values for these amino acids. Serine and threonine can be more accurately determined by extrapolating the values obtained after different times of hydrolysis (see above) to "zero" time. These two residues are usually destroyed to the extent of 10% (Ser) and 5% (Thr) after 24 hours of hydrolysis. If care has been taken to exclude NH_3 (or NH_3-producing substances) from all samples, extrapolation of the NH_3 values to zero time will also give an approximate value for the amide content (Asn + Gln). After acid hydrolysis the analytical value for aspartic acid includes asparagine, and the value for glutamic acid includes glutamine. Other than doing a complete sequence determination, the best way to obtain accurate values for Asp, Asn, Glu, and Gln is to enzymically hydrolyze the protein completely and then analyze with a system that separates the amides from other amino acids (14). The enzymic hydrolysis method can also be used to obtain acid-labile amino acid derivatives such as O-phosphoserine and ε-N-acetyllysine [see Section VII and Glazer et al. (14)].

C.   Derivitization of Cysteine and Methionine

To prevent peptides loss etc. during subsequent sequence studies, it is usually necessary to convert the cysteine residues in histone 3 into a more stable derivative. If the histone 3 is to be used for cyanogen bromide cleavage (Section III,C) we would suggest forming the S-carboxymethyl derivative, but other derivatives might prove better in some cases (*14*). If the histone (any histone) is not to be used for cyanogen bromide cleavage, performic acid oxidation (*14*) may be used to convert any cysteine (or cystine) to cysteic acid and any methionine (and methionine sulfoxide) to methionine sulfone. This has the advantage that during subsequent peptide fractionations, methionine peptides will not separate into two or three fractions due to partial air oxidation.

III.   Cleavage of Histones into Peptides

A.   General

Sequence studies on histones are often complicated by the large numbers of basic residues (23–38%), partial side-chain modifications of certain residues, and sequence heterogeneity at certain sites in some histones (*1–3*). Complete sequence determinations of histones or large peptides from histones will usually require cleavage by at least two separate techniques (see Sections III,B and III,C) to obtain overlapping peptides. However, most other sequence studies (see Section I) and comparative peptide maps (see Section IV) usually require only one method of cleavage, if this method is wisely chosen. Therefore, such methods will be briefly described below.

B.   Specific Enzymic Cleavage of Histones

Although trypsin is one of the most specific proteases known (cleavage is generally limited to peptide bonds in which the carbonyl group is contributed by arginine or lysine), its use for proteolysis of histones is less satisfactory than for other proteins because of the many basic residues which often occur in clusters in histones. Since tryptic hydrolysis of basic residue clusters can lead to the formation of H_2N-Lys-X- (or H_2N-Lys-X-) and -Arg-Y-COOH (or -Lys-Y-COOH; Y = lysine or arginine) bonds, which are cleaved relatively slowly by trypsin, several peptides from the same region of sequence are often produced at lower yields. This often makes peptide fractionations more difficult. For these reasons tryptic hydrolysis of histones has often been restricted to arginyl residues by maleylating the lysine residues by the procedure of Butler and Hartley (*15*). Examples of the application of this procedure to histones are histone 4 (*7, 16*), histone 3 (*17*), and histone 2A (*18–20*). The major peptides obtained in these studies on the calf thymus histones are summarized in Table I together with those peptides obtained by succinylation of histone 2B(30). The advantage of maleylation compared to succinylation is that the maleyl groups can be removed under mildly acidic conditions to allow further tryptic cleavage of isolated peptides at lysine residues.

Of the other enzymes available for specific cleavage of histones only chymotrypsin and thermolysin have been used to any extent. Enzymes such as papain, subtilisin, and pepsin could be used, but usually result in very complex mixtures of peptides due to the lower specificity of these enzymes. However, these enzymes have been used to degrade further the peptides obtained with the more specific enzymes. The use of these enzymes for sequence work is summarized in Boyer (31). A more recently described enzyme from *Staphylococcus aureus* has been used to cleave polypeptides primarily at glutamic acid residues [also aspartic acid residues under appropriate conditions; (32)], but it has not yet been used with histones.

Chymotrypsin ordinarily cleaves at the COOH-terminal side of residues like tryptophan (none in histones), tyrosine, phenylalanine, leucine, methionine, asparagine, glutamine, and histidine. The rates of hydrolysis at these residues are determined to a considerable extent by the types of residues near the susceptible bonds (e.g., acidic residues can greatly decrease the rates of hydrolysis); therefore, several overlapping peptides are often produced in varying yields from the same region of sequence. However, by controlling the conditions of hydrolysis, chymotryptic cleavage of histone 1 was limited largely to the single phenylalanine and two leucine residues to give four major products (22). Chymotryptic cleavage of histone 4 (7) occurred mainly at residue 37 (leucine) to give two major products (37 and 65 residues). Other smaller peptides were produced in lesser yield, but additional time of hydrolysis of added chymotrypsin did not produce significant additional cleavage. This suggested that one of the peptide products was possibly acting as a chymotrypsin inhibitor. A summary of major chymotryptic peptides obtained from histones is given in Table I.

Thermolysin cleaves polypeptides at the NH$_2$-terminal side of residues like leucine, isoleucine, valine, tyrosine, phenylalanine, methionine, and alanine. Examples of the use of thermolysin for proteolysis of histones can be found in the work of Yeoman et al. [(18), histone 2A], Sautiere et al. [(19), histone 2A], and Iwai et al. [(25), histone 2B]. It has also been used to cleave peptides obtained from the other histones by various methods.

C. Specific Chemical Cleavage

The most selective method for cleaving proteins in high yield involves the use of cyanogen bromide to cleave proteins only on the COOH-terminal

TABLE I

PEPTIDES OBTAINED AFTER ENZYMIC OR CHEMICAL CLEAVAGE OF HISTONES[a]

Method of cleavage	Histone 1[b]	Histone 2A[c]	Histone 2B[d]	Histone 3[e]	Histone 4[f]
Trypsin after maleylation	Not done	1-3, 4-11 Not done, 12-17, 18-20, 21-29, 30-32, 33-35, 36-42, 43-71, 72-77, 78-81, 82-88, 89-129 *Other:* 82-129	Not done. Trypsin after succinylation: 1-29, 30-31, 32-33, (34-72)[g], 73-79, 80-86, 87-92, 93-99, 100-125	1-2, 3-8, 9-17, 8-26, 27-40, 41-49, 50-53, 54-63, 64-69, 70-72, 73-83, 84-116, 117-128, 129, 130-131, 132-135 *Other:* 50-52, 53-63, 130-134, 135	1-3, 4-17, 18-19, 20-23, 24-36, 37-39, 40-45, 46-55, 56-67, 68-78, 79-92, 93-95, 96-102
Cyanogen bromide	Not cleaved in many species (Met not present)	Not cleaved in many species (Met not present)	1-59, 60-62, 63-125	1-90, 91-120, 121-135	1-84, 85-102
N-Bromo-succinimide	1-72, 73-223	Not done; expected: 1-39, 40-50, 51-57, 58-129; trout testis[h]: 1-39, (40-50)[g]; 51-125	Not done; expected: 1-37, 38-42, 41-42, 48-83, 84-121, 122-125	Not done; expected: 1-41, 42-54, 55-99, 100-135; chicken erythrocyte[i]: as expected for calf thymus	Not done; expected: 1-51, 52-72, 73-88, 88-98; 99-102

174

Chymotrypsin				
1–44, 45–58, 59–63,	(1–17)[g], 18–23, 24–31,	1–11, 12–19, 20–22,	1–5, (6–39)[g], 40–41,	1–37, 38–102
64–67, 68–72,	32–34, 35–39, 40–85,	23–29, 30–31,	42–48, 49–54, 55–61,	*Other:* 1–22, 23–37,
73–106, 107–223	40–51, 52–57, 58–63,	32–37, 38–40,	62–67, 68–70, 71–78,	38–72, 73–88,
Other: 1–21, 22–44,	64–76, 86–96,	41–42, 43–45,	79–84, 83–90, 91–99,	89–98, 99–102
59–60, 61–63,	125–129, (77–85)[g],	46–59, 60–65,	100–104, 105–109,	
73–77, 84–88,	(97–124)[g]	66–80, 81–83,	110–113,	
84–106		84–90, 91–95,	(114–118)[g],	
		96–100, 101–106,	119–126, (127–135)[g]	
		107–121, 122–125	*Other:* 105–113	

[a] Peptides are indicated by residue numbers (inclusive) and are for calf thymus histones unless otherwise indicated. Sites of peptide cleavage may differ in other species, but will, in general, be more conserved in the arginine-rich histones (B3 and H4) than in the others.
[b] Data for rabbit thymus subfraction 3 unless specified (21–24).
[c] Yeoman et al. (18); Sautiere et al. (19).
[d] Iwai (25).
[e] DeLange et al. (17, 26).
[f] DeLange et al. (7, 27); Ogawa et al. (28).
[g] Peptides not isolated from these regions.
[h] Bailey and Dixon (20).
[i] Brandt and Von Holt (29).

TABLE II
Sites of Amino Acid Changes in Histones from Various Species[a]

Histone 1[b]

Residue	Rabbit-3[b]	Other	Residue	Rabbit-3	Other	Residue	Rabbit-3	Other
14[b]	Glu	D[a](1)[f]	82	Ile	Val(3)	149	Thr	Lys(3)
14	Glu	Pro(2)	84	Leu	Ile(3)	150	Pro	D(3)
20	Lys	Thr(1,2)	85	Gly	Ala(3)	154	D	Ala(3)
21	Lys	Pro(1,2)	86	Leu	Val(3)	159	Lys	Ala(3)
22	D	Val(1,2)	91	Ser	Thr(3)	171	Lys	Ala(3)
25	Ala	Arg(1)	110	Asp	Asn(3)	172	Ala	Thr(3)
28	Pro	Lys(1,2)	114	Ala	D(3)	175	Ser	Lys(3)
29	Gly	Ser(1)	115	Ser	D(3)	181	Ala	Lys(3)
29	Gly	Lys(2)	116	Gly	Val(3)	183	Pro	Ala(3)
30	Ala	Pro(2)	120	Pro	D(3)	184	D	Thr(3)
31	Gly	Ala(2)	123	D	Ala(3)	192	Pro	Ala(3)
32	Ala	Gly(2)	127	Gly	D(3)	205	Ala	Lys(3)
34	Lys	Arg(2)	130	Lys	D(3)	207	Ala	D(3)
38	Ala	Ser(1,2)	133	Lys	Ala(3)	210	Pro	D(3)
57[b]	Asn	Ser(1,2,3)	134	Pro	Lys(3)	216	Pro	D(3)
59	Leu	Val(1,2,3)	135	Ala	Lys(3)	219	Ala	Pro(3)
67	Ala	Ser(3)	136	Gly	Val(3)	220	Gly	D(3)
71[b]	Gly	Ala(1,2)	143	Gly	D(3)	224	Lys	D(3)

Histone 2B[c]			Histone 2A[d]			Histone 3[e]			Histone 4[g]		
Residue	Calf	Other	Residue	Calf	Other	Residue	Calf	Other	Residue	Calf	Other
2	Glu	D(4)	6	Gln	Thr(3)	41[e]	Tyr	Phe(7,8)	60	Val	Ile(7)
4	Ala	D(4)	16	Thr	Ser(5,6)	53	Arg	Lys(7)	63	Thr	Cys(6)
6	Ser	Thr(4)	33	Leu	Phe(6)	90	Met	Ser(7)	77	Lys	Arg(7)
8	Pro	Gly(4)	41[d]	Glu	Asn(6)	96	Cys	Ser(7,9,10)			
8–9	Pro-Ala	Pro-Lys-Ala(4)	78	Ile	D(3)	96	Cys	Ser(2-part)			
9	Ala	D(4)	87	Ile	Val(3)	96	Cys	Ala(40% of 7)			
10	Pro	D(3) Ala(4)	99	Lys	Arg(5)						
12–13	Lys-Gly	Lys-Ala-Gly(4)	99	Lys	Gly(3)						
14	Ser	D(4)	122	Ser	D(3)						
15	Lys	D(4)	123	His	D(3)						
17–18	Ala-Val	Ala-Glx-Lys-Asx-Ile(4)	124	His	D(3)						
20–21	Lys-Ala	Lys-Thr-Ala(3)	128	Gly	Val(3)						
20–21	Lys-Ala	Lys-Thr-D(4)	128–129	Gly-Lys	Val-Ala-Lys(3)						
22	Gln	Gly(3)									
22[c]	Gln	Asx(4)									

[a] The number of changes must be considered minimal, since the sequences of the histones from only a few species have been reported (and some of these are only partial sequences). However, the number of changes for each histone would still seem likely to remain in the order H1 > H2B > H2A > H3 > H4. D = deletion (or an insertion in the comparison sequence).

[b] Rall and Cole (23); Jones et al. (24); R. D. Cole (personal communication); Dixon et al. (39); G. H. Dixon (personal communication). See also DeLange (40). Subfractions of histone 1 differ in sequence at certain positions. The residue numbers are not absolute for any species because of deletions (and insertions). The known sequence of trout H1 begins at residue 56, and the known sequences of rabbit-4 and calf-1, H1 stop at residue 73. The sequence of rabbit-3 H1 beyond residue 143 is based on tryptic and thermolysin peptides and alignment with the sequence of trout H1 (40).

[c] Iwai et al. (25); Elgin et al. (3). The sequence of Drosophila H2B is unknown beyond residue 24 (of calf thymus).

[d] Yeoman et al. (18); Sautiere et al. (19,41); Bailey and Dixon (20). The sequence of sea urchin H2A is unknown beyond residue 42.

[e] DeLange et al. (26,17); Patthy and Smith (42,43); Hooper et al. (44); Brandt and von Holt (45); Brandt et al. (46,47). The sequence of cycad H3 is known only through residue 46.

[f] 1 = rabbit thymus-4; 2 = calf thymus; 3 = trout testis; 4 = Drosophila; 5 = rat; 6 = sea urchin; 7 = pea seedling; 8 = cycad; 9 = carp testis; 10 = chicken erythrocyte.

[g] DeLange et al. (7,16,27); Ogawa et al. (28); Sautiere et al. (48); Strickland et al. (49).

177

side of methionine residues (33). The reaction results in peptides with COOH-terminal homoserine lactone, which is formed from methionine residues by an elimination-cyclization reaction, and is in equilibrium with the noncyclized homoserine form. Cleavage at cysteinyl residues (present usually in histone 3 only) can also be achieved by forming the S-methylcysteine derivative (34) prior to the cyanogen bromide reaction. Proteins subjected to considerable manipulation (during purification, etc.) often undergo partial oxidation to give methionine sulfoxide or even methionine sulfone. These oxidized residues are not cleaved by cyanogen bromide and result in overlapping peptides. Methionine can be regenerated from the sulfoxide form by incubation with thiols (35). The analysis of methionine and its oxidation products, as well as homoserine and homoserine lactone, are summarized in Glazer et al. (14). Met-Ser and Met-Thr bonds are often cleaved only partially, and this produces overlapping peptides containing internal homoserine (Hse) residues in the sequence -Hse-Ser- or -Hse-Thr-. Histones 1 and 2A from many species lack methionine and are not cleaved by cyanogen bromide. Histones 2B, 3, and 4 usually contain 1 or 2 methionines in most species and are cleaved by cyanogen bromide to give easily separable peptides (e.g., see Table I for a summary and references to the use of this reagent with histones).

N-Bromosuccinimide can oxidatively cleave proteins at the COOH-terminal side of tryptophan, tyrosine, and histidine (36). However, since histones lack tryptophan, and since histidine bonds require stringent conditions for cleavage, selective cleavage at tyrosine residues can be obtained in histones. There will usually be oxidative losses of certain residues such as methionine, histidine, and derivatives of cysteine, but the cleavage of histones is almost exclusively at tyrosine residues, generally in yields of 30–60%. Thus, uncleaved overlapping peptides are usually also present when this method is used. This method has been used for histone 1 (21), histone 2A (20), and histone 3 (29), but it should also be applicable to the other histones (see Table I for a summary of peptides obtained from histones 1, 2A, and 3 and for those expected from the other histones).

Most other chemical methods of cleaving polypeptide chains give low yields or are less selective than the two described above. Cleavage at aspartic acid residues with dilute acid (37) has been used with histone 3 (38).

Note that because of amino acid substitutions, sequence studies on histones from other species may not give the same peptides as those listed in Table I. Table II gives most of the amino acid changes which have been found in the five major histones thus far. The table should give a good indication of which peptides (residues) are most likely to be different or identical in various species. Obviously, there will be greater peptide conservation in the arginine-rich histones (H3 and H4) than in the others.

IV. Peptide Mapping and Monitoring of Peptide Cleavage

Peptide maps are most often used to determine how successful a peptide cleavage procedure (see Sections III, B and C) has been and also to compare two or more similar proteins (see Table II for differences in histones). The maps are made with one or more systems which produce good separation of all (or most) of the peptides present after the cleavage procedure. In many cases these same systems serve as indicators of the best methods for fractionating the peptides (Section V), whenever this is of importance.

Since large peptides (e.g., more than 15–25 residues) and small peptides often behave very differently in the mapping systems, a complete map of a mixture of large and small peptides may require an initial separation (e.g., on a column of Sephadex G-25), followed by separate mapping procedures for the two size classes of peptides. In our laboratory we have generally used a column (0.9 or 1.9 cm × 160 cm) of Sephadex G-25 in 30% acetic acid for the initial separation of the large peptides (those eluting at or near the void volume) from the smaller ones (those which elute later). The column effluent is monitored by the ninhydrin method (50) or by the use of a fluorogenic reagent such as o-phthalaldehyde (51). If insoluble peptides are present, they can usually be solubilized (and usually remain soluble in 30% acetic acid) by lyophilizing, dissolving in 98+% formic acid, and diluting to 50% with 30% acetic acid. The maximum sample volume for the 1.9-cm column should be about 4 ml. Alternatively, if all peptides remain soluble in the buffer used for cleavage (e.g., NH_4HCO_3), the column can be run in this solvent and can often be monitored at around 220 nm to detect peptides. However, if an enzyme was used for cleavage of the protein (Section III,B), the enzyme should first be inactivated (e.g., with diisopropylphosphofluoridate for trypsin or chymotrypsin) to prevent further degradation (e.g., chymotryptic-like hydrolysis by trypsin) of the peptides which may elute with the enzyme at the void volume.

We have generally found that the best system for mapping large histone peptides is polyacrylamide gel electrophoresis in acid–urea buffers and 10–20% gels as described by Panyim and Chalkley (13) for separation of histones. Other gel systems such as one run in the presence of sodium dodecyl sulfate (SDS) (52) can also be very useful. Other types of mapping systems, such as paper or ion-exchange chromatography, are often less satisfactory for large histone peptides. However, if another method is desirable, we would recommend the use of carboxymethyl cellulose in 8 M urea with a gradient of increasing salt concentration, possibly coupled to decreasing pH.

The small peptide fraction, generally containing peptides smaller than about 20 residues, can usually be mapped by two-dimensional separations on paper or thin-layer plates (preferably coated with cellulose). We generally use descending paper chromatography (16–20 hours with n-butanol:pyridine:glacial acetic acid:water, 15:10:3:12) in the first dimension, followed by air drying (about 4 hours) and high-voltage electrophoresis at pH 3.6 (2000 V for 60 minutes in pyridine:glacial acetic acid:water, 1:10:289) in the second dimension (long direction of paper). The paper (46 × 57 cm sheets of Whatman 3 MM; origin = 9 cm from long side, 23 cm from short side) is dipped in a ninhydrin solution (0.1% ninhydrin, 5% technical-grade collidine in 95% ethanol) and heated for 30 minutes at 60°C for color development. Various other combinations of chromatography and electrophoresis systems can be used (53). We have found electrophoresis at pH 1.9 (glacial acetic acid: 98 +% formic acid:H_2O, 87:25:888; origin = 9 cm from short end of paper) to be particularly useful (often in one dimension only), but all traces of pyridine (sequanol grade) must be removed from the paper by longer air drying, or a buffer front develops during electrophoresis. Two chromatographic systems can be used if they are sufficiently different. For example, NH_4OH:n-propanol (30:70) may be helpful in the second dimension, but the NH_3 must be completely evaporated and the paper should be stained with a different ninhydrin reagent for proper color development (e.g., 0.1% ninhydrin in 95% ethanol:glacial acetic acid:collidine, 50:15:2). These same procedures can be adapted to thin-layer plates for greater sensitivity. The use of fluorescamine (0.05% in acetone) can be used for greater sensitivity in detecting the peptides on thin-layer plates or paper (54). The reader is referred to the Appendix of Glazer et al. (14) for additional details of mapping techniques, including specific stains for arginine, tyrosine, and histidine peptides.

V. Fractionation of Peptides to Homogeneity

As an initial means of fractionating most histone peptide mixtures, gel filtration (usually on Sephadex G-25 or G-50, but sometimes on G-75 or comparable gels), as described in Section IV, is the method of choice. This has the advantages of high recovery and separation of peptides into groups based mainly on size (tyrosine and phenylalanine residues can often cause a peptide to elute later than expected). Each group of peptides can then be examined by a peptide mapping procedure (Section IV) which is more appropriate for that group to determine the number of peptides present and possibly the best methods for purification.

Although large peptides (e.g., greater than 20 residues) are usually mapped by gel electrophoresis (Section IV), they are usually purified by gel filtration, if possible. By the use of different gels of varying pore size (e.g., Sephadex G-25 and G-50) or different composition (e.g., Sephadex G-50 and Bio-Gel P-6) and by elution with different solvents (e.g., 30% acetic acid and 1 M pyridine; or 30% acetic acid and 1% acetic acid), many large peptides can be purified from each other even though they elute in the same position in the initial gel filtration system. When gel filtration alone is not successful in purifying a desired peptide, ion-exchange chromatography, isofocusing, differential solubility, gel electrophoresis (in slabls or large columns), or even paper electrophoresis (pH 1.9 is generally best) or chromatography can be tried. In the paper systems large peptides (or even very basic smaller ones) often streak or remain at the origin. We have used modified solvent systems (e.g., n-butanol:glacial acetic acid:water, 70–130:60: 75) for paper chromatography of large peptides (7), but on occasion we have obtained multiple zones of the same peptide. If ion-exchange chromatography (e.g., carboxymethyl cellulose) is used, it is usually advisable to include 8 M urea in all buffers to maintain peptide solubility and to sharpen the zones of peptide elution.

Each group of smaller peptides from the initial gel filtration step (see above) can generally be mapped by one or more paper or thin-layer methods (Section IV) to determine the best method(s) of fractionation. Preparative paper electrophoresis or chromatography can be used to prepare several micromoles (on a single sheet of paper) of peptide with the method chosen. This is usually done by streaking the peptide over a 10–30 cm origin and, after the separation has been run, cutting guide strips from the sides (and often the center) to stain with ninhydrin to locate the peptides. Fluorescamine (54) also can be used as a detecting reagent. The unstained peptide strips are then eluted (with 30% acetic acid) and concentrated by rotary evaporation to give a solution (or suspension) which can be checked in other ways (other paper or thin-layer methods, analysis, or end groups) for homogeneity. The migrations of many peptides from histones 3 and 4 in various fractionation systems have been reported (7,16,55).

VI. Sequencing Homogeneous Peptides

A. Analysis and Amide Content

Peptides, which have been isolated by the procedures described in the previous sections, should be hydrolyzed and analyzed as described in Section II,B. The amide content of smaller peptides often can be deduced by determining electrophoretic mobilities (56) or by analysis after complete enzymic hydrolysis (14). Aminopeptidase M is able to degrade most smaller peptides to amino acids, except when aspartic acid residues have β-shifted (57) or NH$_2$-terminal glutamine residues have cyclized (57), or when X-Pro bonds (X = any amino acid) are present. Prolidase (or imidopeptidase, Miles, Elkhart, Indiana) can be added to the aminopeptidase M hydrolysate to cleave any X-Pro dipeptides present (other residues COOH-terminal to the proline are released as free amino acids by aminopeptidase M). Larger peptides may need to be cleaved first to smaller peptides (Sections III,B and III,C) before using aminopeptidase M (either on the resulting peptide mixture or on each separated peptide). The amino acid mixture should be analyzed with a system that separates amides (asparagine and glutamine) from serine (e.g., the 30°–50°C temperature program on a Beckman 120B analyzer) or from each other (58).

B.   Edman Degradation

Most residues in a protein are placed in sequence by use of the Edman degradation; it is, therefore, the single most useful technique in sequence work. Recent advances in the technology of automating this procedure (59) have made it possible to sequence many residues in a very short time. However, since the NH$_2$-terminus is blocked in three of the major histones (α-N-acetylserine in H1, H2A, and H4) and probably in H5 (α-N-acetylthreonine), and since a maximum of 30–50 residues usually can be confidently identified by degrading most large polypeptides, it is usually necessary to cleave the histones (Sections III,B and C) and study individual peptides. The peptide isolations often become the rate-limiting steps in most sequence studies.

In our laboratory we use a Beckman 890C sequenator with Program 030176 (Beckman publication "In Sequence," No. 9, May 1976; Spinco Division, Beckman Instruments, Palo Alto, California) and have been able to sequence 25–50 residues in most proteins or large peptides (using about 200–400 nmol) and 58–86% of the residues in smaller peptides (12–21 residues) which contain arginyl residues (helpful in preventing peptide extraction). Often we can identify 20–25 residues with as little as 80 nmol of a larger peptide or protein. Small peptides (or other peptides lacking arginines) are frequently degraded by a manual method (60). The thiazolinone derivatives obtained from either the Sequenator or manual degradations are converted to the phenylthiohydantoin (PTH) derivatives with 1 N HCl at 80°C (61). Aliquots of the PTH derivatives are examined by gas chromatography (59), thin-layer chromatography (59, 60, 62) and amino acid analysis after regeneration by hydrolysis with 5.7 N HCl at 150°C [±0.1% SnCl$_2$; (63)]. Aliquots of the peptides at various stages of manual degradation are also hydrolyzed and analyzed to check the other results by difference analysis. Other techniques, related to the Edman degradation, which might prove useful are solid-phase degradations (*64*) and the dansyl-Edman procedure for greater sensitivity (*65*). For additional details of the Edman procedure see (*14*) and (*59*). Other enzymic (*66*) and chemical (*67*) methods for NH$_2$-terminal determinations are also available.

C. COOH-Terminal Sequences

Since smaller peptides (and some larger ones) are often extracted into the various solvents during Edman degradations, it frequently is not possible to sequence a peptide completely by the Edman procedure alone. Therefore, it is very helpful to be able to degrade peptides from the COOH-terminal end. Although there is a chemical method for sequentially removing residues from the COOH terminus of a peptide or protein (*68*), enzymic methods using carboxypeptidases are usually employed. Carboxypeptidase B (e.g., Worthington Biochemical, Freehold, New Jersey) is used to remove arginine and lysine residues (*69*), whereas carboxypeptidase A (e.g., Worthington) is used to remove most other residues, although it acts on hydrophobic residues at much faster rates than on other residues (*70*). However, since bonds involving proline, glycine, and acidic residues often are not cleaved at sufficient rates to measure, little or no sequence data can be obtained for some peptides by using these enzymes. Two newer carboxypeptidases from citrus leaf (Rohm and Haas, Darmstadt, West Germany) and yeast (e.g., Pierce Chemical, Rockford, Illinois) have proved useful since they cleave most peptide bonds (including those involving proline, glycine, and acidic residues) at appreciable rates (*71,72*). When a residue cannot be removed by a carboxypeptidase (e.g., glycine by carboxypeptidase A), it frequently can be identified by hydrazinolysis (*73*) of the peptide (protein) which leaves the COOH-terminal residue as the only free amino acid (identified by pH 1.9 electrophoresis or with an analyzer). This method was used to identify the glycine residue (not removed by carboxypeptidases A and B) in the peptide Ac-Ser-Gly-Arg (from NH$_2$-terminus of histone 4) after removing the arginine residue with carboxypeptidase B (*7*). It also allowed identification of the blocking group as acetyl (identified as acetyl hydrazide). We have found that distillation of hydrazine is unnecessary if a new bottle of high-grade reagent (e.g., Pierce, anhydrous) is used each time.

VII. Identification of Modified Residues

Since histones often contain residues which have been modified by posttranslational reactions (*1–3*) it often is necessary to identify the modified residues in the sequence. This is most conveniently done when radioactive labels can be incorporated at the time the modifications take place (74). However, in some situations radioactive labels cannot be used, and this makes it necessary to either (a) purify all peptides and examine each one for the modified residues or (b) isolate particular peptides based on known migrations in certain fractionation systems (7,16).

Certain modified residues (α-N-acetylserine, ϵ-N-acetyllysine, O-phosphoserine, N-phosphohistidine) are not stable during acid hydrolysis and must be released from peptides (or proteins) by hydrolysis with enzymes such as aminopeptidase M (7, 16), Pronase (75), or other enzymes (14). Other modified residues, such as ϵ-N-methyllysines, N-methyllistidines, and N-methylarginines, are stable to acid hydrolysis. The modified residues can be purified by conventional methods (Section V), or they often can be identified in amino acid mixtures. Table III lists for certain systems the migrations and elution positions of some of the most common modified residues in histones. For other modified residues that may be found in histones, particularly those that may be formed during chemical modification experiments *in vitro*, see Glazer *et al.* (14).

Once the identity of a modified residue has been established (see above), its position in the sequence must be determined. If a radioactive label has been used, the position frequently can be determined by the step of Edman degradation at which the radioactivity is removed (74). If the modified residue is not radioactive, at each step of the Edman degradation either the amino acid removed must be checked for the modified residue (possible only if it is stable to acid hydrolysis), or the remainder peptide must be analyzed (this may require enzymic hydrolysis) to determine if the modified residue is still present. Carboxypeptidases (Section VI,C) and other sequence methods also can be used to position the modified residues (7,16).

Note: The nomenclature used here is that suggested at a Ciba Symposium (79). The sequences of trout H1 (80) and trout H2B (81) have been reported recently.

VIII.   Conclusions

The methodology is now available for sequencing any histone fraction completely or in part. However, histone 1 (and perhaps H5) presents special difficulties, and a complete sequence of H1 from a single species has yet to be reported. By appropriate mapping and analytical techniques, differences between two or more similar histone fractions or subfractions can be ob-

TABLE III

Elutions and Migrations of Amino Acid Derivatives from Histones in Various Systems[a]

	Amino acid analyzer		Paper electro- phoresis, pH 1.9[d]	Paper chromatography			Acid hydrolysis[h]	Color with ninhydrin[i]
	System A[b]	System B[c]		System I[e]	System II[f]	System III[g]		
Lys (Ac)[a]	77 minutes Gly = 86 min	ND[a]	0.70 Ala = 1.00	0.69 Val = 1.00	0.77 Val = 1.00	ND	Lys + acetic acid	Like Arg
Lys (CH$_3$)[j]	Shoulder[b] on Lys peak	344 minutes	0.97 Lys = 1.00	0.89 Arg = 1.00	0.90 Arg = 1.00	0.44	No change	Like Lys
Lys (CH$_3$)$_2$[j]	Shoulder[b] on Lys peak	369 minutes	0.93 Lys = 1.00	0.91 Arg = 1.00	0.94 Arg = 1.00	0.76	No change	Like Arg
Lys (CH$_3$)$_3$[j]	Shoulder[b] on Lys peak	387 minutes	0.95 Lys = 1.00	0.65 Arg = 1.00	0.71 Arg = 1.00	0.88	No Change	Like Arg
Ser (PO$_3$H$_2$)	Unretarded position	ND	0.06 Asp = 1.00	0.31 Asp = 1.00	0.48 Glu = 1.00	ND	1–4 hour: Ser + Ser (PO$_3$H$_2$) 24 hour: Ser (low)	Like Arg

[a] Abbreviations are: Lys (Ac) = ϵ-N-acetyllysine; Lys (CH$_3$)$_n$ = ϵ-N-methyllysines; Ser (PO$_3$H$_2$) = O-serine phosphate; ND = not determined. Data are as determined by R. J. DeLange unless otherwise noted. For paper systems the migrations are measured from origin to leading edges of substances.

[b] Using the usual medium column of a Beckman 120B analyzer. Depending on the resin column size, buffer, etc., the shoulder on the Lys peak elutes with Lys or as a separate peak at times (76).

[c] Using a medium column with Durrum DC-2 resin (77); Lys = 297 minutes; His = 413 minutes. See Glazer et al. (14) for other analyzer systems.

[d] High-voltage electrophoresis on Whatman 3 MM paper with pH 1.9 buffer (glacial acetic acid:99 + % formic acid:H$_2$O, 87:25:888).

[e] Descending chromatography on Whatman 3 MM paper with 1-butanol:glacial acetic acid:H$_2$O (200:30:75).

[f] Descending chromatography on Whatman 3 MM paper with 1-butanol:pyridine:glacial acetic acid:H$_2$O (15:10:3:12).

[g] Descending chromatography with m-cresol–phenol–borate buffer (76). Lys = 0.18; Arg = 0.30; His = 0.43.

[h] 5.7 N HCl, 110°C in vacuo.

[i] Using 0.1% ninhydrin in 95% ethanol and 5% technical-grade collidine.

[j] The methylated lysines (and other methylated derivatives) can also be resolved well by two-dimensional thin-layer chromatography (78).

served, and the peptides which differ can be isolated and sequenced to show whether the differences are caused by amino acid substitutions or posttranslational modification (or both).

REFERENCES

1. DeLange, R. J., and Smith, E. L., *Annu. Rev. Biochem.* **40**, 279 (1971).
2. DeLange, R. J., and Smith, E. L., *Ciba Found. Symp.* **28**, 59 (1975); see also *in* "The Proteins" (H. Neurath and R. L. Hill, eds.), 3rd Edition, Vol. 4 Academic Press, New York, in press.
3. Elgin, S. C. R., and Weintraub, H., *Annu. Rev. Biochem.* **44**, 725 (1975).
4. Neelin, J. M., and Butler, G. C., *Can. J. Biochem. Physiol.* **39**, 485 (1961).
5. Wigle, D. T., and Dixon, G. H., *J. Biol. Chem.* **246**, 5636 (1971).
6. Huntley, G. H., and Dixon, G. H., *J. Biol. Chem.* **247**, 4916 (1972).
7. DeLange, R. J., Fambrough, D. M., Smith, E. L., and Bonner, J., *J. Biol. Chem.* **244**, 319 (1969).
8. Johns, E. W., *in* "Histones and Nucleohistones" (D. M. P. Phillips, ed.), p. 1. Plenum, London, 1971.
9. Böhm, E. L., Strickland, M., Thwaits, B. H., Van Der Westhuizen, D. R., and von Holt, C. *FEBS Lett.* **34**, 217 (1973).
10. Bustin, M., and Cole, R. D., *Arch. Biochem. Biophys.* **127**, 457 (1968).
11. Bustin, M., and Cole, R. D., *J. Biol. Chem.* **244**, 5286 (1969).
12. Wangh, L., Ruiz-Carrillo, A., and Allfrey, V. G. *Arch. Biochem. Biophys.* **150**, 44 (1972).
13. Panyim, S., and Chalkley, R., *Arch. Biochem. Biophys.* **130**, 337 (1969).
14. Glazer, A. N., Delange, R. J., and Sigman, D. S., *in* "Laboratory Techniques in Biochemistry and Molecular Biology" (T. S. Work and E. Work, eds.), Vol. 4, Part I: Chemical Modification of Proteins. North Holland Publ. Amsterdam, 1976.
15. Butler, P. J. G., and Hartley, B. S. *in* "Methods in Enzymology," Vol. 25: Enzyme Structure, Part B (C. H. W. Hirs, ed.), p. 191. Academic Press, New York, 1972.
16. DeLange, R. J., Fambrough, D. M., Smith, E. L., and Bonner, J., *J. Biol. Chem.* **244**, 5667 (1969).
17. DeLange, R. J., Hooper, J. A., and Smith E. L., *J. Biol. Chem.* **248**, 3261 (1973).
18. Yeoman, L. C., Olson, M. O. J., Sugano, N., Jordan, J. J., Taylor, C. W., Starbuck, W. C., and Busch, H., *J. Biol. Chem.* **247**, 6018 (1972).
19. Sautiere, P., Tyrou, D., Laine, B., Mizon, J., Ruffin, P., and Biserte, G., *Eur. J. Biochem.* **41**, 563 (1974).
20. Bailey, G. S., and Dixon, G. H., *J. Biol. Chem.* **248**, 5463 (1973).
21. Bustin, M., and Cole, R. D., *J. Biol. Chem.* **244**, 5291 (1969).
22. Bustin, M., and Cole, R. D., *J. Biol. Chem.* **245**, 1458 (1970).
23. Rall, S. C., and Cole, R. D., *J. Biol. Chem.* **246**, 7175 (1971).
24. Jones, G. M. T., Rall, S. C., and Cole, R. D., *J. Biol. Chem.* **249**, 2548 (1974).
25. Iwai, K., Hayashi, H., and Ishikawa, K., *J. Biochem.* **72**, 357 (1972).
26. DeLange, R. J., Hooper, J. A., and Smith, E. L., *Proc. Natl. Acad. Sci. U.S.A.* **69**, 882 (1972).
27. DeLange, R. J., Smith, E. L., Fambrough, D. M., and Bonner, J., *Proc. Natl. Acad. Sci. U.S.A.* **61**, 1145 (1968).
28. Ogawa, Y., Quagliarotti, G., Jordan, J., Taylor, C. W., Starbuck, W. C., and Busch, H., *J. Biol. Chem.* **244**, 4387 (1969).
29. Brandt, W. F., and von Holt, C., *Eur. J. Biochem.* **46**, 407 (1974).
30. Ishikawa, K., Hayashi, H., and Iwai, K., *J. Biochem.* **72**, 229 (1972).

31. Boyer, P. D., ed. "The Enzymes," 3rd ed., Vol. 3. Academic Press, New York, 1971.
32. Houmard, J., and Drapeau, G. R., *Proc. Natl. Acad. Sci. U.S.A.* **69**, 3506 (1972).
33. Gross, E. *in* "Methods in Enzymology," Vol. 11: Enzyme Structure (C. H. W. Hirs, ed.), Academic Press, New York, 1967.
34. Heinrikson, R. L., *J. Biol. Chem.* **246**, 4090 (1971).
35. Jori, G., Galiazzo, G., Marzotto, A., and Scoffone, E., *J. Biol. Chem.* **243**, 4272 (1968).
36. Ramachandran, L. K., and Witkop, B. *in* "Methods in Enzymology," Vol. 11: Enzyme Structure (C. H. W., Hirs, ed.), p. 283. Academic Press, New York, 1967.
37. Schultz, J., *in* "Methods in Enzymology," Vol. 11: Enzyme Structure (C. H. W. Hirs, ed.), p. 255. Academic Press, New York, 1967.
38. Brandt, W. F., and von Holt, C., *FEBS Lett.* **23**, 357 (1972).
39. Dixon, G. H., Candido, E. P. M., Honda, B. M., Louie, A. J., Macleod, A. R., and Sung, M. T., *Ciba Found. Symp.* **28** (*N.S.*) 229 (1975).
40. DeLange, R. J., *in* "Handbook of Biochemistry and Molecular Biology," Vol. 2: Proteins (G. D. Fasman, ed.), 3rd ed., p. 295. CRC Press, Cleveland, Ohio, 1976.
41. Sautiere, P., Wouters-Tyrou, D., Laine, B., and Biserte, G., *Ciba Found. Symp.* **28** (*N.S.*) 77 (1975).
42. Patthy, L., Smith, E. L., and Johnson, J., *J. Biol. Chem.* **248**, 6834 (1973).
43. Patthy, L., and Smith, E. L., *J. Biol. Chem.* **250**, 1919 (1975).
44. Hooper, J. A., Smith, E. L., Sommer, K. R., and Chalkley, R., *J. Biol. Chem.* **248**, 3275 (1973).
45. Brandt, W. F., and von Holt, C., *Eur. J. Biochem.* **46**, 419 (1974).
46. Brandt, W. F., Strickland, W. N., Morgan, M., and von Holt, C., *FEBS Lett.* **40**, 167 (1974).
47. Brandt, W. F., Strickland, W. N., and von Holt, C., *FEBS Lett.* **40**, 349 (1974).
48. Sautiere, P., Tyrou, D., Moschetto, Y., and Biserte, G., *Biochimie* **53**, 479 (1971).
49. Strickland, M., Strickland, W. N., Brandt, W. F., and von Holt, C., *FEBS Lett.* **40**, 346 (1974).
50. Hirs, C. H. W., Moore, S., and Stein, W. H., *J. Biol. Chem.* **219**, 623 (1956).
51. Benson, J. R., and Hare, P. E., *Proc. Natl. Acad. Sci. U.S.A.* **72**, 619 (1975).
52. Laemmli, U.K., *Nature* (*London*) **227**, 680 (1971).
53. Sherod, D., Johnson, G., and Chalkley, R., *J. Biol. Chem.* **249**, 3923 (1974).
54. Furlan, M., and Beck, E. A., *J. Chromatogr.* **101**, 244 (1974).
55. Hooper, J. A., and Smith, E. L., *J. Biol. Chem.* **248**, 3255 (1973).
56. Offord, R. E., *Nature* (*London*) **211**, 591 (1966).
57. Schroeder, W. A., "The Primary Structure of Proteins." Harper and Row, New York, 1968.
58. Benson, J. V., Jr., Gordon, M. J., and Patterson, J. A., *Anal. Biochem.* **18**, 228 (1967).
59. Niall, H. D. *in* "Methods in Enzymology," Vol. 27: Enzyme Structure, Part D (C. H. W. Hirs and S. Timasheff, eds.), p. 942. Academic Press, New York, 1973.
60. Peterson, J. D., Nehrlich, S., Oyer, P. E., and Steiner, D. F., *J. Biol. Chem.* **247**, 4866 (1972).
61. Edman, P., and Begg, G., *Eur. J. Biochem.* **1**, 80 (1967).
62. Inagami, T., and Murakami, K., *Anal. Biochem.* **47**, 501 (1972).
63. Mendez, E., and Lai, C. Y., *Anal. Biochem.* **68**, 47 (1975).
64. Laursen, R. A. *in* "Methods in Enzymology," Vol. 25: Enzyme Structure, Part B (C. H. W. Hirs, ed.), p. 344. Academic Press, New York, 1972.
65. Gray, W. R. *in* "Methods in Enzymology," Vol. 25: Enzyme Structure, Part B (C. H. W. Hirs, ed.), p. 333. Academic Press, New York, 1972.
66. DeLange, R. J., and Smith, E. L., *in* "The Enzymes" (P. D. Boyer, ed.), 3rd ed., Vol. 3, p. 81. Academic Press, New York, 1971.

67. Stark, G. R. *in* "Methods in Enzymology," Vol. 25: Enzyme Structure, Part B (C.H.W. Hirs, ed.), p. 103. Academic Press, New York, 1972.
68. Stark, G. R. *in* "Methods in Enzymology," Vol. 25: Enzyme Structure, Part B (C. H. W. Hirs, ed.), p. 369. Academic Press, New York, 1972.
69. Folk, J. E., *in* "The Enzymes" (P. D. Boyer, ed.), 3rd., Vol. 3, p. 57. Academic Press, New York, 1971.
70. Hartsuck, J. A., and Lipscomb, W. N., *in* "The Enzymes" (P. D. Boyer, ed.), 3rd. ed., Vol. 3, p. 1. Academic Press, New York, 1971.
71. Hayashi, R., Moore, S., and Stein, W. H., *J. Biol. Chem.* **248**, 2296 (1973).
72. Tschesche, H., and Kupfer, S., *Eur. J. Biochem.* **26**, 33 (1972).
73. Fraenkel-Conrat, H., and Tsung, C. M., *in* "Methods in Enzymology," Vol. 11: Enzyme Structure (C. H. W. Hirs, ed.), p. 151. Academic Press, New York, 1967.
74. Candido, E. P. M., and Dixon, G. H., *Proc. Natl. Acad. Sci. U.S.A.* **69**, 2015 (1972).
75. Gershey, E. L., Vidali, G., and Allfrey, V. G., *J. Biol. Chem.* **243**, 5018 (1968).
76. DeLange, R. J., Glazer, A. N., and Smith, E. L., *J. Biol. Chem.* **244**, 1385 (1969).
77. DeLange, R. J., Glazer, A. N., and Smith, E. L., *J. Biol. Chem.* **245**, 3325 (1970).
78. Klagsbrun, M., and Furano, A. V., *Arch. Biochem. Biophys.* **169**, 529 (1975).
79. Bradbury, E. M., *Ciba Found. Symp.* **28** (N.S.) 1 (1975).
80. Macleod, A. R., Wong, N. C. W., and Dixon, G. H., *Eur. J. Biochem.* **78**, 281 (1977).
81. Kootstra, A., and Bailey, G. S., *FEBS Lett.* **68**, 76 (1976).

Chapter 1 2

Determination of the Primary Structures of Histones

M. W. HSIANG AND R. D. COLE

Department of Biochemistry,
University of California, Berkeley,
Berkeley, California

I. Introduction

The remarkable conservation of amino acid sequence in histone H4, comparing plants and animals, might lead the casual reader to conclude there is little point in further sequence determination of histones. This conclusion would be incorrect. None of the other histone classes is as conservative as H4. The strongest contrast is the case of H1 which shows a variation of sequence comparable to that of cytochrome c, but H2A, H2B, and H3 also vary significantly. An argument might well be made that the discovery of deviant amino acid sequences might help to establish correlations between structure and function of histones. Such deviant histone structure might well be sought in primitive eukaryotes which seem to represent the borderline between prokaryotes which lack histones, nuclei, and mitotic chromosomal condensations and higher eukaryotes which are characterized by all three. Which histones exist in these life forms and how deviant are their amino acid sequences are still open questions. In the case of H1, since the recipe of its subfractions varies with the phenotype of their tissue source (*1*), comparative structures might well establish a correlation with function once that function is established.

The fundamentals of amino acid sequence determination are the same for histones as they are for other proteins. There are, however, certain features of histone that superimpose special problems and considerations onto the fundamentals—such features as: their extensive posttranslational modifications; the monotonous presence (extremely high levels) of some of their constituent amino acids and the absence of others, and the possibility of microheterogeneity that might be due to the repetitive genes from which they are derived. In this article we shall cover the general fundamentals as simply as possible since they are reviewed in more detail elsewhere, and we shall give emphasis to the strategies and methods that are pertinent to histones in particular.

The points of emphasis will include methods aimed at determining the levels and sites of posttranslational modifications such as phosphorylation, acetylation, methylation, and adenosine diphosphate (ADP) ribosylation because of the increased attention these modifications are receiving as possible correlates of the dynamic, physiological state of chromatin. A few examples of histone sequence determination have been selected to illustrate the application of methods to histones and especially to show how histones have been specifically fragmented in spite of the problems created by their unusual compositions. Finally, for each class of histone, we recommend a strategy of sequence determination that seems to us the most convenient for workers interested in comparing histones from various sources. Although these recommendations might not be optimal in all cases, and may well become obsolete in the light of further progress in the development of methods, we feel these schemes will provide a helpful guide and encouragement to those interested in the correlation of variations of histone structure with chromatin structure and function.

II. Amino Acid Composition

A. Preparation of Hydrolysates

1. ACID HYDROLYSIS

a. Complete acid hydrolysis. The most general method in preparation of protein hydrolysates for the analysis of amino acid composition is acid hydrolysis. The most commonly used procedure consists of hydrolyzing the protein (approximately 0.2–1%) in 6 *N* or constant-boiling HCl in a sealed, thoroughly evacuated tube at constant temperature of 105–110° for 20–24 hours. Serine, threonine, and frequently (depending on oxidants, etc. in the sample) cysteine and tyrosine will be partially destroyed. Isoleucine and valine and those amino acids in peptide linkage with their carboxyl moieties will also be in low yield, due to the effect of the steric hindrance by the branched side chain on the rate of hydrolysis. Therefore, it is necessary to perform quantitative amino acid analysis on a kinetic series of acid hydrolysates (generally 24, 48, 72, and 96 hours). A crystal of phenol is sometimes added to the hydrolysis tube (2) to protect tyrosine, but this leads to an incorrect analysis for cystine when the latter is present (3).

The correct values for serine, threonine, cystine, and tyrosine may be obtained by extrapolation to zero time; valine and isoleucine contents are usually taken as the level observed at the longest time of hydrolysis. Unless cysteine is converted to a more stable form, such as carboxylmethyl-cyteine, it is partially converted to cystine during acid hydrolysis and is analyzed as cystine after allowing its conversion to go to completion by exposure to air at pH 6.5 for 4 hours (4). Glutamine, asparagine, and modified amino acids which occur in histones, such as ϵ-N-acetyllysine, O-phosphoserine and threonine, and ϵ-N-phospholysine and N^3-phospho-histidine, are extensively or completely converted to the unmodified amino acids under the usual conditions.

Although tryptophan has not yet been observed in histones, it should be noted that it will be completely destroyed by hydrolysis in HCl. When examining histones from unusual organisms, especially those from lower eukaryotes, it would be wise to use hydrolysis in methane–sulfonic acid which gives quantitative recovery of tryptophan (5), especially if the ultraviolet (UV) absorption spectrum indicates the possible presence of tryptophan.

b. Partial acid hydrolysis. Attempts at partial acid hydrolysis of protein and quantitative recovery of phosphorylated serine and threonine as in their native form have been rather unsuccessful. Lipmann and Levene as early as 1932 (6) found 2 N NCl at 100° for 10 hours was not sufficient to hydrolyze all the peptide bonds, and most of the organic phosphate was in small peptides (7). The regular acid hydrolytic condition (6 N HCl, 110°, 20 hours) hydrolyzes almost all the organic phosphate [all of O-phosphoserine and 80% of O-phosphothreonine (8) into inorganic phosphates; also the hydrolytic loss of serine and threonine from O-phosphoserine and O-phosphothreonine is different from that of serine and threonine (8,9). Several conditions such as 2 N HCl, 18–20 hours at 100° (10–12); 6 N HCl, 3 hours at 105° (13); 2 N HCl, 4 hours at 105° (14); 6 N HCl, 80 minutes at 110° (15); and 6 N HCl, 4 hours at 85° (16) have been tried and none of them have been proven adequate for quantitative analysis.

2. TOTAL ENZYMIC HYDROLYSIS

The total hydrolysis of a protein by enzymic methods would circumvent the problems of destruction of acid- or alkali-labile amino acids and their derivatives. Pronase (17) or Viokase [pancreatic preparation by Tower *et al.* (18)] or the combination of papain, leucine aminopeptidase, and prolidase as suggested by Hill and Schmidt (19) have been tried (20) in many proteins but difficulty has been experienced in achieving complete hydrolysis.

Balhorn *et al.* (21) used pronase (24 hours at 37°) and leucine aminopeptidase (96 hours at 37°) on radioactively phosphorylated H1 histone and chromatographically separated O-phosphoserine in a Dowex-50 column. Serine phosphate was shown to be stable to this treatment, but neither the quantitative determination of other amino acids nor the ^{32}P measurement of the complete elution profile was given so that the extent of digestion cannot be estimated.

In the study of acid-labile histone phosphates in histones and chromosomal acidic proteins of regenerating rat liver, Chen et al. (22) used trypsin (37°, 2 hours at pH 8), followed by pronase (24 hours digestion at 37°), and identified acid-labile ϵ-N-phospholysine and 3-phosphohistidine by paper chromatography. There was significant radioactivity at the origin (10%) which might represent incompletely digested peptides.

B.   Amino Acid Analysis

1.   COMMON AMINO ACIDS–AMINO ACID ANALYZER

Usually analyses of protein hydrolysates are carried out by one of the several automated amino acid analyzers available, and the analysis of the common amino acids is straightforward using any of them (23). Since the introduction of such instruments, great attention has been given to reducing the time required for the routine analysis of the usual protein hydrolysate. Among the several factors manipulated to this end was the design of elution programs that essentially eliminated all resolving space (and time) between peaks of the common amino acids. While these programs are obviously advantageous for most analyses of proteins, they are not suitable for analyses that involve unusual or modified amino acids. For histones, then, the otherwise attractive 1-hour analysis on a Durrum Model D-500 analyzer is frequently not as suitable as a longer analysis on an analyzer such as the Beckman 121 that is used in the authors' laboratory.

In the Beckman Model 121 analyzer as well as in others using the ninhydrin color reaction, it is possible to achieve a sensitivity of 10^{-9} moles of each amino acid with a precision of $\pm 3\%$ (24). For histones, then, about 40 to 100 μg of histone are hydrolyzed for analysis in a single-column analyzer. Twice the amount of histone is hydrolyzed for the two-column type of analyzer (one column for acid and neutral amino acids, and a shorter column for basic amino acids).

A new development employs either an Aminanalyzer (American Instrument Company, Silver Springs, Maryland) or another amino acid analyzer modified to use a fluorometric detector in place of a colorimeter (Durrum D-500, Durrum Chemical Corporation). In place of ninhydrin, a fluorescent reagent such as fluorescamine (25) or O-phthaldehyde (26,27) is used. Both reagents are 10 to 100 times as sensitive as ninhydrin in theory and in practice for amino acids other than proline, but their full potential has yet to be realized because of difficulties with proline.

2. SPECIAL PROBLEMS

a. Glutamine and asparagine. These amino acids can be estimated by use of the amino acid analyzer. To separate asparagine and glutamine from the other amino acids (particularly serine) of an enzymic hydrolysate, the usual procedure used for the amino acid analyzer must be modified. Adjusting the elution temperature of 30° is helpful in some systems (*28*). Adequate separation also seems to be obtained by using a lithium–citrate buffer system at 37° for the Beckman 120C analyzer (*29*) or lithium–salt buffers in the Beckman 121M analyzer (*30*).

b. ε-N-Acetyllysine, ε-N-Succinyllysine. In the regular analyzer program ε-N-acetyllysine will be eluted between proline and glycine. For example, in a typical run on a Beckman 120B analyzer, it is eluted at 77 minutes while glycine is eluted at 86 minutes and proline at 70 minutes (*31*). ε-N-succinyllysine will be eluted between glutamic acid and proline (*32*).

c. Methylated lysine, histidine, and arginine. These derivatives can be resolved in the analyzer by use of modified elution programs. An ε-N-methyl derivative of lysine was first observed in histones by Murray (*33*). He used 0.35 *M* sodium citrate buffer at pH 5.28 in a 53-cm column for 8 hours to separate ε-N-methyllysine from lysine and other amino acids; however, mono-, di-, and trimethyllysines eluted out as a single peak (*34*). Paik and Kim (*34*) used a 50-cm column in a standard analyzer, with 0.35 *M* citrate buffer of pH 5.84 at 28° and a flow rate of 30 ml/hour to separate mono- and dimethyllysine and thus to show histone H4 has dimethyllysine as well as the monomethylated derivative. This system was later shown to be adequate to resolve mono-, di-, and trimethyllysines, 3-methylhistidine, 1-methylhistidine (*35*), ω-N-methylarginine, and α-N-methyl-ω-N-methylarginine (*36*). Even a shorter column (0.9 × 36 cm) is sufficient to separate mono-, di-, and trimethyllysines under similar conditions (0.35 *M* citrate buffer, pH 5.837, 30°, 30 ml/hour) (*37,38*). Although considerable time could be saved in a shorter column (10 hours/analysis instead of 14 hours), the separation of methylated histidines and arginines has not been shown in this system, and it is doubtful that methylated derivatives need to be analyzed frequently enough to justify the preparation of this special column, as long as the 50-cm column is available as it is in many amino acid analyzers.

d. O-Phosphoserine and O-phosphothreonine (*12*). These derivatives are not adequately analyzed with the amino acid analyzer since they elute very early near common contaminants and cysteic acid (*39*).

Paper chromatography in 40% ethanol:conc. NH_3 (9:1) gives an R_f of 0.82 for O-phosphoserine (12). In several other systems of paper chromatography, the separation seems to be adequate (10,11,40); however, no R_f values were given.

Paper electrophoresis at acidic pH gives very satisfactory separation (8,11). However, the best quantitative separation is obtained by ion-exchange chromatography at Dowex 50 by 0.01 N HCl at 37° (10) or 0.05 N HCl (12).

    e. ε-N-Phospholysine and N³-phosphohistidine.   Ascending paper chromatography on Whatman No. 3 paper with isopropyl alcohol:ethanol: H_2O: triethylamine (30: 30: 39: 1) gives reasonable separation of acid-labile phosphorylated histidine, arginine, and lysine. Their R_f values are 0.56, 0.68, and 0.75, respectively (22).

III. Selective Cleavage of Polypeptide Chain

A. Introduction

The primary structure of a protein generally cannot be determined simply by sequential degradation from the ends of the whole molecule, and so one must resort to the examination of fragments obtained from the selective cleavage of the polypeptide chain. Commonly one can determine with confidence 20–40 residues, and occasionally 50 residues, from the amino end. Unfortunately, the sequential degradation method from the carboxyl end has not been developed to any stage comparable to Edman degradation. In fact none of the available methods for carboxyl terminal analysis goes well beyond a few residues. Therefore even the sequence of the smallest histone, H4, with 102 residues cannot be determined without its cleavage into smaller peptides. Moreover, since three of the five major histone classes and also the erythrocyte specific histone, H5, have their amino ends blocked by acetylation, these intact histones are not subject to Edman degradation at all, and once again we have to resort to selective cleavage methods.

A good means for selective cleavage of polypeptide chains either by chemical or enzymic methods must meet the following criteria: (1) high specificity, (2) high cleavage yield, and (3) no undesirable side reactions.

In general, when the specificity is high, the number of peptide bonds cleaved is small, and the resulting mixture of peptide fragments is less complex and more amenable to fractionation. Moreover, high specificity is desirable because it produces large fragments for use in automatic sequencers and for overlapping smaller known sequences. Of course, the need for small fragments arises too, particularly when working out those few stubborn pockets of ambiguity that frequently require a major portion of the researcher's time.

B.   Chemical Methods

1.   CYANOGEN BROMIDE (*41*)

To date very few chemical cleavages meet the stated criteria. The most specific and generally applicable method available is the cyanogen bromide (CNBr) cleavage of the peptide chain at the carboxyl end of methionyl residues (*42*). Unfortunately, not all histones possess methionine residues; it is absent in H2A and exceedingly rare in H1 (*43,44*). When it occurs in these proteins, the content is so low that large polypeptide fragments are produced that are advantageous for the automatic sequencer.

The cleavage of histone H4 from calf by cyanogen bromide will be described as an example (*45*). Ninety-five milligrams of calf H4 were treated with 80-fold molar excess of cyanogen bromide (Eastman) at room temperature in 70% formic acid for 20 hours at a protein concentration of 10 mg/ml in the dark. At the end of the reaction, the mixture was diluted 1:10 with deionized water and lyophilized to dryness. There is a side reaction involving cysteine, which is slowly oxidized to cysteic acid; therefore when the cyanogen bromide reaction was applied to H3, the protein was carboxymethylated (*46*), or was oxidized to cystine (*47–49*) before cyanogen bromide reaction. The yield of peptide is in the range of 70–90%.

2.   N-BROMOSUCCINIMIDE (NBS) (*50*)

The reactivity of *N*-bromosuccinimide (*51*) toward the cleavage of peptide bonds is in the order of tryptophan > tyrosine > histidine, and the peptide bond on the carboxyl terminal of histidine is cleaved only after heating at 100° for 1 hour. Since all the histones have tyrosine but no tryptophan, this method has been used widely in histone sequence determination. The yield is generally lower than cyanogen bromide cleavage but can be increased by using excess reagent with the concomitant oxidation of cysteine, if present.

Histone H1 has a single tyrosine residue per molecule (*52,53*). When 1.2 μmol (24 mg) of H1 in 8 ml of 50% acetic acid were reacted with 12 μmol of freshly prepared *N*-bromosuccinimide solution in 50% acetic acid, reaction could be followed at 260 nm. After 3 hours, a second addition (12 μmol) of newly dissolved *N*-bromosuccinimide solution was made and reacted for another 5 hours. After the reaction, acetic acid was flash evaporated and the reaction mixture was fractionated in a Sephadex G-100 column in 0.02 N HCl (*54*). The yields of two fragments were nearly quantitative (*55*). They were better than the yields observed in most other proteins (30–65%) (*56*) because larger amounts of reagent could be used without fear of reaction with histidine. Under these conditions, however, the aspartyl residue split from the tyrosine is partially damaged, blocking a fraction of the large fragment representing the carboxyl region of H1 (*55*). There was also a small amount of H1 that resisted cleavage by *N*-bromosuccinimide, but the absorption spectrum indicated the presence of the spirolactone that is the expected derivative of tyrosine in the cleavage reaction. Sherod *et al.* (*57*) proposed several mechanisms and the structure for this modified but uncleaved H1. Apparently, more of this modified H1 formed under dilute acid conditions (0.1 *N* HOAc) than in 50% acetic acid (*55*).

The three tyrosine residues of H3 were cleaved to different extents by *N*-bromosuccinimide. Tyr–Leu and Tyr–Arg linkages were cleaved to nearly 100% while Tyr–Glx was cleaved to only about 50% completion (*49*).

One of the three peptides derived from *N*-bromosuccinimide cleavage of rainbow trout H2A was not obtained in pure form, but from the yields of the other two peptides the cleavage at two tyrosine residues must have been close to 90% (*58*). Histidine residues in the sequence were modified but not cleaved at room temperature. One of the peptides (NBS-3) of H3, after purification and characterization, was heated in 1.9 ml of H_2O, 0.1 ml of pyridine, and 1.0 ml of acetic acid at 100° for 1 hour to cleave it at the single histidine residue; the yield varied from 60 to 95%.

3.   MILD ACID CLEAVAGE–RELEASE OF ASPARTIC ACID (*59*)

In favorable cases free aspartic acid is released from proteins containing aspartyl or asparaginyl residues at a rate about 100 times faster than any other residues; when proteins are heated at 105° in 0.03 *N* HCl (*60,61*), they sometimes can be cleaved quite selectively. In less favorable circumstances when, for example, an adjacent valine hinders the hydrolysis of aspartyl residues, the rate of hydrolysis might overlap the rates of other labile bonds such as acyl proline (*62,63*) and acyl serine bonds (*64*), and as a consequence the selectivity can be unsatisfactory. A peptide (CN-1, NB-1) of chicken erythrocyte H3 was hydrolyzed for 17 hours at 105° in 0.03 *N* HCl. 90% of the two aspartyl residues were released producing three polypeptide fragments (*49*).

4.   N → O ACYL REARRANGEMENT (*65*)

In concentrated, anhydrous acid an N → O rearrangement takes place at peptide bonds containing the amino group of serine or threonine. The polypeptide chain is then held together by an ester linkage, and since esters are more readily hydrolyzed than peptide bonds, both acid and alkali have been used in selective cleavage of proteins previously exposed to anhydrous acid

(66,67). In general, the main problems are to obtain a complete acyl shift and to prevent reversion from the ester to the peptide form under the conditions used for hydrolysis. The rearrangement at threonine usually is less than 50%, and at serine it is the range of 60 to 90% (65). Although this method is far from ideal in most situations, it has given workable results in a few cases.

C. Enzymic Cleavages

1. GENERAL COMMENTS (68)

There are two categories of proteases; endopeptidases that split internal peptide bonds, and exopeptidases that remove amino- or carboxyl-terminal amino acids or dipeptides. The exopeptidases, such as dipeptidyl amino peptidase, aminopeptidases, and carboxypeptidases, are used to determine the terminal amino acids of proteins and peptides. Of the endopeptidases, trypsin, chymotrypsin, thermolysin, thrombin, and staphylococcal protease possess sufficient specificity to allow them to be used widely for specific cleavage of proteins. Other endopeptidases, such as subtilisin, papain, pepsin, etc. have broader specificity, and therefore they are most commonly used for secondary degradation of fragments that are produced by the more specific enzymes.

Except for thrombin, whose cleavage sites on protein seem to be rather limited, the recognition sites for the other highly specific endopeptidases are fairly abundant in protein (lysine and arginine for trypsin; aromatic residues and leucine for chymotrypsin; the branched chain amino acids, valine, and isoleucine for thermolysin; and glutamic acid for staphylococcal protease). Consequently, the resulting peptide fragments are generally small and their mixtures fairly complex. The subsequent purification of the peptides is frequently the most time-consuming step in sequencing a protein as well as being the least straightforward. Nevertheless, there are two advantages of these enzymic cleavages that make them the most useful techniques for producing, in good yields, a relatively uniform population of peptide fragments, most of which are small enough for easy sequence determination. The first advantage is the mildness of the conditions under which they are used. Generally, room temperature is used near neutral pH to preserve many labile bonds so that asparagine, glutamine, and the modified amino acid residues of histone are kept intact. The second advantage is that their high specificities maximize yields at the primary cleavage sites and minimize cleavages at secondary sites which lower peptide yields and complicate the purification of peptides (69).

The exact conditions for digestion depend upon the enzyme, but (1) pH,

(2) temperature, (3) enzyme to substrate ratio, (4) substrate concentration, and (5) time of incubation are the important factors for consideration.

2. TRYPSIN (70)

The classical studies of Bergmann et al. (71) and Bergmann and Fruton (72) established that trypsin specifically cleaves at the carboxyl end of lysine and arginine. The cleavage rate at arginine is faster than at lysine, and the relative rates of cleavage of susceptible bonds are influenced by the chemical nature of the side chains in the immediate vicinity. Generally, the presence of polar groups close to the susceptible bond decreases the rate of hydrolysis (69). For example, the cleavage takes place more slowly when an acidic residue (aspartic acid, glutamic acid, cysteic acid, S-carboxylmethyl cysteine, or O-phosphoserine and threonine) is adjacent to lysine or arginine. Similarly, an amino terminal lysine, either in the original protein or in one exposed during proteolysis, is cleaved slowly. Most of these observations have been made in the cleavages of histones. However, several Lys–Pro bonds which, along with Arg–Pro bonds, are generally considered totally resistant to tryptic hydrolysis (70) were extensively hydrolyzed in rabbit histone H1 (73). N-Acetyllysine is not cleaved by trypsin, and presumably methylated lysine and arginine do not fit the active site of trypsin [maleylated calf H4 was not cleaved at methyllysine, but it was not certain that methyllysine was free from maleylation (31)]. In porcine H4 (74,75) and rat chloroleukaemic tumor H4 (76) trypsin did not cleave at methyllysine.

The form of the enzyme generally used is the commercially available TPCK-treated trypsin (77). An enzyme to substrate ratio of 1 : 50 (preferably delivered as 1% at zero time and another 1% at 2 hours) at 37° at pH 8 (either in a pH-stat or in 0.1 M ammonium bicarbonate) for 4 hours at a histone concentration of 10 mg/ml is commonly used. Prolonged hydrolysis (~ 20 hours) usually is avoided to reduce nonspecific cleavages. The digestion can be stopped by adjustment of the pH to pH 3 or less or by inactivation of the enzyme with diisopropyl fluorophosphate (DFP) or phenylmethyl sulfonylfluoride (PMSF) at neutral pH. The latter is used when it is desired to preserve acid-labile side-chain modifications such as maleyl groups.

3. MODIFICATIONS OF LYSINE TO RESTRICT TRYPTIC CLEAVAGE

The natural specificity of trypsin is greater than that of most other proteases, and so substrate modifications have been used to exploit it further. New points of cleavage can be created at cysteinyl residues through S-aminoethylation (78,79), or tryptic hydrolysis can be restricted to either arginine or lysine through modifications (70).

Because of the high content of lysine and arginine, a common problem with histones is the restriction of tryptic activity. The modifications of lysine such as maleylation, succinylation, guanidination, and amidation have all been used. The advantages and disadvantages of each modifying agent and some other agents will be discussed in separate sections below.

a. Maleylation (80). Maleylation of calf thymus H4 by DeLange *et al.* *(31)* is described as an example; it was performed essentially as described by Butler *et al (81)*.

To 568 mg of the protein dissolved in 100 ml water at pH 9.0 and 0°, 600 mg of maleic anhydride were added over a 5-minute period while the solution was maintained in the pH-stat between pH 8.5 and 9.5 by the addition of 4 M NaOH. After a total of 15 minutes, the maleylated protein was dialyzed against 0.1 M NaHCO$_3$ for 2 hours and then against two changes of water at pH 8.0 for 3 hours.

After the tryptic digestion, the peptides were demaleylated at pH 3 for 40 hours at 40° in the presence of a few drops of toluene to retard bacterial growth.

The yields of the 13 expected peptides after gel filtration by Sephadex G-25 column chromatography and paper chromatography were around 40–50%. (No indication of cleavage at lysine was observed.) These are high recoveries considering paper chromatography was used, and they indicate that maleylation and demaleylation were nearly quantitative.

Other examples of the restriction of tryptic activity by maleylation of other histones have been published *(48,58,82–88)*.

b. Succinylation (89). Succinylation was used to block lysine or calf thymus H2B *(32,90)*. Calf thymus H2B (44 mg; 3.2 μmol) in 3–4 ml of water was adjusted to pH 8 and kept at 2°. Succinic anhydride, 34-fold excess, was added, and the mixture was maintained at pH 8 in a pH-stat for 3.5 hours. At the end of the reaction, hydroxylamine was added to the reaction mixture to the final concentration of 0.67 M for 10 minutes at pH 6.6 at 25° to desuccinylate the O-succinyl groups introduced. Then the solution was thoroughly dialyzed against water of pH 8 at 4° and lyophilized.

Ishikawa *et al. (32)* confirmed by dinitrophenylation that all the lysine was blocked by the succinylation. Succinyllysine is generally stable during the manipulation of peptide fractionation and purification, but in studying H2A, Bailey and Dixon *(58)* found maleyllysine was partially unstable at pH 6.5 during separation of peptides by paper electrophoresis. Unfortunately, succinylation is irreversible, and there is no way to cleave at lysine residues with trypsin after the completion of the arginine cleavages. Moreover, complications are introduced by O-succinylation of serine *(32)*.

c. Amidination (91). Early attempts to block the lysines of H1 by maleylation were unsuccessful because of incomplete blockage, and trifluoroacetylation failed because the derivative was insoluble *(92)*. Therefore, amidination and guanidination, both of which preserve the positive charge of lysine, were successfully tried in restricting tryptic cleavage.

The amidination of H1 (Hsiang and Cole, unpublished) by methyl acetimidate was carried out as described by Ludwig and Hunter (*91*) in a pH-stat. The completeness of lysine blockage was demonstrated by dinitrophenylation. The reversal of amidination can be performed in several ways (*91*) but all these ways are troubled somewhat by side reactions.

d. Guanidination (93,94). Guanidination of an H1 histone fragment was performed in order to reduce the extractive loss of peptide in the protein sequencer (M. W. Hsiang *et al.*, unpublished). Quantitative guanidination by O-methylisourea was obtained following the method of Kimmel (*93*). One gram of O-methylisourea was dissolved in 3 ml of water, and adjusted to pH 10.5, and then added to 14 mg (1 μmol) of the peptide RTL-3 N1 (*54*) in 1.0 ml of water. The pH was readjusted to 10.3 (slightly higher pH resulted in precipitation and incomplete guanidination), and the solution was stirred at 4° for 96 hours, after which the mixture was dialyzed against several changes of cold water and lyophilized. Homoarginine, the modified form of the lysine, is stable toward acid hydrolysis and can be analyzed in the amino acid analyzer (*94a*).

Guanidination of peptides for use in the automatic sequencer has the disadvantage that the acid-catalyzed cyclization and cleavage step of the Edman reaction at proline residues is greatly hindered by neighboring homoarginine (proline at homoarginine–proline–homoarginine sequence was cleaved to 40–60% completion in one cycle of acid cleavage reaction resulting in extensive overlapping of sequence). Brandt and Von Holt (*95*) also observed a similar effect when proline is located before arginine. A similar limitation probably applies to amidination.

4. CHYMOTRYPSIN

Chymotrypsin is less specific than trypsin, and consequently the peptide mixtures it produces by digestion are more complex. Moreover, the yields of chymotryptic peptides are generally lower than those of tryptic peptides (for calf thymus H4 they are around 10% compared to 50% for tryptic peptides) (*31*).

Chymotrypsin exhibits a preference for peptide bonds involving the carboxyl groups of aromatic amino acids and leucine (*20*), but cleavages may also occur at several other peptide bonds at slower rates. The environment of the susceptible bond is important. No peptide bond linked to the amino nitrogen of proline will be cleaved, and any bond near a negatively charged residue or proximal to the terminus will be hydrolized very slowly. Chymotrypsin hydrolyzes the cluster of 3 lysines (*96*) and 2 lysines (*97*) in cytochrome c. In histones, a few exceptional bonds were cleaved; for example,

Arg–Thr of calf H4 (*31*), Arg–Ser of calf H2A (*87*), and Ser–Ser of trout H2A (*58*). A Lys–Lys bond in calf H2B and the third and fifth bonds in the sequence Lys–Lys–Arg–Lys–Arg–Ser in the same histone were cleaved (*98*), and a similar cleavage at the third bond in the sequence Lys–Lys–Lys–Lys was found in H1 histone (*99*). Because of the relative lack of specificity of chymotrypsin, optimal digestion conditions differ from one substrate to another, and they must be determined for each case individually.

5. THERMOLYSIN

Thermolysin (*100, 101*) cleaves preferentially at the amino side of hydrophobic amino acids (including alanine). If the carboxyl group of the sensitive residue is in peptide linkage with proline, cleavage will not occur at the expected site (*100*).

Thermolysin was first used on histone by DeLange *et al.* (*31*) who found yields of cleavage at pH 7.8 of 10–30%; the cleavages were preferential for branched-chain hydrophobic residues. Hayashi and Iwai (*102*) used thermolysin on H2B and found the enzyme more selective for isoleucine and leucine than for valine and phenylalanine at pH 7.0 but not at higher pH. Poly-L-lysine was not hydrolyzed at a 1:100 enzyme to substrate ratio at 40° for 10 hours. Ando and Suzuki (*103*) also found no cleavage at arginine after prolonged hydrolysis on protamine. However, Rall and Cole (*99*) found that thermolysin at pH 7.8 hydrolyzed a cluster of four basic residues Lys–Lys–Lys–Lys in one H1 histone.

6. THROMBIN (*104*)

Thrombin is known to be highly selective in its substrate specificity (*104*). It hydrolyzes four Arg–Gly bonds of its natural substrate, fibrinogen (*105*) but not the Arg–Gly bond of insulin (*106*), nor the Arg–Gly bond of some other proteins (*107*).

Chapman *et al.* (*108*) found that a bovine thrombin preparation specifically cleaved H1 on several sites. A subsequent study by M. W. Hsiang and R. D. Cole (unpublished) on rabbit H1 showed that four of the cleavages are at the third bond of some particular Lys–Pro–Lys sequences. Cleavages at the carboxyl end of some lysines were also observed at 1 NIH unit of thrombin/mg of H1 at 20 mg/ml for 20 hours at 37° in 50 mM Tris buffer (pH 8.0)–2 mM MgCl$_2$–2mM CaCl$_2$.

7. STAPHYLOCOCCAL PROTEASE

The protease from *Staphylococcus aureus* strain 8325N has been shown to be specific for the carboxyl end of glutamyl residues (*109*), but some other minor cleavages were also observed in the application of this enzyme to the sequence study of neurotoxin (*110*). In an unpublished (Hsiang and Cole) study of histone H1, specific cleavage at the glutamyl bond was observed using a preparation of staphylococcal protease generously provided by Dr. Lars Rydén (University of Uppsala), at an enzyme: substrate weight ratio of 1:200 in 0.1 M ammonium bicarbonate at 37° for 4 hours at a substrate concentration of 1–10 mg/ml.

Another staphylococcal protease from a different strain (strain V8) was reported to hydrolyze aspartyl and glutamyl residues (111, 112). Behrens and Brown (113) observed a strong preference for the glutamyl residues of serum albumin compared to the aspartyl residues. It is available from Miles Laboratory, Elkhart, Indiana (Protease, *S. aureus*, V8 Code 36–900).

IV. Methods of Peptide Fractionation

A. General Considerations (114)

Separation and purification of peptides from an enzyme digest or a cleavage mixture usually are the rate-limiting steps in the sequence determination of proteins. One way to circumvent this difficulty is to choose the cleavage method that gives the least number of fragments and the most complete cleavage at each fragmentation step.

The purification of peptides is usually accomplished by combination of ion-exchange chromatography, gel filtration, paper electrophoresis, and paper chromatography. It is useful to estimate beforehand the number of peptides to expect and to have an approximate notion of their electrophoretic and chromatographic behavior. This can be done with some confidence with histones because they are conserved in sequence.

Gel filtration usually gives quantitative yields, and while it is the simplest technique, its resolving power is somewhat limited. Ion-exchange chromatography and paper electrophoresis have much greater resolving power, but the former is technically tedious and the latter gives losses of about 30% per run. Frequently the best resolution is seen in paper chromatography, but here losses run 50% or even higher. Highly cationic peptides produce streaks on paper and are almost irreversibly bound on strong cation exchangers (e.g., Dowex 50).

B. Gel Filtration

Two types of gel filtration media are widely used for fractionation of peptide mixtures—Sephadex (cross-linked dextrans by Pharmacia Fine Chemicals, Uppsala, Sweden) and Bio-Gel P (polyacrylamide gels in beads by Bio-Rad Laboratories, Richmond, California). The technical bulletins by both companies are most useful in listing the available kinds of Sephadex and Bio-Gel and the effective range of separation of each.

The primary factor in the separation in either Sephadex or Bio-Gel P is hydrodynamic size. Adsorption to Sephadex will retard the peptides that contain hydrophobic residues. Adsorption effects appear to be minimal with Bio-Gel P. The presence of a few carboxyl groups in the Sephadex also retards basic peptides through ionic binding, and the retardation of basic peptides is even more prominent in Bio-Gel P in dilute acid (Hsiang and Cole, unpublished). Elution with high-ionic-strength eluents minimizes this charge effect. Both the charge effect and the adsorption effect could be useful in separation of certain peptides of similar size, but different chemical character.

C. Ion-Exchange Chromatography

Ion-exchange chromatography has been used extensively in fractionation of peptides from histones as well as other proteins. Among the most frequently used ion exchangers are: Dowex-50 (115), Dowex-1 (116) CM-cellulose (117), CM-Sephadex, SE-cellulose (117), SE-Sephadex and DEAE-cellulose (118). Basic peptides are best resolved on cation exchangers while acidic ones generally work best on anion exchangers. There is frequently an advantage to using two different types of resins for mixtures or relatively neutral peptides, especially when a diversity of pKs or a diversity of hydrophobic character can be exploited. A special problem of peptide fragments from histones is that they sometimes contain several cationic groups and so bind almost irreversibly to strong ion exchangers such as Dowex-50 or Bio-Rex 70.

D. Paper Electrophoresis (119)

Paper electrophoresis gives good resolution but because of its limited capacity and modest recovery it is frequently used in the last stages of peptide purification.

For paper electrophoresis, pyridine–acetate buffers at pH 6.5, 3.6, and 1.9 are used most frequently. The resolving power can be extended and adapted to particular problems by the judicious combination of successive runs at different pHs. It is important that the sample be free from salt before application to the paper. Among the three different pHs, 6.5 is the most tolerant toward salt; therefore, pH 6.5 has been used most frequently as the first step of paper electrophoresis. The amount of peptide loaded on the paper affects the quality of the results. Overloading will spread the spots or bands so that they might overlap, and underloading will lead to poor recovery due to the loss of peptides through adsorption. In spreading a band of peptides at the origin, 20 nmol or 0.2 mg/cm is recommended for Whatman No. 1 paper and 100 nmol or 1.0 mg/cm is recommended for Whatman No. 3 MM paper. Substantial departure from this level in either direction is not recommended. Further important details on the method of handling paper and sample application are contained in an article by Harris (*120*), although this article was written for another purpose.

After the electrophoresis, peptides can be located by ninhydrin, fluorescamine, or *O*-phthaldehyde staining of a guide strip, and then the peptide in the main band can be cut and eluted by solvent. Although dilute acetic acid, dilute bicarbonate solution, and other solutions have been recommended for the elution of peptides from the paper strip, we found water generally works effectively.

The range of recovery for most peptides is 60–70% for each dimension of electrophoresis; for very basic peptides it is 30–50%.

E. Paper Chromatography (*119*)

Preparative paper chromatography of peptides is usually performed on Whatman 3 MM paper. The solvents generally used are:

Solvent I: *n*-Butanol:HOAc:H_2O (4:1:5)

Solvent II: *n*-Butanol:pyridine:HOAc:H_2O (90:60:18:72)

The general considerations for separation of peptides on paper as discussed in paper electrophoresis also apply for paper chromatography.

The recovery of peptide is even lower (50–60%) than in the paper electrophoresis.

V. Sequencing Methods

A. Terminal Analysis

1. NH$_2$-TERMINAL ANALYSIS

There are about 40 methods for NH$_2$-terminal determination. Among them, the phenylisothiocyanate method of Edman and the dansyl chloride method of Hartley are currently the most widely used due to their versatility and high sensitivity.

a. Dansyl procedure (*121–123*). One to two nanomoles of peptide are dried down *in vacuo* in a glass test tube (4 mm × 30 mm). Ten microliters of freshly prepared 0.1 *M* sodium bicarbonate are added, and then the tube is taken to dryness in a vacuum desiccator. Next 10 μl of deionized water are added (check pH $\geq$ 8 by pH paper), followed by 10 μl of dansyl chloride in acetone (2.5 mg/ml). The tube is covered with Parafilm and incubated at 37° for 1 hour. After drying the sample in a vacuum desiccator, the dansylated peptide is hydrolyzed in 50 μl of 6 N HCl in a sealed tube (evacuation is unnecessary).

When proline is suspected as the terminal amino acid, 4 hours at 110° ought to be used since it will double the yield of dimethylaminonaphthalene sulfonyl (DNS)–Pro compared to the yield of 20–25% obtained at 16 hours. Peptide bonds at the carboxyl end of valine and isoleucine are slowly hydrolyzed; therefore the yields of DNS–valine and DNS–isoleucine are low, and DNS–dipeptide might be observed unless the time of hydrolysis is greatly extended.

The hydrolysate is thoroughly dried down *in vacuo* and then dissolved in 3 μl of 50% pyridine to be applied equally to both sides of a thin-layer polyamide sheet (7.5 × 7.5 cm) (Cheng Chin Trading Co., Ltd.; distributed in the United States by Gallard-Schlesinger, Carle Place, New York). To one side is applied 1 μl of a mixture of DNS–amino acids (DNS–proline, DNS–isoleucine, DNS–phenylalanine, DNS–glycine, DNS–glutamic acid, DNS–serine, and DNS–arginine, at 0.1 nmol/μl of each per μl of 50% pyridine).

The plate is chromatographed in the first dimension (ascending) for 13 minutes in 1.5% formic acid (v/v) and then thoroughly dried. Then it is run in benzene: glacial acetic acid (9:1, v/v), in the second dimension. Most of the DNS–amino acids are resolved in this system. A further chromatography in ethyl acetate–glacial acetic acid–methanol (20:1:1, v/v) in the same direction as the second dimension will improve the resolution of DNS–aspartic acid, DNS–glutamic acid, DNS–serine, DNS–threonine, and DNS–histidine.

Unknown DNS–amino acids are located by reference to the standard DNS–amino acids on the opposite side of the sheet under a long wavelength (365 nm) ultraviolet (UV) lamp; short wavelengths (254 nm) are also effective in exciting fluorescence, but will promote photo-oxidation of DNS–amino acids.

The NH$_2$-terminus of as little as 0.02 nmol of peptide can be identified by the dansyl procedure. Unfortunately, it is not as satisfactory for large proteins because protein precipitation during the labeling decreases accessibility of the end group by the reagent. Sodium dodecyl sulfate (SDS) can solve some of the problems of insolubility (*124*). Still large amounts of DNS–OH are found after hydrolysis and the fluorescent DNS–ϵ-lysine, DNS–O-tyrosine, and DNS–NH$_2$ may obscure the unknown DNS–amino acid.

b. Edman procedure (125,126). The peptide or protein (2–10 nmol) is dried in a thick-walled screw-capped 12-ml conical centrifuge tube. Then 200 μl of 50% pyridine is added, followed by 100 μl of 5% phenylisothiocyanate in redistilled pyridine. The tube is immediately flushed with nitrogen and covered with Saran wrap; the screw-cap is replaced before incubation in a 45° oven for 1 hour to allow the coupling reaction to take place. A clear, pale, yellowish solution of phenylthiocarbamylated protein is obtained at the end of the incubation. The mixture is dried in a heated (60°) desiccator over P_2O_5 and moist sodium hydroxide pellets *in vacuo* to remove excess reagent and solvent. After the contents of the tube are completely dried, about 0.5 ml of anhydrous trifluoroacetic acid (TFA) is added to cover the residue, and the tube is again flushed with N_2, covered with Saran wrap, and capped. After mixing on a Vortex mixer, it is incubated in a 45° oven for 45 minutes. At the end of cyclization the open tube is placed in a desiccator for drying *in vacuo* over NaOH (vacuum applied gradually). Next 150 μl of deionized distilled water and 1.5 ml of spectro grade *n*-butyl acetate are added and mixed. Two phases are separated by centrifugation, and the extraction procedure is repeated twice.

The aqueous phase which contains the original protein or peptide minus the NH_2-terminal residue is dried down in a vacuum desiccator. An aliquot of this residual peptide can be taken for amino acid analysis as in "subtractive Edman method" or for dansyl NH_2-terminal determination in the "dansyl Edman method" for sequence determination.

The organic phase which contains the thiazolinone is pooled in a 5-ml conical tube and dried down in a heated water bath (40°) under a stream of nitrogen. When the contents are completely dried, 0.2 ml of 1 N HCl is added, followed by N_2 flushing, and the tube is covered tightly with a silicon rubber stopper. The contents are mixed and heated in an oil bath at 80° for 10 minutes to convert the unstable 2-anilino-5-thiazolinone derivatives of amino acids into their stable isomers, phenylthiohydantoins (PTHs). The contents are cooled to room temperature and extracted twice with 0.7 ml of ethyl acetate (sequencer grade). The aqueous phase which contains PTH–arginine, PTH–histidine, and PTH–cysteic acid is dried down *in vacuo*. The organic phase which contains other PTH–amino acids is dried.

c. Methods of identification of PTH–amino acids. PTH–amino acids can be directly identified by many procedures. The advantage of the successive chemical degradation by the Edman procedure and especially the development of the automatic sequencer (*127*) have prompted improvements in the analysis of PTH–amino acids. Paper chromatography (*128*), thin-layer chromatography (TLC) (*129*), gas chromatography (GC) (130), and high-pressure liquid chromatography (HPLC) (*131*) have been used for direct identification of the PTH–amino acids. Thin-layer chromatography, which does not require any expensive instrument, has been most commonly used, but only qualitatively. Paper chromatography has been superseded by TLC techniques, but it still plays an important role in quantitative determination (126). GC has been used extensively in recent years because of its advantages on sensitivity and speed. However, PTH–arginine and PTH–cysteic acid cannot be identified by GC, and difficulties with PTH–serine, –threonine, –asparagine, –glutamine, –histidine, and –lysine are still not completely solved by silylation (132). The separation of PTH–amino acids by HPLC is the most recent addition (131). Although the resolution achieved so far is not as good as that attained by GC (132), new improvements are expected. HPLC may well become the method of choice in the future because: (1) it is sensitive to 10 pmol; (2) all PTH–amino acids are stable during the analysis and have similar extinction coefficients; (3) all the PTHs are eluted from a single column; (4) no further derivatization of PTHs is required; (5) it is speedy ($\sim$40 minutes for each analysis).

Mass spectrometry has been used for identification of PTHs (133).

Back hydrolysis by either HI (134), HCl (135), or NaOH (136) to regenerate the parent amino acid and identification in the amino acid analyzer have been used in conjunction with GC. The principal difficulties of this method are that some amino acids are destroyed and do not produce the parent amino acids. Still this is the most sensitive method of quantitating PTH–lysine, –arginine, –histidine, and –cysteic acid when HPLC is not available.

(i) *Thin-layer chromatography.* TLC on both silica gel and polyamide sheet has been used. Edman (126) advocates silica gel TLC, and it has been used widely. Three solvent systems are generally: system D—xylene/formamide; system E—n-butyl acetate/propionic acid/formamide; and system H—ethylene chloride/acetic acid. The identification of closely spaced spots on silica gel requires practice; two-dimensional TLC on polyamide sheets, however, seems more easily reproducible (137). Cheng-Chin polyamide sheets are cut into 5 × 5 cm squares. Samples in methanol are applied on both sides and 1 μl of PTH–amino acid markers (PTH–proline, –isoleucine, –phenylalanine, –glycine, –glutamic acid, –serine, –arginine, and –tryptophan in 0.5 nmol/μl methanol) is also applied on one side. The plate is chromatographed for about 10 minutes in the first dimension (ascending) with toluene:pentane:HOAc (60:30:35) containing 250 mg/liter of 2-(4-t-butylphenyl)-5-(4'-biphenyl)-1,3,4-oxidiazole as fluorescent indicator, after 15-minute equilibration against the solvent vapors. After thorough drying in a stream of hot air, the plate is chromatographed with 35% HO Ac for about 15 minutes in the second dimension, perpendicular to the first solvent. Short wavelength UV light will reveal the PTHs on the dried plate. The identification of the unknown is made by reference to the markers in a manner similar to that used with DNS–amino acids in TLC (Section V,A,1,c).

(ii) *Gas chromatography.* The separation of PTHs in GC has been extensively reviewed by Pisano (*132*). The Beckman GC-45 and GC-65 are most convenient and give very satisfactory separation of PTHs, but unfortunately these instruments are no longer produced. Other models can be used for the GC of PTHs after modification to include glass columns, on-column injection, and hydrogen-flame ionization detector. The He flow rate and temperature program may need some adjustment for optimal separation.

(iii) *High-pressure liquid chromatography.* The separation of PTHs by HPCL is still being refined. Already the result is quite impressive, except for PTH–proline, –valine, and –methionine; the other PTHs are well resolved. To date, and as far as we know, the best separation of PTHs by HPLC was performed in a reversed phase column (Waters Assoc., Milford, Massachusetts; μ Bondapak C_{18} 4 mm ID $\times$ 30 cm) with a gradient from 14 to 56% MeOH in 0.1 M NaOAc (pH 4.3) for those PTHs extractable by ethyl-acetate and 25% MeOH isocratic in the same buffer for the aqueous-soluble PTHs. At a total flow rate of 2.2 ml/minute, the run took 27 minutes for the former and 14 minutes for the latter at room temperature.

(iv) *Back hydrolysis.* Parent amino acids of some PTH derivatives can be regenerated by hydrolysis in 0.1 N NaOH for 12 hours at 120° (*136*), or in 57% HI for 20 hours at 127° (autoclaving at 21-pound pressure) (*134*), or in 6 N HCl with $SnCl_2$ (*135*). The yields of most of the PTHs are in the range of 70–90% when oxygen is thoroughly excluded. However, instead of returning the parental amino acids, certain PTHs give other products upon hydrolysis in acid. No amino acid is recovered from PTH–methionine. PTH–serine and PTH–cystine are converted by acid to alanine in quantitative yield; PTH–tryptophan gives small yields of glycine and alanine, confusing the interpretation of the sequence. PTH–asparagine and PTH–glutamine and other labile modified amino acids (acetyllysine or phospho-amino acids) are also converted in obvious ways.

Hydrolysis in NaOH alone (*136*) or in the presence of sodium dithionite (*134*) will recover methionine and tryptophan as well as most of the other amino acids in reasonable yield. However, the yield of identifiable product from threonine, serine, cysteine, glutamine, and arginine is low.

d. Other methods. Sanger's FDNB (fluorodinitrobenzene) procedure (*138*) has its unique position in the history of biochemistry, but with the development of the dansyl procedure, which is two orders of magnitude more sensitive, it has fallen into disuse even though it is generally more conveniently quantitated than the dansyl method. However, amino terminus of calf thymus H2B was found to be proline (*139*) and that of histone H3 was found to be alanine (*140*) by Sanger's procedure.

Enzymes such as leucine aminopeptidase or aminopeptidase M are not widely used in the NH_2-terminal analysis of histones, in part because many histones are acetylated at the NH_2 terminus.

2. CO$_2$H-TERMINAL ANALYSIS (123)

In contrast to NH$_2$-terminal determination, chemical methods available for CO$_2$H-terminal determination have severe shortcomings. Although the results of terminal determination by enzymic methods are often ambiguous due to the relative specificities of the enzymes, carboxypeptidases A, B, and C and lately carboxypeptidase Y are widely used to complement the results of the chemical analysis.

a. Hydrazinolysis (141, 142). The peptide (0.1 to 0.5 μmol) and 50 mg of dry Amberlite CG-50 are lyophilized to dryness over P$_2$O$_5$ in a thick-walled ignition tube and 0.5 ml anhydrous hydrazine from a previously unopened bottle (Matheson, Coleman, and Bell Norwood, Ohio, 97% +) is added. The sample is frozen; then while the tube is under vacuum (aspirator) the contents are thawed and refrozen before sealing. The tube is incubated at 80° for 24 hours. Then it is opened and dried over H$_2$SO$_4$ and P$_2$O$_5$ under low vacuum (aspirator) for 6 hours and then under high vacuum (Welch dual-seal pump). The resin is washed once with 3 ml of H$_2$O and twice with 1 ml. The washes are pooled, concentrated, and examined for free amino acids by pH 1.9 electrophoresis and amino acid analyzer.

b. Tritium-labeling by racemization (123, 143, 144). Protein or peptide (50 to 100 nmol) and an internal standard such as bovine insulin or hen egg lysozyme (10 to 100 nmol) are placed in a test tube (0.8 × 10 cm). Then 5 μl of T$_2$O (100 mCi/ml, 500 μCi) and 10 μl of redistilled pyridine are added. If the solution is not clear, proportional amounts of both reagents are added until the solution clarifies. Next 10 μl of acetic anhydride are added with cooling in an ice bath, and the tube is sealed with Parafilm and kept at 0° for 5 minutes and then at 20° for an additional 15 minutes. Another 20 μl of pyridine and 20 μl of acetic anhydride are added at 20°, and the reaction mixture is kept at 0° for 5 minutes and then at 20° for 1 hour. After a further addition of 5 μl of T$_2$O, it is left at 20° for 1 hour to decompose excess acetic anhydride. After completion of the tritiation reaction, the highly radioactive T$_2$O, pyridine, and acetic anhydride are carefully vacuum-distilled into a trap and discarded.

The residue is redissolved in 0.1 ml of 10% acetic acid to remove the exchangeable tritium and the distillation is repeated. The washing procedure is repeated 5 times to remove the weakly bound tritium completely.

The residual protein is hydrolyzed with 6 N HCl, and the resulting tritiated amino acid is characterized. A convenient way to analyze the amino acid uses paper electrophoresis on Whatman No. 1 paper at pH 6.4 to resolve the products into acidic amino acids, basic amino acids, and a neutral band. Subsequently the neutral amino acids are separated in a second electrophoresis at pH 1.8, while the basic amino acids are separated by paper chromatography with pyridine: acetone: conc. NH_4OH: H_2O (50: 20: 5: 50, v/v/v/v). The radioactivity of the paper strips can be expected to show about 300 to 1000 cpm/0.2 nmol of protein in a scintillation counter if T_2O of 5 Ci/ml is used.

Since the mechanism of the tritium labeling is through oxazolone formation, carboxy-terminal proline and hydroxyl proline will not be labeled. Serine and threonine incorporate radioactivity poorly, probably due to side reactions involving dehydration. The extent of tritium incorporation also varies among the amino acids. For quantitative estimation in CO_2H-terminal analysis, correction factors derived from relative ratios of tritium incorporation into various acetyl amino acids can be used (123).

c. *Carboxypeptidases* (145,146). Carboxypeptidases A, B, and C are widely used in sequence determination (145). Commonly, carboxypeptidase A is used in conjunction with carboxypeptidase B since the latter releases the basic amino acids. Even when this combination is used, however, the exopeptidase reaction stops at proline or hydroxylproline and slows down at glycine, aspartic acid, glutamic acid, cysteic acid, and S-carboxymethylcysteine. Carboxypeptidase C can release proline as well as the other amino acids, but the release of glycine is very slow; also the peptide bonds Gly–Pro, Pro–Gly, and Pro–Pro are resistant to hydrolysis. Lately carboxypeptidase Y from yeast has become available commercially (146; Pierce Chemical Co., Rockford, Illinois, #20212). This enzyme releases proline, S-carboxymethylcysteine, and homoserine as well as other amino acids, but the release of glycine and sometimes aspartic is slow. One advantage of this enzyme is its stability in 4 *M* urea and its partial retention of activity in 6 *M* urea; this allows it to be applied to some sequence determinations which are not possible with other carboxypeptidases.

3. BLOCKED AMINO TERMINI

Two types of NH_2-terminal blocking groups, *N*-acetyl and pyroglutamyl groups, have been encountered in the sequence determination of histones. The formylated NH_2-terminus is rarely observed in eukaryotic proteins (147) and has not been found in histones. No method of NH_2-terminal determination can be directly applied to such blocked peptides or proteins.

a. *N-Acetyl group.* H1, H2A, H4 and H5 all have their NH_2-termini acetylated (148,149). The acetylation of chicken erythrocyte H5 was found to be partial (150; Sautiere, unpublished), but in calf thymus H1, H2A, and H4 and in rabbit thymus H1 the NH_2-terminals are completely acetylated (Hsiang and Cole, unpublished).

The acetyl group can be identified as its hydrazide derivative by hydrazinolysis of the acetyl peptide (see Section V,A,2, a). Since the amino group of the peptide is blocked, it will not react with ninhydrin stain; a chlorination stain (*151*) should be used instead. Acetyl hydrazide can be directly identified by paper chromatography (*123*) or further characterized as DNP–acetyl hydrazide or DNS–acetyl hyrazide (for dansylation see Section V,A,1,*a*) and identified by two-dimensional thin-layer chromatography on a silica gel plate (*123*).

Conversion into methyl acetate by methanolysis of acetyl peptides and identification in GC has also been tried (*123*).

Probably the best method for identifying the acetyl group and also determining the sequence of the acetyl peptide is mass spectrometry (*152*).

b. Pyroglutamyl group. Heating peptides or proteins with NH_2-terminal glutamic acid and glutamine at neutral or slightly acidic condition tends to promote cyclization to a pyroglutamyl group (*153*). Even contact with the acidic form of Dowex 50 ion-exchange will accelerate the cyclization (*154*).

Ikenaka *et al.* (*155*) showed that incubation in 1 *N* NaOH at 24° for 40 hours and 72 hours opened the pyroglutamyl ring of Pyrglu-Pro-Ile-OH to 36% and 50%, respectively. Unfortunately, when the same condition was tried on the protein, peptide bonds were also cleaved to some extent concomitantly.

Pyrrolidone carboxylyl peptidase was purified by Doolittle and Armentrout (*156*). This enzyme specifically hydrolyzes NH_2-terminal pyroglutamyl linkages with the exception of Pyrglu-Pro and Pyrglu-Lys (*157,158*). Incubation of the blocked peptide with this enzyme opens the blocked peptide to usual NH_2-terminal analysis. Unfortunately, it is not yet commercially available.

B. Manual Sequence Determination

1. Dansyl–Edman Method (*121,159*)

This is the recommended procedure for sequence determination of small peptides (e.g., most tryptic peptides) where the difference in solubility of thiazolinone and residual peptide is too small for satisfactory separation, or for small amounts of peptide where the brightly fluorescent DNS–amino acid could be detected down to 0.02 nmol on TLC. In this procedure the parent peptide and those peptides obtained after each Edman degradation (see Section V,A,1,*b*) are analyzed for their NH_2-terminals by the dansyl procedure (see Section V,A,1,*a*).

2. Subtractive Edman Degradation (*160,161*)

Subtractive Edman degradation is performed by determination of the amino acid composition of the residual peptide after each step of the degradation (Section V,A,1,*b*). The method is quantitative and as sensitive as the amino acid analyzer available. Since the residual peptides are analyzed, the state of purity and the yield of the peptide at each stage of degradation can be obtained. However, a fraction of the peptide is destroyed at each stage, and at this date, the sensitivity of the best analyzer is much lower than that of methods for direct measurement of PTH–amino acids. Therefore, the subtractive Edman procedure will continue to be useful mainly in special situations, for example, when the NH_2-terminal residue (PTH–arginine, –histidine, and –cysteic acid) yields a PTH that is not well extracted or when a poor yield of end group is obtained in hydrolysis (as in the cases of PTH–serine and –threonine). Since the subtractive Edman procedure detects NH_2-terminal groups by the loss of one residue from the composition of the residual peptide, it tends to give ambiguous results when the amino acid in question occurs more than about 5 times in the peptide. Consequently proteins and large peptides usually cannot be sequenced by this method.

3. EDMAN DEGRADATION AND DIRECT IDENTIFICATION OF PTH–AMINO ACIDS (126)

The degradation is performed essentially as described in Section V,A,1,*b* and identification of PTHs is carried out as described in Section V,A,1,*c*. Sequencer-grade reagents or the best reagents available are used. Since the completion at every reaction is essential for a successful experiment, the complete solubilization of protein or peptide in the coupling medium is necessary. The medium must be buffered around pH 9.5, usually by *N*-dimethylallylamine. After the coupling reaction, excess reagents and by-products are extracted by benzene.

The amount of peptide used in the degradation is scaled up according to the aim of the experiment and the availability of the peptide. Without resorting to the use of radioactively labeled Edman reagent or peptide, 20 nmol is minimal for the sequencial determination of 10 residues, and more will be needed for particularly troublesome sequences. The amount of peptide needed increases much faster than an increase in the number of residues to be determined. However, due to oxidative desulfurization of the thiocarbamylated terminal group, sequential determination beyond about 15 residues is difficult without carrying the reaction in a nitrogen hood. The automatic sequencer has the advantage of carrying the reaction in nitrogen atmosphere.

C. Automatic Degradation-Sequencer (162)

Two types of automatic sequencers are commerciably available. In one reactions and extractions take place in a spinning cup, as designed by Edman and Begg (127), and this type includes the Beckman sequencer and the Jeol sequencer. Another type of instrument is the sequemat (Sequemat Inc.,

Watertown, Massachusetts) in which the peptide is coupled to a resin which then can be washed with reaction media and extraction solvents (*163*). The instrument is less expensive than the spinning-cup type, but it seems better suited to smaller peptides which can be coupled to the resin in reasonable yields. The spinning cup works best for large peptides that are not easily lost into extraction solvents, but recent attempts have been made to increase retention of small peptides in the cup by their attachment to glass beads (*164*) or the spinning cup itself (*165*).

VI. Alignment of Peptide Fragments after Their Sequences Have Been Determined

A. Overlapping of Peptides from Different Fragmentation Procedures

This is the most commonly used method in the final stage of sequence determination. Peptides from at least two different types of cleavages are needed, and usually three or more are needed.

B. Ordering the Peptide Fragments by Homology

An outstanding example of this method for the alignment of peptides was the elucidation of California gray whale cytochrome c sequence after the horse heart cytochrome was determined (*166*). Four of the major classes of histones—H2A, H2B, H3, and H4—are even more conserved than cytochrome c; therefore, this method has particular importance in the sequence determination of histones.

In arranging by homology, the most favorable enzymic or chemical fragmentation would be chosen, and the peptide fragments would be purified and sequenced. Fragments could then be arranged to maximize the similarity between the sequence of the histone under study and the known sequence of a histone of the same class. Therefore, the complete sequence of a protein often could be determined on the basis of a single digestion, if the homology between the two histones is great. Part of trout H2A was determined by this method (*58*), as were carp H3 (*84*) and pea seedling H4 (*82*).

C. Ordering the Peptide Fragments by Identification of Their NH₂ and CO₂H Terminals

Theoretically, the sequence of a protein can be determined without overlapping information if the protein and each of a succession of cleavage products are fragmented into three daughter peptides. The advantage of this approach is vividly exemplified in the sequence determination of H3 histone. DeLange *et al.* (*167*) isolated and purified 48 peptides (18 tryptic peptides from maleylated H3, 12 thermolytic or tryptic peptides from secondary cleavage of the previous peptides, 15 chymotryptic peptides, and 3 cyanogen bromide fragments) to delineate the sequence of calf thymus H3 by overlapping, while Brandt and Von Holt (*95*) determined the whole sequence of chicken H3 after isolation and characterization of only 11 peptides, 5 of which were for the confirmation of the terminal identifications. Since peptide purification is the most time-consuming part of sequence determination, limiting the total number of peptides to be purified at each stage (and overall) greatly facilitates the work. Although ordering peptides by identification of terminals seems like the method of choice, most proteins do not have such favorable numbers and locations of the crucial amino acids so that ordering of peptides can be completed by this approach alone. A successful sequencing usually will use a combination of these approaches.

VII.   Location of Polymorphism and Posttranslational Modification

There is polymorphism in at least some histone sequences which is the consequence of multiple genes (we shall not consider the polymorphism due to strain differences in slaughter house materials, etc.). Manual sequencing methods by fragmentation and purification of small peptides may overlook polymorphism unless the amino acid interchange results in a charge difference—e.g., arginine, glutamic interchange at chicken H5 (*168*)—or the minor sequence contributes a significant amount [e.g., 40% in the case of pea histone H3 (*48*)]. A striking example of the difficulty in detecting polymorphism is the case of calf thymus histone H3 (*167*). Although two distinct forms of calf thymus H3 were known (*169*) and the minor component in fact contributes one-third of the total H3, the minor sequence was not established until 3 years after the major one (*46*).

The best method to detect and establish polymorphic sequences is automatic degradation in sequencer with its quantitative determination of PTHs. Obviously the sequencer avoids the escape of minor peptides. A computer is helpful in analyzing data to establish the extent of polymorphism at each position. The shortcoming of this method is that when polymorphism is observed in more than one position in a histone preparation (rat thymus H2A; *170*) the actual number of sequences present could range anywhere between the largest number of residues found at any single position to the maximal number of combinations of all the polymorphisms observed. The confirmation of any particular sequence would then still depend on the isolation of particular peptides.

Many posttranslational modifications of histones have been observed, such as acetylation of lysine (45, 171); mono-, di- and trimethylation of lysine (33,34,37); methylation of histidine (35) and arginine (36); phosphorylation of serine (172), threonine (173), and lysine and histidine (22); and adenosine 5'-phosphate (ADP)–ribosylation probably on glutamic (174), and arginine (175). Determination of the total amino acid sequence and hence the sites of modification was done first on calf H4 (31). Usually a fractional amount of the histone is modified, and very small amounts of histone are available because studies of the correlation of histone modification and physiological state involve synchronous cell cultures (176). Therefore, very sensitive means are necessary to detect and localize the modification. Radioactively labeled precursors for modification have been used. Peptides of histone were isolated, and modified peptides were characterized, such as in calf thymus H2B phosphorylation by kinase (177). In rat liver, the amino terminal sequence of the peptide and the exact position of cAMP-dependent phosphorylation site were established by digestion with leucine aminopeptidase (178). The exact position and extent of modification at each site can also be pinpointed by Edman degradation and measurement of the label at each cycle, as in the case of acetylation in trout H4 (179) and trout H2A (180). In trout H2B and H3, since amino termini are not blocked, Candido and Dixon (181) sequenced 22 and 25 residues respectively in the automatic sequencer and localized and measured the extent of modification at each acetylation site. This method is also readily applicable to the location of methylation sites of lysine, histidine, and arginine, acetylation of lysine, and phosphorylation of serine and threonine. However, phosphorylated lysine and histidine are acid-labile, and probably the modified group will not withstand the acid-cleavage step of the Edman reaction. Similarly the ADP–ribosyl group on glutamic acid is labile at alkaline pH and probably will be released during repeated Edman cycles at pH 9.5.

VIII.   Examples of Histone Sequence Determination

A.   H1 Histone

Bustin and Cole (54) bisected a rabbit thymus H1 histone (RTL3) by NBS into an NH_2-terminal fragment containing about one-third of the residues and a CO_2H-terminal fragment containing two-thirds. Rall and Cole (99) sequenced the NH_2-terminal NBS–peptide of 72 residues by overlapping tryptic, thermolytic, and chymotryptic peptides of the unmodified H1. The sequence of RTL3 was further extended to residue 106 by overlapping tryptic, chymotryptic, and thermolytic peptides (73) obtained from the COOH-terminal fragment produced by NBS. The structures of the tryptic peptides from residue 107–213 were determined, but overlapping them proved difficult. Attempts to overlap the sequence by extensive thermolytic digestion yielded more than 40 peptides (Largeman and Cole, unpublished), almost all of which gave ambiguous results because they terminated in the extremely common lysine and alanine residues. Automatic degradations were carried out in the sequencer on the carboxyl half and its thrombin peptides and a few large thermolytic peptides to overlap and complete the sequence of RTL3 (Hsiang and Cole, unpublished).

B.   H2A Histone

Hayashi and Iwai (182) isolated tryptic peptides of unmodified H2A from calf thymus and sequenced most of them by the subtractive Edman procedure and aminopeptidase M digestion.

Busch and his co-workers determined the total sequence of calf thymus H2A by overlapping the tryptic peptides of maleylated H2A, and the thermolytic, tryptic, and peptic peptides of unmodified H2A. Both subtractive Edman and dansyl–Edman procedures were used for the sequence determination (85–87).

Bailey and Dixon (58) isolated and sequenced 26 tryptic peptides of either succinylated or maleylated H2A from immature testes of rainbow trout. Nine chymotryptic and two of the three expected NBS peptides were also purified and sequenced. The subtractive Edman procedure, dansyl–Edman procedure, and Beckman automatic sequencer were used, with PTH–amino acids from sequencer by GC or by back hydrolysis and amino acid analysis. Eleven of the 13 tryptic peptides were arranged unambiguously from the overlapping data, while two tryptic peptides could only be placed in order by homology with the H2A sequence known for calf thymus. In addition to confirming the sequence of calf H2A proposed by Yeoman et al. (85), Sautiére et al. (88) have examined histone H2A from rat thymus. In the latter histone, Sautiére (170) observed molecular polymorphism with both serine and threonine present at residue 16 and arginine and lysine at residue 99. Similar polymorphism at position 99 was observed in calf thymus H2A (P. Sautiere, unpublished; see also 183).

C. H2B Histone

Iwai and his co-workers isolated 23 tryptic peptides from unmodified calf H2B, 13 tryptic peptides from succinylated H2B (*32*), and 27 chymotryptic peptides and 35 thermolytic peptides (*102*) from unmodified H2B. Sequence determination was mainly performed by the subtractive Edman procedure and occasionally confirmed by back hydrolysis of the extracted thiazolinone derivatives of amino acids or DNS or DNP methods for the NH_2-terminus. Most of the sequence was confirmed by Hnilica *et al.* (*184*).

Candido and Dixon (*181*) using a Beckman automatic sequencer determined the NH_2-terminal 22 residues of H2B from trout testis and located four sites of acetylation at lysyl residues 5, 10, 13, and 18.

S. C. R. Elgin *et al.* (unpublished) sequenced the amino-terminal sequence 25 residues of H2B from *Drosophila* in an automatic sequencer (*183*).

D. H3 Histone

DeLange *et al.* (*167*) determined the complete amino acid sequence of calf thymus H3. Eighteen tryptic peptides of maleylated H3 were isolated and sequenced by substractive Edman procedure and occasionally confirmed by DNS–Edman (*83*). Fifteen chymotryptic peptides from *S*-carboxymethylated H3 were purified and partially sequenced in similar fashion by Hooper and Smith (*47*). Although these 15 chymotryptic peptides could not account for the whole sequence of H3, with the overlapping information from cyanogen bromide fragments of *S*-carboxymethylated H3 and thermolytic and peptic peptides, the total sequence of calf thymus H3 was obtained (*184a*). Olson *et al.* (*185*) confirmed the sequence. H3 from carp also was determined by Hooper *et al.* (*84*) by examining cyanogen bromide and tryptic, chymotryptic, and thermolytic peptides for their compositions and by partial sequencing or identification of terminals; the alignment of peptides depended on homology with calf thymus H3. Only one replacement (serine replaced cysteine) in composition was detected and was pinpointed to the single sequence replacement at residue 96. Lysines 9 and 27 were also found to be methylated to different extents (0 to 3 methyl groups) but no ϵ-*N*-acetyllysine was found in carp, in contrast to calf thymus H3, which has two sites (positions 14, 23) acetylated.

Patthy and Smith (*48*) found pea seedling H3 is polymorphic in sequence. Residue 96 was found to be 60% alanine and 40% serine. After sequence determination of tryptic, chymotryptic, and thermolytic peptides of cyanogen bromide fragments, they found that the primary structure of H3 is highly conserved; only four amino acids were changed from the calf thymus sequence. While the same sites (lysines 9 and 27) as in calf thymus were found to be methylated, these lysine residues in pea seeding were modified only to the mono- and dimethyl forms. No ϵ-acetyllysine was detected.

Patthy and Smith (46) also confirmed the polymorphism in H3 from calf thymus which was not detected in the previous sequencing work. Residue 96 was found to be cysteine in one form and serine in the other.

Brandt and Von Holt (49,95) used the Beckman automatic sequencer exclusively for the determination of H3 sequence from chicken erythrocytes. PTHs were identified by GC or converted to amino acids by back hydrolysis. The very favorable distribution of crucial amino acids enabled them to complete the total sequence determination of 135 residues without the use of overlapping peptide fragments. Sequencer experiments were performed on complete H3 to identify 48 residues. CNBr cleavage resulted in three peptides (residue 1 to 90, 91 to 120, 121 to 135). Two smaller CNBr peptides (30 and 15 residues, respectively) were completely sequenced by sequencer; the 30-residue piece was also confirmed by the composition of two peptides obtained after NBS cleavage. The largest of the CNBr peptides (90 residues) was cleaved into three peptides by NBS cleavage at two tyrosine residues located at 41 and 54. NBS peptides of residue 42 to 54 and 55 to 90 were completely sequenced by the sequencer. The sequence of the peptide (55–90) was confirmed by the peptides isolated from acid cleavage at aspartyl residues. A summary of their scheme is given in Section IX, D.

Lys 36 as well as Lys 9 and 27 that had been found in calf are methylated. Lys 27 is nearly completely methylated, while Lys 9 and 36 are methylated to a small extent. Dimethylation predominates over mono- and trimethylation. ϵ-N-Acetyllysine is present but was not assigned a location, although known to be in the first 40 residues of the molecule (95).

Brand et al. (186) determined the first 40 to 50 residues of amino acids from H3 of calf, chicken, shark, sea urchin, mollusk, and a plant (cycad pollen). Except for replacement of tyrosine by phenylalanine at residue 39 for the plant, the primary structures are identical in the first 48 residues from all these organisms, but the extent of modification by methylation does vary from organism to organism.

Brandt et al. (187) also determined the complete sequence of H3 from shark erythrocytes using the same scheme they used for chicken erythrocyte H3. They found shark H3 to be identical to chicken and carp H3 in sequence; however, the extent of modification seems to be different again. Methylations were assigned to lysines at positions 9, 27, and 36. They were not able to assign the acetyl group precisely, but it was localized in the first 41 residues, and the level of acetylation is rather minor.

E. H4 Histone

The arginine-rich histone H4 was the first histone molecule for which the complete amino acid sequence was determined. DeLange *et al. (188)* cleaved H4 at the single methionine with cyanogen bromide and isolated an 18-residue carboxyl terminal peptide. The sequences of such peptides from pea embryo and calf thymus were found to be identical. DeLange *et al. (31)* further isolated 13 tryptic peptides from maleylated calf H4 and 8 chymotryptic peptides of unmodified calf H4 and sequenced all these peptides by manual Edman degradation. Residues released during degradation were characterized by combinations of direct identification of PTHs with paper chromatography in different solvent systems and of amino acids after back hydrolysis of the PTHs.

For the sequence determination of pea seedling H4, DeLange *et al. (31,82)* purified and sequenced only 13 tryptic peptides of the maleylated H4 and settled the total sequence by arranging the peptide fragments according to homology to calf H4.

Ogawa and his co-workers *(189)* and Sautiére *et al. (190)* independently sequenced the calf thymus H4 and confirmed the sequence obtained by DeLange *et al. (45)* through overlapping of tryptic, chymotryptic, thermolytic, and cyanogen bromide peptides of the unmodified H4. They also characterized the carboxyl-terminal 18 residues from lymphosarcoma and Novikoff hepatoma and found that they are identical to the comparable sequence from calf thymus and pea seedling *(191)*.

Sautiére and his co-workers confirmed the sequence of calf thymus H4 *(74,192)* and also sequenced the porcine thymus H4 *(75)* by isolating 18 tryptic, 5 chymotryptic, and cyanogen bromide peptides of the unmodified H4 and found the sequence identical to calf thymus H4.

Strickland *et al. (193)* pinpointed the difference between the sequences of sea urchin H4 and calf thymus H4 as a single replacement of threonine by cysteine at position 73; this was done by a study of several peptides released in cyanogen bromide cleavage and mild acid hydrolysis.

F. H5 Histone

Tryptic and thermolytic peptides of chicken H5 were isolated and characterized by Greenaway *(194)*. Garel *et al. (195)* established the sequence of the amino-terminal 70 residues by overlapping tryptic and chymotryptic peptides, NBS fragments, and CNBr fragments. Sautiére *et al. (196)* further extended the sequence for 25 residues around the only phenylalanine of the molecule.

The sequences of the amino-terminal 16 residues of H5 from chicken, quail, goose, and duck and 19 residues from pigeon were determined by automatic sequencer (*150*).

IX.   Conclusion: Recommended Schemes for Histone Sequencing

The recommended schemes are based on the assumption that an automatic sequencer is available and the determination of the total sequence of the histone molecule is the goal. When a particular section of the molecule is the focus of interest for localization of modification, different approaches might be more advantageous.

A.   Histone H1

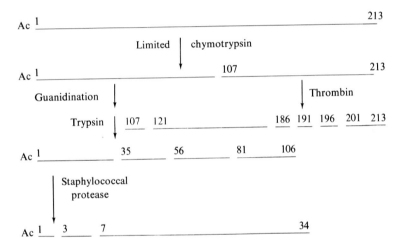

B. Histone H2A

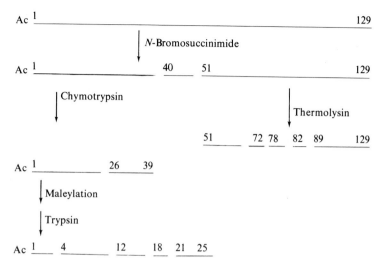

C. Histone H2B

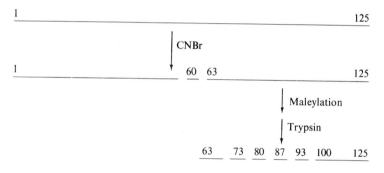

D.   Histone H3

This is the same scheme Brandt and Von Holt (*49, 95, 197*) used.

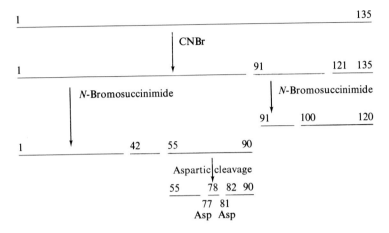

E.   Histone H4

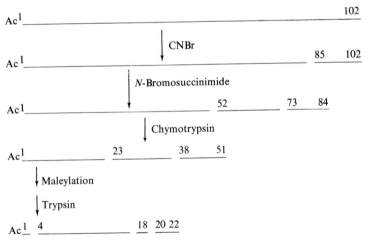

References

1. Bustin, M., and Cole, R. D., *J. Biol. Chem.* **243**, 4500 (1968).
2. Benisek, W. F., Raftery, M. A., and Cole, R. D., *Biochemistry* **6**, 3780 (1967).
3. Africa, B., and Carpenter, F. H. *Biochemistry* **9**, 1962 (1970).
4. Moore, S., and Stein, W. H., *in* "Methods in Enzymology" (S. P. Colowick and N. O. Kaplan, eds.), Vol. 6, p. 819. Academic Press, New York. (1963).
5. Simpson, R. J., Neuberger, M. R., and Liu, T. Y., *J. Biol. Chem.* **251**, 1936 (1976).
6. Lipmann, F., and Levene, P. A., *J. Biol. Chem.* **98**, 109 (1932).
7. Levene, P. A., and Hill, D. W., *J. Biol. Chem.* **101**, 711 (1933).
8. Allerton, S. E., and Perlman, S. E., *J. Biol. Chem.* **240**, 3892 (1965).
9. Schaffer, N. K., and Balakir, R. A., *Anal. Biochem.* **14**, 167 (1966).
10. de Verdier, C.-H., *Acta Chem. Scand.* **7**, 196 (1953).
11. Engström, L., *Biochem. Biophys. Acta* **52**, 49 (1961).
12. Schaffer, N. K., *in* "Methods in Enzymology," Vol. 11: Enzyme Structure (C. H. W. Hirs, ed.), p. 702. Academic Press, New York, (1967).
13. Ingles, C. J., and Dixon, G. H., *Proc. Natl. Acad. Sci. U.S.A.* **58**, 1011 (1967).
14. Jergil, B., and Dixon, G. H., *J. Biol. Chem.* **245**, 425 (1970).
15. Shepherd, G. R., Noland, B. J., and Robert, C. N., *Biochem. Biophys. Acta* **199**, 265 (1970).
16. Hohmann, P., Tobey, R. A., and Gurley, L. R., *Biochem. Biophys. Res. Commun.* **63**, 126 (1975).
17. Nomoto, M., Narahashi, Y., and Murakami, M., *J. Biochem. (Tokyo)* **48**, 593 (1960).
18. Tower, D. B., Peters, E. L., and Wherrett, J. R., *J. Biol. Chem.* **237**, 1861 (1962).
19. Hill, R. L., and Schmidt, W. R., *J. Biol. Chem.* **237**, 389 (1962).
20. Hill, R. L., *Adv. Protein Chem.* **20**, 37 (1965).
21. Balhorn, R., Rieke, W. O., and Chalkley, R., *Biochemistry* **10**, 3952 (1971).
22. Chen, C.-C., Smith, D. L., Bruegger, B. B., Halpern, R. M., and Smith, R. A., *Biochemistry* **13**, 3785 (1974).
23. Hare, P. E., *in* "Protein Sequence Determination" (S. B. Needleman, ed.), 2nd ed., p. 204. Springer-Verlag, Berlin and New York, (1975).
24. Hamilton, P. B., *in* "Methods in Enzymology," Vol. 11: Enzyme Structure (C. H. W. Hirs, ed.), p. 15. Academic Press, New York, (1967).
25. Udenfriend, S., Stein, S., Bohlen, P., Dairman, W., Leimgrabew, W., and Weigele, M., *Science* **178**, 871 (1972).
26. Roth, M., *Anal. Chem.* **43**, 880 (1971).
27. Roth, M., and Hampaii, A., *J. Chromatogr.* **83**, 353 (1973).
28. Spackman, D. H., Stein, W. H., and Moore, S., *Anal. Chem.* **30**, 1190 (1958).
29. Benson, J. V., Gordon, M. J., and Patterson, J. A., *Anal. Biochem.* **18**, 228. (1967).
30. Lee, P., *Biochem. Med.* **10**, 107 (1974).
31. DeLange, R. J., Fambrough, D. M., Smith, E. L., and Bonner, J., *J. Biol. Chem.* **244**, 319 (1969).
32. Ishikawa, K., Hayashi, H., and Iwai, K., *J. Biochem. (Tokyo)* **72**, 299 (1972).
33. Murray, K., *Biochemistry* **3**, 10 (1964).
34. Paik, W. K., and Kim, S., *Biochem. Biophys. Res. Commun.* **27**, 479 (1967).
35. Gershey, E. L., Haslett, G. W., Vidali, G., and Allfrey, V. G., *J. Biol. Chem.* **244**, 4871 (1969).
36. Paik, W. K., and Kim, S., *Arch. Biochem. Biophys.* **134**, 632 (1969).
37. Hempel, K., Lange, H. W., and Birkofer, L., *Naturwissenschaften* **55**, 37. (1968).
38. Hempel, K., Lange, H. W., and Birkofer, L., *Z. Physiol. Chem.* **349**, 603. (1968).
39. Moore, S., and Stein, W. H., *J. Biol. Chem.* **211**, 893 (1954).

40. Schaffer, N. K., May, S. C., Jr., and Summerton, W. H., *J. Biol. Chem.* **202**, 67 (1953).
41. Gross, E., *in* "Methods in Enzymology," Vol. 11: Enzyme Structures (C. H. W. Hirs, ed.), p. 238. Academic Press, New York, (1967).
42. Gross, E., and Witkop, B., *J. Am. Chem. Soc.* **83**, 1510 (1961).
43. Kinkade, J. M., *J. Biol. Chem.* **244**, 3375 (1969).
44. Medvedeva, M. N., Huschtscha, L. I., and Medvedev, Z. A., *FEBS Lett.* **53**, 253 (1975).
45. DeLange, R. J., Fambrough, D. M., Smith, E. L., and Bonner, J., *Proc. Natl. Acad. Sci. U.S.A.* **61**, 1145 (1968).
46. Patthy, L., and Smith, E. L., *J. Biol. Chem.* **250**, 1919 (1975).
47. Hooper, J. A., and Smith, E. L., *J. Biol. Chem.* **248**, 3255 (1973).
48. Patthy, L., and Smith, E. L., *J. Biol. Chem.* **248**, 6834 (1973).
49. Brandt, W. F., and von Holt, C. *Eur. J. Biochem.* **46**, 407 (1974).
50. Ramachandran, L. K., and Witkop, B., *in* "Methods in Enzymology," Vol. 11: Enzyme Structure (C. H. W. Hirs, ed.) p. 283. Academic Press, New York, (1967).
51. Witkop, B., *Adv. Protein Chem.* **16**, 221 (1961).
52. Kinkade, J. M., Jr., and Cole, R. D., *J. Biol. Chem.* **241**, 5790 (1966).
53. Hnilica, L. S., *Prog. Nucl. Acid Res. Mol. Biol.* **7**, 48 (1967).
54. Bustin, M., and Cole, R. D., *J. Biol. Chem.* **244**, 5291 (1969).
55. Rall, S. C., and Cole, R. D., *J. Am. Chem. Soc.* **92**, 1800 (1970).
56. Spande, T. F., Witkop, B., Degani, Y., and Patchornik, A., *Adv. Protein Chem.* **25**, 97 (1970).
57. Sherod, D., Johnson, G., and Chalkley, R., *J. Biol. Chem.* **249**, 3923 (1974).
58. Bailey, G. S., and Dixon, G. H. *J. Biol. Chem.* **248**, 5463 (1973).
59. Schultz, J. *in* "Methods in Enzymology," Vol. 11: Enzyme Structure (C. H. W. Hirs, Ed.), p. 255. Academic Press, New York, (1967).
60. Partridge, S. M., and Davis, H. F., *Nature* (*London*) **165**, 62 (1950).
61. Schultz, J., Allison, H., and Grice, M., *Biochemistry* **1**, 694 (1962).
62. Piszkiewicz, D., Landon, M., and Smith, E, L., *Biochem. Biophys. Res. Commun.* **40**, 1173 (1970).
63. Payne, J. W., Jakes, R., and Hartley, B. S., *Biochem. J.* **117**, 757 (1970).
64. Harris, J. I., Cole, R. D., and Pon, N. G. *Biochem. J.* **62**, 154 (1956).
65. Iwai, K., and Ando, T., *in* "Methods in Enzymology," Vol. 11: Enzyme Structure (C. H. W. Hirs, ed.) p. 263. Academic Press, New York, (1967).
66. Lenard, J., and Hess, G. P., *J. Biol. Chem.* **239**, 3275 (1964).
67. Smyth, D. G., and Stark, G. R., *Anal. Biochem.* **14**, 152 (1966).
68. Caufield, R. E., and Aⁱfinson, C. B., *in* "The Proteins" (H. Neurath, ed.), Vol. 1, p. 311. Academic Press, New York, (1963).
69. Plapp, B. V., Raftery, M. A., and Cole, R. D., *J. Biol. Chem.* **242**, 265 (1967).
70. Smyth, D. G., *in* "Methods in Enzymology," Vol. 11: Enzyme Structure (C. H. W. Hirs, ed.), p. 214. Academic Press, New York, (1967).
71. Bergmann, M., Fruton, J. S., and Pollock, H., *J. Biol. Chem.* **127**, 643 (1939).
72. Bergmann, M., and Fruton, J. S. *Adv. Enzymol. Relat. Subj. Biochem.* **1**, 63 (1941).
73. Jones, G. M. T., Rall, S. C., and Cole, R. D., *J. Biol. Chem.* **249**, 2548 (1974).
74. Sautiére, P., Breynaert, M., Moschetto, Y., and Biserte, G., *C. R. Hebd. Sciences Acad. Sci.* **271**, 364 (1970).
75. Sautiére, P., Lambelin-Breynaert, M., Moschetto, Y., and Biserte, G., *Biochimie* **53**, 711 (1971).
76. Sautiére, P., Tyrou, D., Moschetto, Y., and Biserte, G., *Biochimie* **53**, 479 (1971).
77. Carpenter, F. H., *in* "Methods in Enzymology," Vol. 11: Enzyme Structure (C. H. W. Hirs, ed.), p. 237. Academic Press, New York, (1967).

78. Lindley, H., *Nature (London)* **178**, 647 (1956).
79. Cole, R. D., *in* "Methods in Enzymology," Vol. 11: Enzyme Structure (C. H. W. Hirs, ed.), p. 315. Academic Press, New York, (1967).
80. Butler, P. J. G., and Hartley, B. S., *in* "Methods in Enzymology," Vol. 25: Enzyme Structure, Part B (C. H. W. Hirs, ed.), p. 191. Academic Press, New York, (1972).
81. Butler, P. J. G., Harris, J. I., Hartley, B. S., and Leberman, R., *Biochem. J.* **103**, 78P (1967).
82. DeLange, R. J., Fambrough, D. M., Smith, E. L., and Bonner, J., *J. Biol. Chem.* **244**, 5669 (1969).
83. DeLange, R. J., and Smith, E. L., *J. Biol. Chem.* **248**, 3248 (1973).
84. Hooper, J. A., Smith, E. L., Sommer, K. R., and Chalkley, R. *J. Biol. Chem.* **248**, 3275 (1973).
85. Yeoman, L. C., Olson, M. O. J., Sugano, N., Jordan, J., Taylor, C. W., Starbuck, W. C., and Busch, H., *J. Biol. Chem.* **247**, 6018 (1972).
86. Sugano, N., Olson, M. O. J., Yeoman, L. C., Johnson, B. R., Taylor, C. W., Starbuck, W. C., and Busch, H., *J. Biol. Chem.* **247**, 3589 (1972).
87. Olson, M. O. J., Sugano, N., Yeoman, L. C., Johnson, B. R., Jordan, J., Taylor, C. W., Starbuck, W. C., and Busch, H., *Physiol. Chem. Physics* **4**, 10 (1972).
88. Sautiére, P., Tyrou, D., Laine, B., Mizon, J., Ruffin, P., and Biserte, G., *Eur. J. Biochem.* **41**, 563 (1974).
89. Klotz, I. M., *in* "Methods in Enzymology," Vol. 11: Enzyme Structure (C. H. W. Hirs, ed.), p. 576. Academic Press, New York, (1967).
90. Iwai, K., Ishikawa, K., and Hayashi, H., *Nature (London)* **226**, 1056 (1970).
91. Ludwig, M. L., and Hunter, M. J., *in* "Methods in Enzymology," Vol. 11: Enzyme Structure (C. H. W. Hirs, ed.), p. 595. Academic Press, New York, 1967.
92. Rall, S. C., Ph.D. Thesis Univ. of California, Berkeley, 1971.
93. Kimmel, J. R., *in* "Methods in Enzymology," Vol. 11: Enzyme Structure (C. H. W. Hirs, ed.), p. 584. Academic Press, New York, 1967.
94. Habeeb, A. F. S. A., *in* "Methods in Enzymology," Vol. 25: Enzyme Structure, Part B (C. H. W. Hirs, ed.), p. 558. Academic Press, New York, 1972.
94a. Kimmel, J. R., *in* "Methods in Enzymology," Vol. 11: Enzyme Structure (C. H. W. Hirs, ed.), p. 589. Academic Press, New York, 1967.
95. Brandt, W. F., and Von Holt, C., *Eur. J. Biochem.* **46**, 419 (1974).
96. Margoliash, E., and Smith, E. L., *J. Biol. Chem.* **237**, 2151 (1962).
97. Matsubara, H., and Smith, E. L., *J. Biol. Chem.* **238**, 2732 (1963).
98. Iwai, K., Hayashi, H., and Ishikawa, K., *J. Biochem. (Tokyo)* **72**, 357 (1972).
99. Rall, S. C., and Cole, R. D., *J. Biol. Chem.* **240**, 7175 (1971).
100. Matsubara, H., *in* "Methods in Enzymology," Vol. 19: Proteolytic Enzymes (G. Perlmann, ed.), p. 642. Academic Press, New York, 1970.
101. Matsubara, H., and Feder, J., *in* "The Enzymes" (P. Boyer, ed.), Vol. 3, p. 721. Academic Press, New York, 1971.
102. Hayashi, H., and Iwai, K., *J. Biochem. (Tokyo)* **72**, 327 (1972).
103. Ando, T., and Suzuki, Ko, *Biochim. Biophys. Acta* **140**, 375 (1967).
104. Baughman, D. J. *in* "Methods in Enzymology," Vol. 19: Proteolytic Enzymes (G. Perlmann, ed.), p. 145. Academic Press, New York, 1970.
105. Blombäck, B., *in* "Blood Clotting Enzymology" (W. H. Seegers, ed.), p. 143. Academic Press, New York, 1967.
106. Bailey, K., Bettelheim, F. R., Loraud, L., and Middlebrook, W. R., *Nature (London)* **167**, 233 (1951).
107. Gladner, J. A., *in* "Fibrinogen" (K. Laki, ed.), p. 87. Dekker, New York, 1968.

108. Chapman, G. E., Hartman, P. G., and Bradbury, E. M., *Eur. J. Biochem.* **61**, 69 (1976).
109. Rydén, A., Rydén, L., and Philipson, L., *Eur. J. Biochem.* **44**, 105 (1974).
110. Halpert, J., and Eaker, D., *J. Biol. Chem.* **250**, 6990 (1975).
111. Drapeau, G. R., Boily, Y., and Houmard, J., *J. Biol. Chem.* **247**, 6720 (1972).
112. Houmard, J., and Drapeau, G. R., *Proc. Natl. Acad. Sci. U.S.A.* **69**, 3506 (1972).
113. Behrens, P. Q., and Brown, J. R., *Anal. Biochem.* **18**, 228 (1976).
114. Kasper, C. B., *in* "Protein Sequence Determination" (S. B. Needleman, ed.), 2nd ed., p. 114. Springer-Verlag, Berlin and New York, 1975.
115. Schroeder, W. A. *in* "Methods in Enzymology," Vol. 25: Enzyme Structure, Part B (C. H. W. Hirs, ed.), p. 203. Academic Press, New York, 1972.
116. Schroeder, W. A. *in* "Methods in Enzymology," Vol. 25: Enzyme Structure, Part B (C. H. W. Hirs, ed.), p. 214. Academic Press, New York, 1972.
117. Peterson, E. A., and Sober, H. A., *in* "Methods in Enzymology," Vol. 5: Preparation and Assay of Enzymes (S. P. Colowick and N. O. Kaplan, eds.) p. 3. Academic Press, New York, 1962.
118. Roy, D. and Konigsberg, W., *in* "Methods in Enzymology," Vol. 25: Enzyme Structure, Part B (C. H. W. Hirs, ed.), p. 221. Academic Press, New York, 1972.
119. Bennett, J. C., *in* "Methods in Enzymology," Vol. 11: Enzyme Structure (C. H. W. Hirs, ed.), p. 330. Academic Press, New York, 1967.
120. Harris, J. I., *in* "Methods in Enzymology," Vol. 11: Enzyme Structure (C. H. W. Hirs, ed.), p. 395. Academic Press, New York, 1967.
121. Gray, W. R., *in* "Methods in Enzymology," Vol. 11: Enzyme Structure (C. H. W. Hirs, ed.), p. 139, Academic Press, New York, 1967.
122. Woods, K. R., and Wang, K. T., *Biochim. Biophys. Acta* **133**, 369 (1967).
123. Narita, K., Matsuo, H., and Nakajima, T., *in* "Protein Sequence Determination" (S. B. Needleman, ed.), 2nd ed., p. 30. Springer-Verlag, Berlin and New York, 1975.
124. Wiener, A., Platt, T., and Weber, K., *J. Biol. Chem.* **247**, 3242 (1972).
125. Gray, W. R., *in* "Methods in Enzymology," Vol. 11: Enzyme Structure (C. H. W. Hirs, ed.), p. 469. Academic Press, New York, 1967.
126. Edman, P., and Henschen, A., *in* "Protein Sequence Determination" (S. B. Needleman, ed.), 2nd ed., p. 232. Springer-Verlag, Berlin and New York, 1975.
127. Edman, P., and Begg, G., *Eur. J. Biochem.* **1**, 80 (1967).
128. Schroeder, W. A., *in* "Methods in Enzymology," Vol. 11: Enzyme Structure (C. H. W. Hirs, ed.), p. 445. Academic Press, New York, 1967.
129. Cherbulier, E., Baehler, B., and Rabinowitz, A., *J. Helv. Chim. Acta* **43**, 1871 (1960).
130. Pisano, J. J., van den Heuvel, W. J. A., and Horning, E. C., *Biochem. Biophys. Res. Commun.* **7**, 82 (1962).
131. Zimmerman, C. L., Pisano, J. J., and Appella, E., *Biochem. Biophys. Res. Commun.* **55**, 1220 (1973).
132. Pisano, J. J., *in* "Protein Sequence Determination" (S. B. Needleman, ed.), 2nd ed., p. 280. Springer-Verlag, Berlin and New York, 1975.
133. Stephanov, V. M., Vullfson, N. S., Puchow, V. A., and Zyakum, A. M., *J. Gen. Chem. USSR Engl. Transl.* **34**, 3771 (1964).
134. Smithies, O., Gibson, D., Fanning, E. M., Goodfliesh, R. M., Gilman, J. G., and Ballantyne, D. L., *Biochemistry* **10**, 4912 (1971).
135. Mendez, K., and Lai, M., *Anal. Biochem.* **68**, 47 (1975).
136. Africa, B., and Carpenter, F. H., *Biochem. Biophys. Res. Commun.* **24**, 113 (1966).
137. Summers, M. R., Smythers, G. W., and Oroszlan, S., *Anal. Biochem.* **53**, 624 (1973).
138. Sanger, F., *Biochem. J.* **39**, 507 (1945).
139. Phillips, D. M. P., and Johns, E. W., *Biochem. J.* **72**, 538 (1959).

140. Johns, Z. W., Phillips, D. M. P., Simson, P., and Butler, J. A. V., *Biochem. J.* 77, 631 (1960).
141. Fraenkel-Conrat, H., and Tsung, C. M., *in* "Methods in Enzymology," Vol. 11: Enzyme Structure (C. H. W. Hirs, ed.), p. 151. Academic Press, New York, 1967.
142. Schroeder, W. A., *in* "Methods in Enzymology," Vol. 25: Enzyme Structure, Part B (C. H. W. Hirs, ed.), p. 138. Academic Press, New York, 1972.
143. Matsuo, H., Fujimoto, Y., and Tatsuno, T., *Biochim. Biophys. Res. Acta* 22, 69 (1966).
144. Matsuo, H., and Narita, K., *in* "Protein Sequence Determination" (Needleman, S. B., ed.), 2nd ed., p. 104. Springer-Verlag, Berlin and New York, 1975.
145. Ambler, R. P., *in* "Methods in Enzymology," Vol. 25: Enzyme Structure, Part B (C. H. W. Hirs, ed.), p. 143. Academic Press, New York, 1972.
146. Hayashi, R., Moore, S., and Stein, W. H., *J. Biol. Chem.* 248, 2296 (1973).
147. Fujika, H., Brawnitzer, G., and Rudloff, V., *Hoppe-Seyler's Z. Physiol. Chem.* 351, 901 (1970).
148. Phillips, D. M. P., *Biochem. J.* 87, 258 (1963).
149. Phillips, D. M. P., and Johns, E. W., *Biochem. J.* 94, 127 (1965).
150. Seligy, V., Roy, C., Dove, M., and Yaguchi, M., *Biochem. Biophys. Res. Commun.* 71, 196 (1976).
151. Mazur, R. H., Ellis, B. W., and Cammarata, P. S., *J. Biol. Chem.* 237, 1619 (1962).
152. Gray, W. R., and del Valle, U. E., *Biochemistry* 9, 2134 (1970).
153. Neuberger, A., and Sanger, F., *Biochem. J.* 36, 662 (1942).
154. Narita, K. *Biochim. Biophys. Acta* 30, 352 (1958).
155. Ikenaka, T., Bammerlin, H., Kaufmann, H., and Schmid, K., *J. Biol. Chem.* 241, 5560 (1966).
156. Doolittle, R. F., and Armentrout, R. W., *Biochemistry* 7, 516 (1968).
157. Doolittle, R. F., *in* "Methods in Enzymology," Vol. 19: Proteolytic Enzymes (G. Perlmann, ed.), p. 555. Academic Press, New York, 1970.
158. Doolittle, R. F., *in* "Methods in Enzymology," Vol. 25: Enzyme Structure, Part B (C. H. W. Hirs, ed.), p. 231. Academic Press, New York, 1972.
159. Gray, W. R., *in* "Methods in Enzymology," Vol. 25: Enzyme Structure (C. H. W. Hirs, ed.), p. 333. Academic Press, New York, 1972.
160. Konigsberg, W., *in* "Methods in Enzymology," Vol. 11: Enzyme Structure (C. H. W. Hirs, ed.), p. 461. Academic Press, New York, 1967.
161. Konigsberg, W., *in* "Methods in Enzymology," Vol. 25: Enzyme Structure, Part B (C. H. W. Hirs, ed.), p. 326. Academic Press, New York, 1972.
162. Niall, H. D., *in* "Methods in Enzymology," Vol. 27: Enzyme Structure, Part D (S. H. W. Hirs and S. Timasheff, eds.), p. 942. Academic Press, New York, 1973.
163. Laursen, R. A., *in* "Methods in Enzymology," Vol. 25: Enzyme Structure, Part B (C. H. W. Hirs, ed.), p. 344. Academic Press, New York, 1972.
164. Spinco Division of Beckman Instruments, Inc. Palo Alto, California. *Sequence* 8, 1 (1975).
165. Spinco Division of Beckman Instruments, Inc. Palo Alto, California *Sequence* 6, 1 (1975).
166. Goldstone, A. D., and Smith, E. L., *J. Biol. Chem.* 241, 4480 (1966).
167. DeLange, R. J., Hooper, J. A., and Smith, E. L., *Proc. Natl. Acad. Sci. U.S.A.* 69, 882 (1972).
168. Greenaway, P. J., and Murray, K., *Nature (London) New Biol.* 229, 233 (1971).
169. Mazluff, W. F., Jr., Sanders, L. A., Miller, D. M., and McCarthy, K. S., *J. Biol. Chem.* 247, 2026 (1972).
170. Sautiére, P., *Ciba Found. Symp.* 28, 77 (1975).
171. Gershey, E. L., Vidali, G., and Allfrey, V. G., *J. Biol. Chem.* 243, 5018 (1968).
172. Ord, M. G., and Stocken, L. A., *Biochem. J.* 98, 5P (1966).

173. Kleinsmith, L. J., Allfrey, V. G., and Mirsky, A. E., *Proc. Natl. Acad. Sci. U.S.A.* **55**, 1182 (1966).

174. Dixon, G. H., Wong, N., and Poirier, G. G., *Fed. Proc. Fed. Am. Soc. Exp. Biol.* **35**, 1342 (1976).

175. Rohrer, H., Zillig, W., and Mailhammer, R., *Eur. J. Biochem.* **60**, 227 (1975).

176. Jackson, V., Shires, A., Tauphaichit, N., and Chalkley, R., *J. Mol. Biol.* **104**, 471 (1976).

177. Farago, A., Romhanyi, T., Antoni, F., and Takats, A., *Nature (London)* **254**, 88 (1975).

178. Langan, T., *Proc. Natl. Acad. Sci. U.S.A.* **64**, 1276 (1969).

179. Candido, E. P. M., and Dixon, G. H., *J. Biol. Chem.* **246**, 3182 (1971).

180. Candido, E. P. M., and Dixon, G. H., *J. Biol. Chem.* **247**, 3868 (1972).

181. Candido, E. P. M., and Dixon, G. H., *Proc. Natl. Acad. Sci. U.S.A.* **69**, 2015 (1972).

182. Hayashi, H., and Iwai, K., *J. Biochem.* (Tokyo) **70**, 543 (1971).

183. Elgin, S. C. R., and Weintraub, H., *Annu. Rev. Biochem.* **44**, 725 (1975).

184. Hnilica, L. S., Kappler, H. A., and Jordan, J. J., *Experientia* **26**, 353 (1970).

184a. DeLange, R. J., Hooper, J. A., and Smith, E. L., *J. Biol. Chem.* **248**, 3261 (1973).

185. Olson, M. O. J., Jordan, J., and Busch, H., *Biochem. Biophys. Res. Commun.* **46**, 50 (1972).

186. Brandt, W. F., Strickland, W. N., Morgan, M., and Von Holt, C., *FEBS Lett.* **40**, 167 (1974).

187. Brandt, W. F., Strickland, W. N., and Von Holt, C., *FEBS Lett.* **40**, 349 (1974).

188. DeLange, R. J., Fambrough, D. M., Smith, E. L., and Bonner, J., *J. Biol. Chem.* **243**, 5906 (1968).

189. Ogawa, Y., Quagliaratti, G., Jordan, J., Taylor, C. W., Starbuck, W. C., and Busch, H., *J. Biol. Chem.* **244**, 4387 (1969).

190. Sautiére, P., Starbuck, W. C., Roth, C., and Busch, H., *J. Biol. Chem.* **243**, 5899 (1968).

191. Desai, L., Ogawa, Y., Mauritzeu, E. M., Taylor, C. W., and Starbuck, W. C., *Biochim. Biophys. Acta* **181**, 146 (1969).

192. Sautiére, P., Moschetto, Y., Dautrevaux, M., and Biserte, G., *Eur. J. Biochem.* **12**, 222 (1970).

193. Strickland, M., Strickland, W. N., Brandt, W. F., and Von Holt, C., *FEBS Lett.* **40**, 346 (1974).

194. Greenaway, P. J., *Biochem. J.* **124**, 319 (1971).

195. Garel, A., Mazen, A., Champagne, M., Sautiére P., Kmiecik, D., Loy, O., and Biserte, G., *FEBS Lett.* **50**, 195 (1975).

196. Sautiére, P., Kmiecik, D., Loy, O., Briand, G., Biserte, G., Garel, A., and Champagne, M., *FEBS Lett.* **50**, 200 (1975).

197. Brandt, W. F., and Von Holt, C., *FEBS Lett.* **23**, 357 (1972).

Part D. Physical Properties of DNA– Nuclear Protein Complexes

Chapter 13

Preparation of Chromatin and Chromosomes for Electron Microscopy

HANS RIS

Department of Zoology,
University of Wisconsin,
Madison, Wisconsin

In this chapter I shall describe some of the methods used by our group to prepare chromatin and chromosomes for electron microscopy.

I. Fixation

A. General Remarks

Cells are exposed to a large number of more or less drastic treatments from the time of death to the final electron micrograph. Therefore, the microscopist, when interpreting the image, must constantly worry about how much is left of the specimen. Chromatin consists largely of DNA, histones, and variable amounts of nonhistone proteins and RNA. This complex must be fixed by linking the components together so that ideally nothing is lost and no relocations occur during subsequent treatments. Two classes of fixatives have become established in chromosome research: the aldehydes (formaldehyde and glutaraldehyde) and acetic acid-containing fixatives.

1. ALDEHYDE FIXATION

The interaction of aldehydes with nucleohistones has recently been investigated by direct chemical analysis. The most detailed study was published by Chalkley and Hunter (*1*). They considered nucleohistones "fixed" if the histones were no longer extractable by acids (0.2 N sulfuric acid).

This is a useful requirement for analysis, but in practice we can assume that fixation is achieved if neither DNA nor histones are extracted or relocated by any of the ensuing treatments. In isolated calf thymus chromatin the histones became insoluble in acid after 2 hours of treatment with 2% formaldehyde. Fixation is due largely to cross-linking of all five histone types to DNA. This can be reversed by prolonged dialysis against water, which renders histones acid extractable again. The action of glutaraldehyde (0.6%) is different. It forms irreversible cross-links between the histone molecules themselves. Aldehydes not only bind histones and DNA together, but also preserve the basic organization of nucleohistone. For example, X-ray diffraction patterns of oriented nucleohistone fibers have revealed a periodic structure with repeats of 11.0, 5.5, 3.8 nm (2). This regular structure is in part due to the close packing of the 11-nm histone beads in the basic nucleohistone fiber and is preserved by both formaldehyde (2) and glutaraldehyde (3). Aldehyde fixation therefore would seem to be the method of choice for preserving chromatin structure.

2. ACETIC ACID-CONTAINING FIXATIVES

For certain preparations aldehyde fixation is not feasible. For example, if one wants to fix chromosomes *in situ* and afterward prepare them by the classical cytogenetic squash techniques, aldehydes are useless because they fix the cytoplasm into a rubbery gel that prevents clean separation of chromosomes. A 3:1 mixture of ethanol (or methanol) and glacial acetic acid has long been used by light microscopists for chromosome squash preparations. This fixative is also useful for electron microscopy even though it extracts a certain amount of histones. Dick and Johns (4) have analyzed the effect of ethanol–acetic acid (3:1) on isolated calf thymus chromatin. They report that it extracts 7–8% of total histone, predominantly H4 and some H3 and H2A. The remaining histones, however, can still be extracted with 0.25 N HCl. Brody (5) studied the extraction of histones by methanol–acetic acid (3:1) from cultured mouse L cells. He confirmed the greater resistance of H1 and H2B to extraction but found that only about 20% of acid-extractable proteins are left in the cells after three washes with fixative. Although this fixative therefore is not suitable for a study of nucleohistone structure, it is still extremely useful for investigating the arrangement of the native (20 nm) chromatin fiber in chromosomes, since we find that the 20-nm chromatin fiber is clearly preserved even though somewhat thinner than when fixed in aldehydes (Figs. 1 and 2).

For squash preparations a mixture of formalin and acetic acid may give somewhat better preservation while still allowing dispersal of the cytoplasm between slide and coverslip (6). No information is available about loss of histones in his fixative. For examples of chromosomes fixed with 45% acetic acid containing 4% formaldehyde, see Ris (7). Since methanol–acetic acid

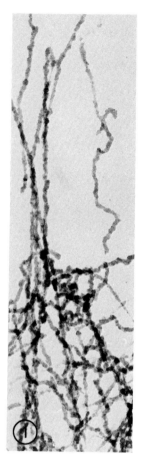

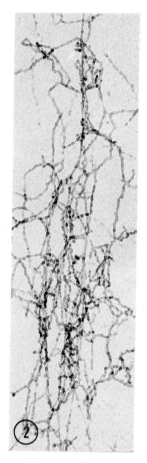

FIG. 1. *Triturus viridescens* erythrocytes spread on 0.1 mM PIPES buffer (pH 6.8) fixed in 4% paraformaldehyde, stained in 7.5% uranyl magnesium acetate, critical-point dried. × 50,000.

FIG. 2. *Triturus viridescens* erythrocytes spread on 1 mM PIPES buffer (pH 6.8), fixed in 3:1 methanol-acetic acid, stained in 7.5% uranyl magnesium acetate, critical point dried. × 50,000.

fixation gives better squash preparations, it is preferable for chromosome studies unless one wants to preserve other structures, such as nucleoli and nuclear membrane.

B. Preparation of Fixatives

1. FORMALDEHYDE

Since commercial formalin contains added preservatives, it is better to prepare fixing solutions from paraformaldehyde powder (a formaldehyde polymer). An 8% stock solution is prepared by dissolving 2 gm of para-formaldehyde in 25 ml of water by heating to 60–70 °C. One to three drops of 1 N NaOH are added, and the solution is stirred until it becomes clear. After cooling the pH is adjusted to neutrality with 0.1 N HCl. This stock solution can be kept at 4 °C for several days.

2. GLUTARALDEHYDE

Glutaraldehyde fixatives should be prepared fresh from concentrated solutions stored in ampoules under nitrogen. These can be obtained from electron microscopy supply firms.

II.   Chromatin and Chromosomes in Sections

A.   General Remarks

Analysis of chromatin structure *in situ* must generally be done in sections of cells. Since chromatin fibers are usually closely packed and complexly folded, thin sections of well-fixed cells are not very informative. The situation is somewhat improved by swelling the chromatin in hypotonic fixatives. The fibers are then separated from each other, and measurement of fiber thickness is greatly facilitated (Fig. 3). In the usual thin sections it is, however, quite impossible to get information on the spatial arrangement of chromatin fibers. Recently, high-voltage electron microscopes (one million eV) have become available for biomedical research providing a resolution of 20–50 Å with sections up to several micrometers thick. As section thickness increases, longer segments of chromatin fibers are contained in a section and the three-dimensional arrangement becomes more evident, especially in stereoscopic photographs. The practical limit to section thickness is in the overlap of structures, which makes analysis difficult. There are two million-volt electron microscope facilities available to biomedical investigators in the United States—one located at the University of Wisconsin, Madison, the second at the University of Colorado, Boulder. For information, contact Dr S. Stimler, Biotechnology Resources Branch, National Institutes of Health, Bethesda, Maryland.

B.   Procedures

1. HYPOTONIC FIXATION

Pieces of tissue 1 mm³ are immersed in 4% paraformaldehyde in water for 1 hour at room temperature. The fixative is washed out by several

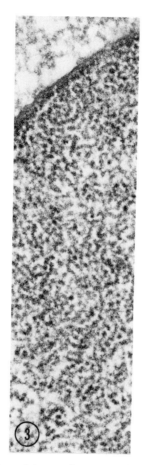

Fig. 3. Thin section of *Rana pipiens* erythrocyte nucleus fixed in 4% paraformaldehyde in water, washed in water (4°C) for 18 hours, postfixed in 1% aqueous OsO₄, and embedded in Epon–Araldite. × 50,000.

changes in water at 4°C over a period of 2 hours. This is followed by post-fixation in 1% OsO_4 in water for 1 hour. The tissue is then rinsed in several changes of water and stained for 2 hours in 0.5% aqueous uranyl magnesium acetate. After washing in 50% ethanol, the tissue is dehydrated in 70%, 80%, 95%, and absolute ethanol and embedded in Epon–Araldite (8). Thin sections are stained for 2 hours with 7.5% aqueous uranyl magnesium acetate (9) and for 15 minutes with lead citrate (10).

2. THICK SECTIONS

Sections 0.5 μm thick or thicker are prepared with glass knives and picked up on Formvar-covered single-hole (1 × 2 mm) copper grids. It is often use-

ful to have several consecutive sections on a grid. For ease of alignment of
the ribbons, they are picked up from the trough with Formvar-covered loops
(6 mm in diameter) made from thin copper wires. Under the dissecting
microscope the films are transferred to grids sitting on top of a peg (4 mm
in diameter) by sliding the loops over the peg. The Formvar films are prepared by dipping a glass slide into a solution of 0.5% Formvar in ethylene
dichloride. When dry, the film is cut with a needle into 8-mm squares on the
slide. The film squares are floated off on water, picked up with the wire loops,
and dried.

Consecutive thick sections prepared with glass knives often have breaks
parallel to the knife edge on the top surface. These can be avoided by making
a few thin sections in between the thick sections (11). Thick sections must be
stained for longer periods than the usual thin sections to allow for penetration of the stain. The 1-μm sections are stained in aqueous 7.5% uranyl
magnesium acetate for 4 hours at 50°C (in a humid chamber). They are then
further stained in lead citrate for 20 minutes. After rinsing and thorough
drying, a thin layer of carbon is evaporated on the side of the grid not carrying the sections. The grids are carboned only after staining to allow the
stain to penetrate from both surfaces into the sections.

III. Preparation of Isolated Chromatin Suspension for Electron Microscopy

A.   Positive Staining and Drying from Amyl Acetate

This is a quick method giving adequate preservation for checking on a
chromatin preparation. A drop of the chromatin suspension is placed on a
Formvar–carbon–coated grid (200 or 400 mesh) which has been rendered
hydrophilic by glow discharge in the evaporator (12). After standing for a
few minutes, the grid is floated with specimen side down on 4% formaldehyde and fixed for 5 minutes. To obtain partially oriented chromatin fibers
a hydrophilic grid may be streaked over the surface of a chromatin preparation and then fixed by floating on formaldehyde. After fixation the grids are
rinsed in distilled water and positively stained either with uranyl magnesium
acetate (UMA) or phosphotungstate (PTA) as follows.

1.   URANYL STAINING

Grids are immersed in 7.5% aqueous solution of UMA for 5 minutes,
rinsed in three changes of 50% ethanol, dehydrated through 70%, 80%, 95%,
and absolute ethanol, and placed into amyl acetate.

2. PTA STAIN

Grids are dehydrated through 70%, 80%, and 95% ethanol and immersed for 5 minutes in 1% PTA in absolute ethanol. They are then rinsed in absolute ethanol and placed into amyl acetate.

For drying from amyl acetate, a grid is picked up with forceps and excess amyl acetate on the grid and between the tongs of the forceps is quickly removed with filter paper. The grids are stored in a desiccator.

B. Positive Staining and Drying with the Critical-Point Method

1. GENERAL REMARKS

The critical-point method (13,14) is the best procedure for preserving specimens that must be dried for electronmicroscopy (whole mounts). Because of the large surface-tension forces, drying in air from water destroys most structure (Fig. 4). Drying from liquids with low surface tension may cause less distortion, but the critical-point method avoids surface-tension distortion altogether and is always preferable (Figs. 1 and 2). The specimen is immersed in a liquid with a critical temperature in a range that will not damage the structure. The critical temperature is the temperature above which the liquid has expanded into a gas and will not condense at any pressure. As the temperature rises the pressure also increases, until at the critical pressure the density of liquid and gas are identical and surface tension is zero. At this critical point the liquid expands into a gas without any interfacial tension distorting the specimen. Two liquids are used today for critical-point drying: carbon dioxide and Freon 13. Carbon dioxide has a critical temperature of 36.5°C and a pressure of 1080 psi. The critical temperature of Freon 13 is 28.9°C, and the pressure is 561 psi. Liquid CO_2, which can be obtained in siphon tanks, is so much cheaper and usually cleaner, that it is generally preferred. There are several instruments now commercially available for critical-point drying. The instrument should have the following features: safety pressure-release valve, lighted pressure chamber with glass viewing port, heater with thermostat, device to cool the chamber, temperature and pressure gauges.

2. PROCEDURE

Chromatin is placed on grids, fixed, and stained and dehydrated as described in Section III,A. Amyl acetate is used as a transitional fluid since it mixes well with ethanol and CO_2. The pressure chamber of the critical-point apparatus is filled with enough amyl acetate to cover the grids, and the grid holder is quickly transferred into the chamber. The chamber is cooled to about 40°C, and the amyl acetate is then replaced with liquid CO_2. It is

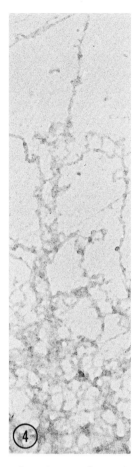

FIG. 4. *Triturus viridescens* erythrocytes spread on water, fixed in 4% paraformaldehyde, stained with 7.5% uranyl magnesium acetate, dried in air from water. The chromatin fibers are smeared out owing to air drying. × 50,000.

important to wash out all amyl acetate by changing the CO_2 several times and soaking the grids in liquid CO_2 for about 15 minutes followed by several further CO_2 replacements. The specimen chamber is then heated to approximately 45°C and allowed to equilibrate until the chamber looks uniformly clear and without schlieren effects. The gas is then released slowly to avoid any recondensation. Absolute acetone also mixes with liquid CO_2, and thus dehydration in acetone (in steps of 5 or 10% beginning with 30%) has also been used for critical-point drying. The grids should be stored in a desiccator.

C. Negative Staining of Chromatin

Negative staining has the advantage that dehydration through organic solvents can be avoided. The specimen is covered with an aqueous solution of an electron-opaque salt (e.g., 1 or 2% uranyl acetate or 1–2% sodium phosphotungstate), which dries into an amorphous glasslike structure. The specimen then appears bright embedded in the dark electron-scattering salt, which outlines the surface structure of the specimen. If the layer of negative stain is thick enough, it protects the structure against drying artifacts. Electron diffraction shows that negatively stained catalase, for instance, still

Fig. 5. *Triturus viridescens* erythrocytes spread on water, fixed in 4% paraformaldehyde, embedded in a thick layer of 2% uranyl acetate (negative staining). Photographed with the 1 million volt electron microscope at the University of Wisconsin, Madison. × 80,000.

contains regular structure to about 8 Å whereas all periodicity is lost after drying in air. (*15*). Negative stain deep enough to embed the 200 Å chromatin fiber is too opaque for the 100 kV electron beam. With the 1 MeV microscope, however, chromatin fibers can be observed in thick layers of uranyl acetate or PTA (*7*) (see Fig. 5).

For the negative stain to spread and dry evenly over the grid, the film surface must be hydrophilic. A simple method to make films wettable is to expose them to a high-voltage glow discharge in the evaporator (*12*). The chromatin suspension is applied as in Section III,A to 400-mesh copper grids with Formvar–carbon films freshly cleaned in a glow discharge. After fixation in aldehydes the grids are washed in distilled water and covered with a drop of 1 or 2% aqueous uranyl acetate, or 1–2% PTA adjusted to pH 7. The excess stain is removed with filter paper. The amount of stain left to dry must be empirically determined for each case. It must be emphasized that chromatin has to be fixed with aldehydes before negative staining, otherwise histones will dissociate from the DNA as the salt concentration increases during drying.

D. Metal Shadowing of Chromatin Fibers

The surface structure of chromatin fibers can be made visible by metal shadowing. Carbon–platinum gives a fine-grained deposit and is conveniently evaporated from carbon–platinum pellets held between recessed carbon rods obtained from suppliers of electron microscope accessories. Grids with critical point-dried chromatin are mounted in the evaporator about 15 cm from the carbon rods with an evaporation angle of 10–45° (for an example, see Fig. 6).

IV. Spreading of Chromatin and Chromosomes on a Hypophase (Langmuir Trough)

A. General Remarks

In 1963 Gall (*17*) showed that chromatin could be prepared for electron microscopy by a modification of the DNA–protein film technique of Kleinschmidt and Zahn (*18*). This has since become the most extensively used method for electron microscopy of chromatin. Cells and nuclei are lysed osmotically by placing them on a clean water surface, or they may be mechanically disrupted before spreading. Cellular proteins form a monolayer on the water surface, and chromatin remains attached to this film

FIG. 6. Chicken erythrocyte chromatin prepared according to Olins and Olins (16), centrifuged in 1% formaldehyde onto a grid, critical-point dried, and shadowed with carbon–platinum. The chromatin beads after shadowing are about 12 nm in diameter. × 120,000.

hanging into the water phase. In contrast to the Kleinschmidt technique, where DNA is actually trapped in the surface protein film, chromatin fibers are not in the surface film but only attached to it. The method thus represents a rapid release of chromatin into "low ionic strength medium."

B. Procedures

Any flat dish with a hydrophobic surface can be used as a trough for spreading. Glass must be covered with a thin layer of paraffin to make it water repellent. By far the most convenient are metal troughs covered with a layer of black Teflon. The liquid surface is cleaned by sweeping several times with a Teflon-covered bar (or paraffined glass rod). The bar is rubbed clean each time with filter paper. This cleaning of the water surface is facilitated by using two bars, one being left on the trough as the other one is swept through. When the surface is clean the bars are left on the trough, limiting the area on which the film will be formed. To make the expanding surface film visible, talcum powder is sprinkled over the area between the bars. A 2-ounce rubber bulb is partly filled with talcum powder. A 200-mesh electron microscope grid is then taped tightly over the opening. By gently squeezing the bulb, talcum can be dusted over the trough. By moving the bars apart, one can test whether the surface is clean: if a contaminating film is present, the particles remain fixed; if the surface is clean they spread out following the bar. The material can be applied to the trough in a number of ways: (a) a small drop is touched to the surface from a pipette or from the tip of an acid-cleaned glass rod; (b) an acid-cleaned glass rod is held vertically into the trough and the material is allowed to run down the rod onto the surface; (c) an acid-cleaned glass slide is placed at a shallow angle into the trough, and a drop of material is allowed to run down the slide onto the surface. If a surface film is formed, the talcum particles will be pushed apart, thus delimiting the periphery of the film and making it easily visible.

The surface film is picked up by touching a Formvar–carbon coated grid to the surface of the trough. The grid is then floated on the fixative, rinsed in water, and stained and dried with the critical point procedure described in Section III.

1. Interphase Chromatin

a. Native fiber of inactive chromatin. Inactive interphase chromatin consists of branched fibers about 20 nm thick. They are visualized by spreading cells on distilled water or nonmetal binding buffers (*19*) (for an example see Fig. 1). Nucleated erythrocytes of vertebrates are a good source, since cells and nuclei rupture osmotically when a drop of blood is touched to the water surface. More resistant cells must be broken before spreading. This is accomplished, for instance, by grinding cells with the tip of a glass rod and touching the rod to the trough (*20*) or by grinding cells between two slides and dipping the slides into the trough (*21*).

b. The elementary fiber of inactive chromatin. When chromatin is treated with chelating agents, it unravels into fibers about 10 nm thick. This seems to be the smallest fiber of inactive chromatin in which all histones are still present and in their original relationship. These elementary fibers can be obtained by spreading cells on chelating agents, e.g., 5 mM sodium citrate, EDTA, or EGTA.

  c. Demonstration of the histone beads in the elementary fiber. Recent biochemical and physical studies have established that histones assemble into regular beadlike structures about 10 nm in diameter. These combine with DNA, with about 200 base-pairs wrapping around each bead. These beads were first visualized with the electron microscope by Olins and Olins (*16,22*) and by Woodcock (*23*), using the technique developed by Miller and Bakken (*24*) for active chromatin. It was suggested that chromatin beads were not seen earlier because surface spreading and critical-point drying destroyed the structure. I have recently found, however, that chromatin beads can easily be demonstrated in surface-spread and critical-point-dried chromatin by slight stretching of the fibers after aldehyde fixation (Fig. 6). The beads are usually not seen because they are closely packed in the 10-nm fiber. They become visible if fixed chromatin is centrifuged onto a grid in a plastic chamber as in Miller's technique (*24*).

  The construction of such a chamber is shown in Fig. 7. It consists of a Lucite disk about 5 mm thick and with a diameter that fits loosely into the tube of a swinging-bucket centrifuge. A hole with a diameter of 4–5 mm is drilled through the center, and a coverslip is glued with Epon on one side to make a well. Any excess glue on the inside is removed so that the bottom and sides of the well are smooth and clean. The well is partly filled with 4% formaldehyde; a Formvar–carbon-coated grid is placed at the bottom of the well, avoiding air bubbles, and the well is then filled with the chromatin suspension or with nuclei swollen in 0.2 M KCl (see Chapter 5, this volume). Chromatin spread on a hypophase is picked up on a grid and fixed in aldehyde. It is then placed in the well in water. A coverslip is placed over the well, taking care that no air bubbles are formed. After centrifugation at 2000–3000 g for 5–10 minutes the grid is either positively or negatively stained, then dried with the critical-point method. After drying it may be

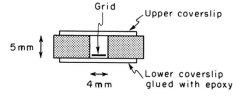

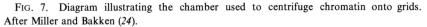

FIG. 7. Diagram illustrating the chamber used to centrifuge chromatin onto grids. After Miller and Bakken (*24*).

shadowed with carbon–platinum as described earlier. An example of chicken erythrocyte chromatin prepared in this way is shown in Fig. 6.

The chromatin beads can also be revealed in surface-spread and critical-point-dried chromatin by extracting the very lysine-rich histones (H1 and H5) before fixation, for instance by immersion for 30 minutes in 0.6 M NaCl. These histones are bound to the DNA bridge between beads. After they are removed, the beads become separated about 15 nm, as this DNA extends at low ionic strength (as in 2% formaldehyde) and can be visualized by positive or negative staining or heavy-metal shadowing (Ris, unpublished observations).

2. Mitotic Chromosomes

Mitotic chromosomes are usually obtained from cell cultures after treatment with colchicine to accumulate metaphases or from tissue with a high mitotic index, such as testes. Cultured cells are swollen by incubation in hypotonic solutions (diluted medium, or half-strength Hanks salt solution), concentrated by centrifugation and placed on the water surface. Pieces of tissue are crushed with the tip of a glass rod or by grinding between two slides, and the debris is spread on the trough. Further treatment is identical to that of interphase chromatin.

V.   The Squashing Technique for Electron Microscopy

A.   General Remarks

The chromosomes prepared by spreading cells on a water surface are generally considerably distorted. This method is therefore not recommended where the arrangement of chromatin fibers in higher order structures is to be preserved. For better preservation chromosomes should be fixed *in situ* first and then separated from the rest of the cell (see Figs. 8 and 9). The acetic acid squash technique has accomplished this for light microscopy. To be useful for electron microscopic studies, this technique must fulfill two conditions: the methanol–acetic acid fixation must preserve the chromatin fiber, and the material must be transferred from a slide to the electron microscope grid without air-drying or other distortions. We have found that, despite some loss of histones, the methanol–acetic acid fixation preserves the structure of the chromatin fiber quite well (Fig. 2). A method to transfer the squashed chromosomes from slide to grid was first published in 1961 (*25*).

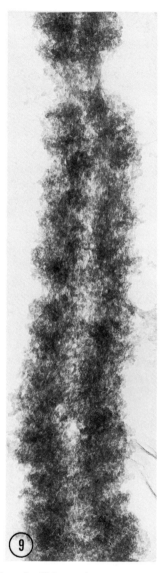

FIG. 8. Chinese hamster metaphase chromosome spread on water, fixed in 4% paraformaldehyde, acetone dehydrated, and critical-point dried. Photographed with the 1 million volt electron microscope at the University of Wisconsin, Madison. × 20,000.

FIG. 9 Chinese hamster metaphase chromosome, fixed *in situ* with methanol–acetic acid 3:1, squashed in 45% acetic acid, and transferred to a grid as described in text. Critical-point dried. Photographed with the 1 million volt electron microscope at the University of Wisconsin, Madison. × 20,000.

B.   Procedure

The procedure is the same for interphase chromatin, mitotic or meiotic chromosomes, or polytene chromosomes.

1.   Fixation

The material is fixed in methanol–acetic acid (3:1) for 24 hours with several changes of fixative.

2.   Squashing

Cell suspensions are centrifuged and resuspended in a small amount of 45% acetic acid. A drop of this suspension is placed on a slide, covered with a No. 2 coverslip, and squashed as described below.

Small pieces of tissue (e.g., testis) are removed from the fixative and placed on a slide into a drop of 45% acetic acid. With dissecting needles any extraneous material, such as unwanted cell types, connective tissue, etc., is removed. Ideally only free cells containing the desired chromosomes should remain. The drop of liquid on the slide must be just large enough to fill the volume beneath the coverslip. A square No. 2 coverslip is placed on the drop by first touching one edge on the slide and then slowly lowering the coverslip to prevent air bubble formation. The coverslip must be evenly flat on the slide, since any thicker object holding up the coverslip will prevent even squashing. The amount of liquid under coverslip is crucial. If there is too much, all material will be squirted from beneath the coverslip over the edges. If there is not enough, the proper vertical movement of the coverslip during "tapping," which disperses the cells, is prevented. The liquid can be adjusted by adding acetic acid or removing it with filter paper.

The dispersal of the cells is best controlled with a modified vibro tool. Any engraving tool with a vertically vibrating tip can be used by replacing the steel tip with a pointed plastic tip. While one edge of the coverslip is held firmly so that it cannot slip sidewise, the vibrating tip is gently touched to the coverslip over the tissue. By frequent checking with the phase microscope one can control the degree of dispersal of cells and chromosomes. Once the chromosomes are free of cytoplasmic debris the acetic acid is left to evaporate until all movement of particles under the coverslip has stopped.

3.   Transfer of Specimens to Electron Microscope Grids

To remove the coverslip the slide is frozen by dipping into liquid nitrogen. The coverslip is flipped off with a single-edge razor blade, and the slide is placed quickly into distilled water. After staining in 7.5% aqueous uranyl magnesium acetate for 30 minutes, the slide is rinsed in two changes of 50% ethanol and dehydrated in 70%, 80%, 95%, and 100% ethanol (2 minutes each). The slides are then soaked for 5 minutes in amyl acetate. Next the material is embedded in a film of Parlodion. The slide is quickly wiped dry except for the area containing the specimens, which is now flooded with a solution of 2% Parlodion in amyl acetate. Care must be taken that the specimen never becomes dry during this procedure. The Parlodion is then allowed to dry in a dust-free place. The slide must be kept horizontal so that the film dries into a uniformly thick layer. Enough Parlodion must be added so that the specimens are completely embedded in the film. After drying, the slide can be searched with the phase microscope to select suitable specimens. A Zeiss diamond object marker is then used to cut a circle about 2 mm in diameter around the selected specimens. Under the dissecting microscope this circular disk is then removed from the glass slide and placed on a 75-mesh copper grid covered with a 0.5% Formvar film in the following way: A drop of water is placed over the disk. With forceps and needle the rim of the disk is gently loosened from the glass. The disk is then lifted to the water surface and picked up from underneath with a Formvar-coated grid. After drying in a 60°C oven for 5 minutes, the grid is immersed into amyl acetate with the specimen uppermost. After several hours, preferably overnight, the Parlodion will have dissolved, leaving the specimen undistorted on the Formvar film. The grids are now dried with the critical-point procedure, and finally a thin layer of carbon is evaporated on the underside of the grid.

For stereomicrographs of metaphase chromosomes prepared in this way, see Ris (*7a*).

REFERENCES

1. Chalkley, R., and Hunter, C. *Proc. Natl. Acad. Sci. U.S.A.* **72**, 1304. (1975).
2. Pardon, J. F., Richards, B. M., and Cotter, R. J. *Cold Spring Harbor Symp. Quant. Biol.* **38**, 75. (1974).
3. Olins, D. E., and Olins, A. L. *J. Cell Biol.* **53**, 715.
4. Dick, C., and Johns, E. W. *Exp. Cell. Res.* **51**, 626. (1968).
5. Brody, T. *Exp. Cell Res.* **85**, 255. (1974).
6. Holmquist, G., and Steffensen, D. M. *J. Cell Biol.* **59**, 147a. (1973).
7. Ris, H. *Struct. Funct. Chromatin, Ciba Found. Symp.* [*N.S.*] **28**, (1975).
7a. Ris, H. "Proceedings of the Sixth European Congress on Electron Microscopy," Volume 2, pp. 21–25. Tal International Publishing Co., Israel.
8. Mollenhauer, H. H. *Stain Technol.* **39**. 111. (1964).
9. Frasca, J. M., and Parks, V. R. *J. Cell Biol.* **25**, 157. (1963).
10. Reynolds, E. S. *J. Cell Biol.* **17**, 208. (1963).
11. Favard, P., and Carasso, N. *J. Microsc.* (*Oxford*) **97**, 59. (1972).
12. Reissig, M., and Orrell, S. A. *J. Ultrastruct. Res.* **32**, 107. (1970).
13. Anderson, T. F. *Phys. Tech. Biol. Res.* **3**, 178. (1956).
14. Cohen, A. L., Marlow, D. P., and Garner, G. E. *J. Miscrosc.* (*Paris*) **7**, 331. (1968).
15. Taylor, K. A., and Glaeser, R. M. *Science* **186**, 1036. (1974).
16. Olins, A. L., and Olins, D. E. *Science* **183**, 330. (1974).

17. Gall, J. G. *Science* **139**, 120. (1963).

18. Kleinschmidt, A., and Zahn, R. K. *Naturforsch., Teil B* **14**, 770. (1959).

19. Good, N. E., Wignet, G. D., Winter, W., Connolly, T. N., Isawa, S., and Singh, R. M. *Biochemistry* **5**, 467. (1966).

20. DuPraw, E. J. *Proc. Natl. Acad. Sci. U.S.A.* **53**, 161. (1965).

21. Wolfe, S. L., and John, B. *Chromosoma* **17**, 85. (1965).

22. Olins, A. L., and Olins, D. E. *J. Cell. Biol.* **59**, 252a. (1973).

23. Woodcock, C. L. F. *J. Cell Biol.* **59**, 368a. (1973).

24. Miller, O., and Bakken, A. H. *Acta Endocrinol. (Copenhagen), Suppl.* **168**, 155. (1972).

25. Ris, H. *Can. J. Genet. Cytol.* **3**, 95. (1961).

Chapter 14

Flow-Cytometric Analysis of Chromatin

DAVID BLOCH,[1] CHI-TSEH FU,[1] AND EUGENE CHIN [2]

I. Introduction

Flow-cytometry is the rapid automated measurement of cell fluorescence. Pulse-height analysis of the induced fluorescence of individual cells in a stream, as they flash by a detector unit at rates of up to 300,000 cells per minute, permits determination of distributions of amounts of fluorescent stained materials. The basic elements of the instrument are the flow cell, which contains and shapes the stream; an intense light source, usually a continuous-beam argon laser; a detector unit consisting of one or more photomultipliers with their power supply; and a pulse-height analyzer consisting of signal amplifiers, analog digital converters, a memory, a data retrieval system, and a variety of accessories the need for which varies with the complexity of the instrument. Volume sensors, circuitry to accommodate multiparameter analysis, a cell sorter, and interfacing with computer systems afford a versatility that holds promise of such accomplishments as automated karyotyping and cell cycle analysis in the near future.

Numerous articles and reviews have been written during the past several years on various aspects of the method. Some selected works are listed in Table 1, and the reader is referred to these for a more detailed account of the development, rationale, and some specific uses of the technique (*1–24*). The present work will cover these aspects in summary form, giving more attention to some of the practical considerations in analyzing chromatin by flow-cytometry (FCM).

The method consists of preparation of the cells, measurement, and analysis of the data. Each of these steps requires attention to problems that are unique to the cell or tissue being studied and to the method of staining used.

[1] Botany Department and Cell Research Institute, The University of Texas at Austin, Austin, Texas and the Cell Biology Section, National Center for Toxicological Research, Jefferson, Arkansas.

[2] Department of Zoology. The University of Texas at Austin, Austin, Texas.

TABLE I

SOME REFERENCES OF GENERAL AND HISTORICAL INTEREST

Historical	
UV absorption and 90° scatter	*1*
Light scatter	*2*
Fluorescence, pulse photometer[a]	*3*
Fluorescence, FMF	*4*
Two-color fluorescence	*5*
Fluorescence	*6*
General reviews	*7–10*
Multiparameter measurement	
Two-color fluorescence	*5, 11*
Fluorescence and light scatter	*12*
Scatter, fluorescence anisotropy	*13*
Computerized analysis of distributions	
Stage frequencies and S period profiles	*14, 15, 15a*
Gating[b]	*16*
Residuals[c]	*17*
Cell sorting	*7, 13, 18, 19*
Chromosome analysis	*20–22*
Future directions	*23, 24*

[a] The pulse photometer is a European version of the flow-cytometer using a more conventional light source with a more efficient light-collecting system.

[b] Gating is the selection of cells based on measurements of one parameter for analysis in a second parameter.

[c] Residuals are departures from theoretical values. Their use here is in the resolution of populations into subpopulations, based on finding best fit of observed curve to the sum of overlapping normal distributions of assumed or derived subpopulations.

II. Cell Preparation

Preparation of cells entails tissue dispersal, in some cases fixation of the separated cells, then staining and differentiation. Few generalizations can be made. The recommended procedures in each step depend on the tissue, its ease of dispersal, and the type of measurement.

A. Cell Dispersal

1. NONADHERENT CELLS

The sample for analysis consists of separated cells in fluid suspension. Some cell types, such as erythrocytes, sperm cells, some plant endosperm developing microspores, ascites tumor cells, and other cultured cells in fluid suspension lend themselves easily to flow analysis because of their nonadherent nature. Lymphocytes from the spleen and thymus and bone marrow cells would also fall into this category. These require no special treatment for their dispersal, although washing in saline prior to fixation may aid in preventing aggregation during subsequent processing.

Ehrlich ascites cells are routinely washed in Earle's (25) physiological saline. Several washes, each followed by centrifugation at 500 g for 5 minutes, has the additional effect of freeing the tumor cells from the lighter erythrocytes. Suspension of the cells in 100 times their packed volume of cold 10% formalin (3.7–4.0% formaldehyde) previously brought to pH 7.0 provides fixation. Chicken erythrocytes have been fixed similarly after dilution of whole blood in an equal volume of acid citrate–dextrose (26) or by fixing heparinized whole blood (ca. 40 units of heparin to 10 ml of whole blood), mixing equal parts with 20% neutralized formalin.

Dispersal of Chinese hamster ovary (CHO) cells from monolayer culture has been accomplished by mild digestion with trypsin and pancreatic DNase (9), according to the following regime. (The function of the DNase is to prevent cell aggregation that might otherwise occur in the presence of DNA liberated from damaged cells.)

1. Wash in cold saline (6 mM glucose–0.137 M NaCl–5 mM KCl–1 mM Na$_2$HPO$_4$–1 mM KH$_2$PO$_4$)

2. Incubate for 10 minutes at 37°C in dispersal medium (above saline containing 0.5 mM EDTA and 0.1 mg of trypsin per milliliter).

3. Add an equal volume of saline containing also 6 mM MgSO$_4$, 1 mM CaCl$_2$, 0.2 mg/ml of soybean trypsin inhibitor, 0.01 mg/ml of DNase I, and 1 mg/ml of bovine serum albumin.

4. Work the suspension up and down in a transfer pipette and fix.

Sperm cells have been washed with Hanks (27) or Tyrode (28) solution before formalin (29). Rabbit leukocytes have been prepared for analysis by suspending in fixing fluid directly from buffy-coat suspensions dispersed in saline (S. Olson and D. Bloch, unpublished). Chicken lymphocytes have been obtained by perfusing spleen with Hanks solution, mincing the organ, then collecting the liberated cells and fixing in formalin (D. Bloch, unpublished).

2. Solid Tissues

Few studies have been attempted on cells dispersed from more typical tissues, where requirements for dispersal would be more stringent. The art of obtaining single-cell suspensions has been highly developed in the case of a few selected tissues.

Table II lists some tissues that should lend themselves successfully to

TABLE II

TECHNIQUES IN TISSUE DISPERSAL

Methods	Tissue	References
Mechanical disruption	Human cervical	*30–32*
	Human cervical and mouse squamous cell carcinoma	*33*
Above plus dissociating agents	Human cervical	*31*
Above plus enzymic digestion		
Trypsin	Monkey kidney	*34*
Collagenase	Embryonic heart	*35*
Collagenase	Rat adipose	*36*
Trypsin	Rat thyroid	*37*
	Mouse, various	*38*
Trypsin, subtilisin	Embryonic chick retina	*39*
Collagenase	Mouse mammary	*40*
Trypsin, collagenase	Rat ovarian	*41*
Trypsin	Rat adrenals	*42*
Various	Human cervical	*31*
Trypsin, collagenase	Human cervical, mouse squamous cell carcinoma	*43*
Collagen, hyaluronidase	Rodent liver	*44*
Thermolysin, Pronase, trypsin	Rat lung	*45* *45*
Trypsin, chymotrypsin, papain	Various	*46*
Trypsin, DNase	Mouse testes	*47,47a*
Pronase, DNase	Gastric mucosa	*48*
General articles		*32,46,47a,49,50*

such analysis because of the effectiveness of the dispersal methods (*30–50*).

3. ISOLATED NUCLEI

Tissues that present problems in dispersal may be more easily analyzed for chromatin content (either DNA or protein) by the use of isolated nuclei. Profiles of propidium diiodide (PI)-stained and acriflavine–Feulgen-stained ascites nuclei, described in Section II,B., suggest the results that might be obtained when nuclear isolation techniques are applied to more tenacious tissues. Caution should be exercised in such investigations, since a complex tissue with various cell types may not yield nuclei in the same proportion as occurs *in vivo*. The lymphocyte nuclei in ascites cell preparations tent to be lost to the supernatant during nuclear isolation, because of their small size. Also, the effects of condensation of nuclear chromatin on FCM analysis may be exaggerated in isolated nuclei (see Section IV).

The following method of nuclear isolation of Zweidler (see *51*) has been used on Ehrlich ascites cells, CHO cultured cells, and rat spleen cells, among others.

Washed cell preparations are suspended in a hypotonic buffer [50 μM glycine—20 mM Tris-maleate (pH 7.4)–5 mM KCl–1 mM MgCl$_2$–1 mM CaCl, 5 mM 2-mercaptoethanol] and broken by homogenization at 4°C for 2 minutes at 60% maximum voltage in a Waring Blendor fitted with a foam-proof plastic adaptor. The suspension is centrifuged at 800 g for 10 minutes.[1] Homogenization and centrifugation are repeated until the number of whole cells is reduced to an acceptable level and cytoplasmic adhesion is minimal. The supernatant at this point is nearly clear. The nuclear pellets can be stored in 50% glycerol in the above buffer at −20°C. DTT, 3 mM, is substituted for the 2-mercaptoethanol in the storage buffer. The results of analysis of such nuclei are shown in Fig. 1.

Aids and cautions: Extremely clean nuclei from some sources, for example, Ehrlich ascites cells, tend to clump, especially when pelleted, complicating flow analysis. On the other hand, nuclear suspensions should be free of whole cells. These conflicting needs require that homogenization be moderated to achieve the best balance. Sticky nuclei may require sedimentation at low speeds or at unit gravity. It should also be noted that PI-treated nuclei may be diluted to suitable concentrations and analyzed without washing out unbound PI.

B. Fixation and Staining of Chromatin in Cells and Nuclei

Fixation, or prestaining treatment where there is no fixation, is dictated by the requirements for staining. Fixation will be considered further in the context of the specific staining protocols. Details and generalities of fixation are discussed by Crissman *et al.* (*9*).

1. STAINING OF DNA

a. Acriflavine–Feulgen Staining. Feulgen staining of DNA using Schiff's reagent made with fluorochromes has been employed for many years. Kasten (*52*) listed a variety of Schiff's reagents, including several made with

---

[1] Addition of sucrose following homogenization was called for in the original procedure, but has been eliminated as unessential. The nuclei are more easily dispersed in the absence of sucrose.

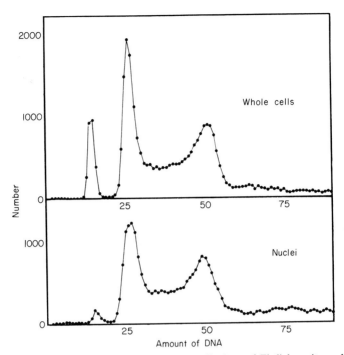

Fɪɢ. 1. Comparison of DNA fluorescence distributions of Ehrlich ascites cells and of nuclei isolated therefrom. Upper panel: Fluorescence profile of acriflavine–Feulgen-stained cells. Lower panel: Profile of nuclei isolated as described in Section II,A. The left-hand peak belongs to contaminating lymphocytes (or lymphocyte nuclei), and the right-hand peaks with the intervening valley, belong to the G_{-1}, G_{-2}, and S-period ascites cells and nuclei. The ascites cells are hypotetraphoid. From P. Seitz, unpublished observations.

highly fluorescent dyes. The initial, and still most commonly used fluorescent Feulgen for FCM is acriflavine–Feulgen (8). The binding of this stain to apurinic acid does not require the presence of sulfite, hence appears to be different from that of the traditional magenta-based Feulgen (53), although neither of the reactions is well understood. In any event sulfite is used, needed or not, and the acriflavine–Feulgen reaction provides the expected profiles on FCM analysis with a variety of cell types (8). The following protocol taken from Kraemer (8) describes the steps in Feulgen staining.

1. Fix dispersed cells overnight in formalin as described in Section II,A,1.

2. Centrifuge for 5 minutes at 3000 rpm and resuspend sample containing about 10^7 to 10^8 cells (about 0.05 ml of packed cells) in 5 ml of buffered saline or water.

3. Resuspend pellets after each change by vortexing in 0.1 ml of residual fluid before adding the next solution. Each solution change is affected by decanting, vortexing, then adding the new solution.

4. Rinse in water. Let stand 3 minutes.

5. Hydrolyze in 4 N HCl at room temperature for 20 minutes.[2]
6. Rinse in water. Let stand 3 minutes.
7. Stain 30 minutes in the dark in acriflavine–Schiff's reagent.[3]
Stain is prepared by dissolving 20 mg of acriflavine hydrochloride in 100 ml of 0.5% potassium metabisulfite–0.05 N HCl, letting stand for 3 minutes, then filtering with pressure through a 0.2 μm Millipore filter.
8. Differentiate several times, 2 minutes each, in 1% concentrated HCl–70% ethanol.
9. Rinse 3 times in distilled water.
10. Store in refrigerator until ready for analysis.
11. The final suspension should be made by dispersing the cells in a small volume. A final dilution[4] can be made before analysis, as described, depending on the cell yield, and convenience. The cells can usually be dispersed easily before analysis by forcing the suspension up and down a transfer pipette a few times, holding the end of the pipette against the bottom of a plastic beaker.

The stain is quite stable; although it is desirable to perform the analysis soon after staining, preparations stored for several months in the refrigerator have been used routinely for tuning the instrument. Such preparations can also be used to monitor instrument stability (see Section IV,D.)

The acriflavine–Feulgen excitation maximum, and that of EB and PI-stained DNA, and protein-conjugated fluorescein are close to the 488 line of the argon laser, which is the wavelength giving the maximum output. Characteristics of these and other spectra are summarized in Table III (5, 9, 55–66).

Potential candidates for fluorescent Feulgen staining for FCM are auramine O, proflavine, coriphosphine, among others.

b. Staining with Intercalating Dyes. EB and PI require similar conditions for preparation, staining, and analysis. The PI may have a slight advantage. The emission maximum is reported to be closer to the red end of the spectrum, permitting better resolution of the fluorescence signals from DNA and protein when the two are double-stained (the latter with fluorescein derivatives) for multiparameter analysis (9). However, the values of emission maxima reported for EB range both above and below those for PI.

PI intercalates into double-stranded nucleic acids, and specific DNA staining requires removal of RNA. Cells that are low in RNA have been

---

[2] Duijndam and Van Duijn (54) recommend hydrolysis for longer times, to minimize the differences in response of naked DNA and protein-bound DNA to hydrolysis and subsequent staining.

[3] In our experience, exposing chicken erythrocytes to fluorescent light during staining gave a slightly lowered coefficient of variation (C.V.).

[4] The fluorescence signal can be increased and the C.V. lowered by swelling the stained cells with 1% SDS for at least 15 min, then analyzing in 1% SDS.

TABLE III

Some Stains Used for the Flow-Microfluorometric Analysis of Chromatin, and the Positions of Their Excitation and Emission Peaks

	Stain	Excitation	Emission	References
DNA	Acridine orange	338 (free), 565 (bound)	505, 530 (DNA, bound 660 (RNA, bound)	55
	Acriflavine-Feulgen	460	515	56
	Auramine O	430	530	57
	Ethidium bromide	475 (free)	580–610 (bound)	58
			580 (bound	9
		525 (bound)	612 (bound)	59
			645 (bound)	60
	Hoechst 33258	338 (free)	505 (free)	61
		356 (bound)	465 (bound)	62,63
	Mithramycin	395	535	64
	Propidium diiodide	472–480 (free)	610 (bound)	9; C. T. Fu and D. P. Bloch, unpublished
RNA	Acridine orange	(see above)		
Protein	Conjugated fluorescein	475	527–530	57,65
	DANS (1-dimethyl-5-aminonaphthalene sulfonyl chloride		520	5
Histone	Eosin Y	518–520 (free)	540–545 (free and bound, pH 3.1), 560–580 (bound, pH 6.9)	66
			550 (protamine bound)	57; C. T. Fu and D. P. Bloch, unpublished

stained with PI without RNA removal (see below), the advantages of simplicity and speed being of value in monitoring results during experiments, or in making rapid diagnosis.

Staining with PI (modified from Crissman and Steinkamp, *11*) is done as follows:

1. Fix in cold 70% ethanol, 0.5 hour.
2. Rinse in water.
3. Hydrolyze in preboiled ribonuclease A, 1 mg/ml, at 37°C for 30 minutes [or 6% saturation Ba(OH)$_1$ in 70% ethanol, 37°C, 30 minutes (*11a*)].
4. Rinse in water (preceded by 70% ethanol, if Ba hydrolysis is used).
5. Stain in 0.25% PI in 1/100 × saline-citrate (SSC) (1 × SCC is 0.14 M NaCl–0.014 M sodium citrate).

6. Rinse twice in water.

7. Store in refrigerator until analysis.

The PI-stained cells, like the Feulgen-stained cells, are quite stable and can be analyzed at leisure. Figure 8 (see Section V,B) shows the fluorescence profile of PI-stained Ehrlich ascites cells.

Alternatively, lymphocytic cells or other cells with scanty cytoplasm or low amounts of RNA can be stained rapidly in PI by swelling in hypotonic solution. The staining solution is 0.05 mg/ml PI–0.1% sodium citrate. No fixation is used. Ribonuclease treatment is unnecessary for such cells. The cells are stained for 5 minutes, then analyzed directly (67).

Isolated ascites nuclei may be treated similarly (E. Chin, unpublished). We have stained nuclei suspended in isolation buffer (Section II,A,3) by adding equal volumes of the suspension to 0.1 mg/of PI per milliliter in water. After 5 minutes, the nuclei are washed twice in water, then analyzed. Ribonuclease treatment is avoided because of its tendency to cause nuclei to clump.

c. Thermal Denaturation of Chromatin.   The fluorescence induced by intercalation of dyes such as EB and PI is sensitive to changes in the secondary structure of DNA. Fluorescence is lost on denaturation by heat or on loss of structure consequent to enzymic degradation. The loss of the characteristic yellow-green fluorescence of acridine orange-stained DNA, manifested by a shift in the ratio of red to yellow fluorescence, following enzymic degradation or heating was described by Alvarez (68), Campbell and Gledhill (69) and Darzynkiewicz et al. (70). Thermal denaturation can be studied by the use of such dyes to stain DNA after prior treatment, as above, or the progress of denaturation can be monitored by determining loss of fluorescence from previously stained cells during the treatment leading to the change in DNA structure. The latter approach is more direct, although it is subject to the effect of the dye in stabilizing the double-stranded DNA.

Figure 2 shows loss of PI-induced fluorescence of Ehrlich ascites cells on heating the stained cells. These cells were heated in glycerolated low-salt solutions, rapidly quenched, diluted in water, and analyzed. The effect of the dye in stabilizing the DNA is considerable. Most of the melting occurs above 100°C. Half the fluorescence remains even above 120°C. Formamide destabilizes the chromatin. Its immediate effect is loss of about 80% of the fluorescence at low temperature, followed by more typical melting between 60° and 80°C. The initial loss is characteristic of the protein-associated DNA of chromatin, and is not seen on addition of formamide to PI-stained free DNA.

d. DNA Staining by Mithramycin. The fluorescent antitumor antibiotic mithramycin provides another rapid and convenient method for FCM analysis (64). Staining is accomplished in a one-step procedure. Unfixed or

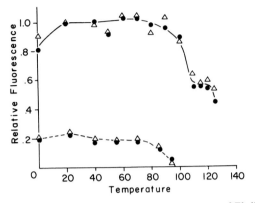

F<small>IG</small>. 2. Loss of propidium diiodide-induced (*PI*) fluorescence of Ehrlich ascites cells on heating. The cells were stained, heated in 1/100 × SSC in 20% glycerol, quenched in ice water, diluted, and analyzed. The △'s and ●'s represent peak values of G_1 and G_2 cells relative to the fluorescence at room temperature, used as the standard condition. The lower curve shows fluorescence after treatment with 40% formamide, then heating in this medium, relative to the same standard conditions. (Profile of the PI-stained cells is shown in Fig. 8.)

ethanol-fixed cells are treated for 20 minutes with a staining solution containing 20% ethanol–15 mM MgCl$_2$–100 μg of mithramycin per milliliter. Lower concentrations of dye can be used, but the staining time is increased. The coefficients of variation are governed by the combined effects of dye concentration and processing time. Two related antibiotics, chromomycin A$_3$ and olivomycin, have also been successfully employed for specific DNA staining (*65*). Mithramycin has also been used in combination with EB, yielding very low coefficients of variation (*65a*).

2. S<small>TAINING OF</small> RNA

DNA and RNA can be stained with acridine orange (AO), giving metachromatic fluorescence which provides the basis for discriminating between the two. AO intercalates into the DNA, giving a yellow-green fluorescence, and stacks along single-stranded RNA, yielding a red fluorescent complex (*55*) (Table III). The following staining protocol is from Braunstein *et al.* (*71*), who used AO for two-parameter analysis of lymphocytes.

The cells are suspended in phosphate-buffered saline, centrifuged, resuspended in a small volume of the saline, and mixed with 9 volumes of 1:1 ethanol–acetone. Fixation is for 3–24 hours. The fixed cells are centrifuged, resuspended in a small volume of 10^{-5} M EDTA–10^{-5} M NaH$_2$PO$_4$ (pH 6.0) to denature the RNA, and an equal volume of 3 × 10^{-5} M AO in phosphate-buffered saline is added. The cells are then analyzed, using a 530 nm and a 660 nm interference filter for DNA and RNA fluorescence, respectively.

3. General Staining of Proteins

Two fluorescent stains that have been used for protein staining are fluorescein isothiocyanate (FITC) (*11*) and 1-dimethylaminonaphthalene 5-sulfonyl chloride (DANS) (*5*). Both were used in conjunction with intercalating dyes as double stains for DNA and protein. For FITC staining, cells are fixed in cold 70% ethanol for 30 minutes rinsed in water (or prestained with the intercalating dye after removal of RNA—see above), then stained for 30 minutes in the cold with FITC (0.03 mg/ml)–0.5 *M* sodium carbonate. The cells are rinsed in water and analyzed. For DANS, the cells are fixed at −30°C in 96% ethanol, then stained for 30 minutes at room temperature in 1 part per 10,000 of the fluorochrome in acetone, then rinsed (or double stained for DNA) and analyzed.

Eosin Y can also be used as a general protein stain by staining at pH lower than that required for histone specificity (*72*; see below).

4. Staining of Histones

Eosin Y has been used in absorption microspectrophotometry to measure the histone content at nuclear and chromosomal levels (*73*). A modified method for fluorescence measurement by FCM uses staining at pH of over 10. The requirement for the higher pH for staining specificity might be due to the more sensitive and spatially undiscriminating FCM (*72*).

The following procedure has been used successfully for Ehrlich ascites cells, Chinese hamster ovary (cultured), Chinese hamster testis, and rat testis cells.

1. Fix the dispersed cells in cold 10% buffered formalin overnight.
2. Centrifuge the fixed sample at 500 *g* for 5 minutes.
3. Wash the dispersed pellet twice in distilled water at room temperature.
4. Hydrolyze in 5% trichloroacetic acid at 95°C for 25 minutes.
5. Rinse with four changes, 25 minutes each, of 70% ethanol.
6. Stain 70 minutes in eosin Y at room temperature. Stain is freshly prepared by mixing 83 ml of distilled water, 0.726 gm of Tris crystals, and 86 mg of the dye.
7. Differentiate the stained cells with three changes, 3 minutes each, of freshly prepared 50 m*M* Tris buffer (pH 10.25), 0°–4°C.
8. Differentiate further in two changes of cold 70% ethanol, 1 minute each.
9. Resuspend cells in cold 70% ethanol, made pH 3.8 HCl, then let stand for 5 minutes at room temperature.
10. Sonicate to disperse cells only if necessary.

Measurements should be made within 1 hour. The resolution and staining intensity decrease with time. Nonspecific staining can often be corrected by repeating the differentiation cycle.

Eosin Y staining of Ehrlich ascites cells shows yellow-green fluorescence under the UV illumination of the fluorescence microscope. Fluorescence is primarily in the nucleus, but some is also seen in the cytoplasm. Laser-activated fluorescence, at 488 nm, appears to be more specific than UV-activated. The eosin fluorescence distribution of ascites cells is shown in Fig. 8. The G_2 peak is twice that of the G_1. The flattened valley prevalent in DNA-stained cells is not apparent due to the higher C.V. for histone fluorescence (72).

Double staining of DNA and histone for two parameter analysis has given promising results with rat testis cells. The optimal conditions may not yet have been achieved, but the procedure is useful for resolving cells at different stages of spermatogenesis (Fig. 9).

5. STAINING OF NUCLEAR ANTIGENS

The potential of FMF analysis of nuclear antigens has hardly been tapped. Nuclei and chromosomes can be fluorescent stained, directly or indirectly, using antibodies against a number of nuclear constituents. Antibodies against adenosine and cytosine (74) bind, respectively, to AT-rich (75) and GC-rich sequences of denatured DNA (76). Antibodies against double- and single-stranded DNA, against DNA (reviewed by Stoller, 77), and against bromodeoxyuridine-containing DNA (78) have been obtained. Antibodies against histones ($79,80$), virus-specific cell-coded T antigens and other viral antigens (81), tumor antigens (81), nonhistone chromosomal proteins (82), lupus erythematosis serum (83), and serum from cancer patients (81), among others, have been used in fluorescent microscopy studies of nuclei and chromatin. The methods of labeling antibodies and conjugating with antigenic proteins are routine, and should be applicable to FCM analysis with little modification.

FCM analysis using fluorescent-labeled proteins has been carried out with fluorescein-labeled concanavalin A (11) and labeled serum from animals infected with hog cholera virus, a cytoplasmic RNA virus (84). Both the above were demonstration experiments, the Con A being used in conjunction with PI staining of DNA in a two-parameter analysis. However, little work has been done by way of FCM analysis of nuclear antigens.

6. STAINING OF ISOLATED CHROMOSOMES

Various modifications of the following procedure, from Wray and Stubblefield (85), have been used in the FCM analysis of individual chromosomes (20).

Suspend mitotic cells (obtained by Colcemid enrichment and selection by trypsinization) in cold buffer consisting of 1.0 M hexylene glycol (2-

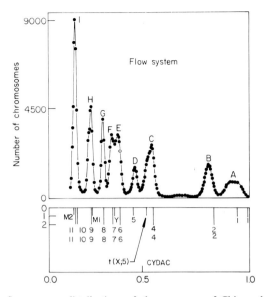

Fɪɢ. 3.   DNA fluorescence distributions of chromosomes of Chinese hamster M3-1 line cultured cells. Upper panel: A computer-fitted curve connecting experimental points. Lower panel: The amount of DNA in individual chromosomes as determined by image scanning of squashed preparations with the CYDAC. The positions of the bars indicates DNA amount, and lengths; the number 1 chromosomes show a difference in DNA content of the two homologs. t(X;5) represents a translocation. From Gray *et al.* (22).

methyl-2, 4-pentanediol)–0.5 mM CaCl$_2$–1.0 mM piperazine N,N'-bis-2-ethanesulfonic acid monosodium salt (PIPES) (pH 6.5). Incubate for 10 minutes at 37°C. Shear cells by extruding several times through a 22-gauge hypodermic needle (85). Carrano (20) carries out the staining by adding 10^{-4} M ethidium bromide to the cell suspension, using 0.075 M KCl in PIPES buffer to swell the cells. The final concentration of chromosomes in the suspension is about 2 × 10^7 per milliliter. The results of analysis of isolated chromosomes are shown in Fig. 3.

III.   Measurement

The following considerations are important in measurement: alignment of the instrument; choice of exciting wavelengths and barrier filters to give optimum response in terms of signal to noise; the flow rate and cell concentration; the refractive index of the medium under certain circumstances;

and overall sensitivity of the instrument; see Mullaney *et al.* (*7*) for further discussion.

A.   Alignment, Centration, Focusing

Most of the conditions enumerated below are applicable to the flow-cytometer developed by the Los Alamos group, as described by Kraemer *et al.* (*8*); but will apply generally to other instruments. Figure 4 summarizes

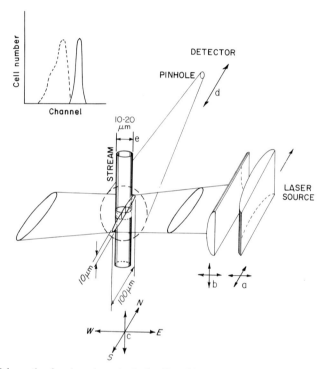

FIG. 4.   Schematic showing the principal adjustable components of a typical flow-cyto-meter and the effects of the adjustments. (The drawing is not to scale.) The dimensionless sets of arrows indicate movements of the vertical cylindrical lens.(a), horizontal cylindrical lens (b), flow cell (c), and pinhole (d). Movement of a and b in an east-west direction focus the laser beam. Movement of a in a north-south direction, and of b in an up-down direction position the beam. Movement of the flow cell in an east-west direction positions the stream with respect to the pinhole; in a north-south direction, it positions the stream with respect to the beam. Movement of the pinhole brings it into the plane of focus of the stream and beam. Change in the relative hydrostatic pressures of the sample and sheath reservoirs (not shown) change the diameter of the sample stream (e). The inset figure shows the decrease in the channel number and left-hand skewing of the profile when the stream is off-center in a north-south direction.

these conditions. Alignment requires the optical axes of the laser and detector to be 90° to one another and to lie in the same horizontal plane. The positions of these two elements are usually fixed, once established, and require no additional adjustment. The flow cell, fixed in orientation but movable along three mutually perpendicular axes, is positioned at the intersection of the optic axes of the laser beam and the detector. Two cylindrical lenses, one vertical and the other horizontal, are interposed between the laser and flow cell. These are positioned to give a horizontal beam elliptical in cross section, with its focus centered with respect to the flow cell.

The laser beam is visible in the flow cell because of Raman fluorescent scattering. The sample stream can be made visible when aligning the instrument by running through a dilute solution of fluorescein.

A pinhole in front of the photomultiplier, in the detector housing, is illuminated by positioning a small light behind it. Repositioning the flow cell, and a slight vertical adjustment of the horizontal cylindrical lens, if necessary, can bring the beam and the center of the flow cell, and the pinhole, to the same horizontal plane. Moving the flow cell in a direction parallel to the laser beam will bring the intersection of the stream and beam to the center of the illuminated pinhole. Horizontal positioning of the flow cell in a direction perpendicular to the laser beam to give maximum intensity at the intersection of the stream and beam, gives approximate centration in this direction. When the intersection is brought into sharp focus with the viewing microscope, it will appear as a sharp bright line, whose ends mark the width of the sample stream. The viewing microscope is focused on the pinhole, and the flow cell is moved in and out so that the intersection is in sharp focus. If necessary, the vertical cylindrical lens is also moved laterally to illuminate the stream. If, because of parallax, the intersection appears in two different regions of the disk of the pinhole as viewed through the two eyepieces of the microscope, the pinhole needs to be brought toward or away from the viewer. Changing the position of the pinhole, followed by compensatory movements of the flow cell and beam, (north and south in Fig. 4) will either improve the situation or make it worse. When the stream and beam occupy the same central position in the disk of the pinhole, as viewed with both eyes, the proper centration has been attained.

Figure 4 shows the relationship between the flow cell, stream, beam, and pinhole, the directions of the centerable and focusable elements, and their effects.

An old sample of acriflavine–Feulgen or PI stained cells provides a handy test system for further alignment and tuning. The original distribution profile of such cells is maintained for long periods with no evidence of deterioration if the cells are kept refrigerated and in the dark. The stream carrying the cells is usually about 20–40 μm wide. The lens-shaped cross section of the beam is about 100 μm wide in a horizontal direction and 10 μm high. The stream can be moved in and out in a horizontal direction, 90° to the laser beam, to give a maximum width of the illuminated stream. Optimal centration in this direction will give maximum response on analysis (higher channel numbers at peak position). A symptom of off-center alignment is a fluorescence distribution profile that has a shoulder, or is skewed to the lower channels, because of the passage of some cells through the narrower regions of the lens-shaped beam.

The width of the stream is varied by changing the relative rate of flow of the sample and sheath stream. Flow is regulated by vacuum or pressure, the former in the case of the instrument being described, combined with gravity feed. The relative rates of the sheath and stream flows are governed by the relative heights of the sample and sheath reservoirs. During long runs, automatic feed-in devices can be used to maintain these levels. The sample stream should be small enough so that it does not extend into the regions of the lens-shaped beam, where some cells would receive less illumination. The disadvantage of a stream that is too narrow is the decreased flow rate and long times necessary to accumulate data, especially from dilute suspensions of cells. Large streams are more likely to give coincidence, two or more cells traversing the beam at the same time.

Coincidence and adherent cells show up as peaks at the "4N," "6N," etc., positions, in a diminishing series of peak heights (e.g., Figs. 6 and 7). Whether this artifact is due to coincidence or cell adherence can be determined by diluting the sample. If the spurious peaks do not disappear, gentle shearing by forcing the suspension up and down several times through a transfer pipette while holding the end against the bottom of the sample beaker should disperse the cells. Light sonication should serve this purpose if the shearing does not work. When adherence is the result of packing during centrifugation, these gentle means of dispersion usually suffice.

Once the optimum tuning of the instrument has been attained at this level, uniform fluorescent spheres can be used to test its performance. These spheres can be obtained from Becton-Dickinson, Mountain view, Ca. Goals to strive for are high sensitivity, proportionality, a normal distribution, and a low coefficient of variation (C.V.). Achievement of the last usually ensures the others. A C.V. of 1.4% using the spheres may be attained, 2% being a more typical lower figure. The C.V. can be approximated quickly using the following formula (*29*)

$$\text{C.V.} = \frac{\text{width in channels} \times 0.426}{\text{peak channel number}} \times 100\%$$

B.   Choice of Exciting Wavelengths and Detection Wavelengths

In general, the lower and voltage required for detection, the lower the C.V. Given alignment and tuning to give the maximum response, selection of exciting wavelengths (the choices limited by the characteristics of the laser) and of barrier filters is made to give optimum signal-to-noise ratio. Ordinarily, this means maximum responsiveness of the instrument, i.e., detection at low sensitivity. Compromises are necessitated, especially in multiparameter analysis, where detection of two signals whose energies overlap may require the use of filters that pass wavelengths off the maximum of the emission spectra, and whose activation by a single-wavelength exciting light might not yield peak efficiency for either of the signals.

Table III shows recommended exciting wavelengths and barrier filters for various fluorochromes and conditions of measurement.

C.   Choice of Refractive Index of Ambient Media

Where results indicate heterogeneity of a population that is expected to consist of a homogeneous population of cells, and error due to refractility is suspected, a run should be made in a medium whose refractive index matches that of the cell. This requires permeating cells of the sample and substituting the entire flow system with the desired media (see Section IV).

IV.   Sources of Error

Given a properly tuned and aligned instrument and conditions of staining that produce low coefficients of variation and a typical profile, remaining sources of error have their origins in the physical characteristics of the cells themselves and of the instrument. These include cell shape, refractive index, and degree of condensation. Birefringence might also have an effect, since the laser light is polarized. The first three have all been encountered in flattened mammalian sperm cells. The effects of the last have not yet been demonstrated, and, in fact, no affect was seen to occur even where birefringence does exist—although FCM has been used to study anisotropy (13).

A.   Effects of Shape, Orientation, and Refractility in Sperm Cells

Flat sperm give a distribution with a sharp peak at the lower range of values and an extended tail ranging in values to over twice that of the main

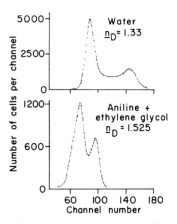

FIG. 5. Effects of changing refractive index on the fluorescence profile of flattened mammalian sperm. Upper curve: Acriflavine–Feulgen-stained boar sperm measured in aqueous medium. Lower curve: Sample run in an ethylene glycol–aniline mixture. From Gledhill *et al.* (*29*).

peak in some cases (*29*). The tail of the curve comprises roughly one half of the cells (Fig. 5). The higher values are due to the combined effects of orientation and refractility. Cell sorting has shown that both the peak cells and the tail cells regenerate the original profile when rerun after sorting. If the refractive index of the medium containing the sperm is increased, by substituting a mixture of aniline and ethylene glycol,[5] the tail retracts toward the peak as the refractive index approaches that of the dehydrated cell. Gledhill *et al.* (*29*) found that neither round nor cylindrical sperm show an orientation effect and that the effect with the flat sperm could be simulated in a static system by measuring the fluorescence of sperm whose profiles are variously oriented toward the detector. They concluded that the tail is due to the combined effects of the cells displaying a broad surface to the light source, giving maximum collecting efficiency, with a "piping" of light to the detector whenever the sperm present an edge to the detector. Matching the refractive index diminishes this effect, thus both refractility and orientation (hence shape) of the sperm contribute to the error.

[5]The matching medium is made by adding aniline to ethylene glycol to give a mixture that yields least contrast of cells, when examined with a phase microscope. The refractive index will usually range around 1.54 to 1.56, depending on cell type, fixation, and staining treatment. (The aniline should be almost colorless. Freshly distilled aniline can be stabilized by adding Zn powder after distillation.) A ratio of aniline to ethylene glycol of 3.4 gave a workable solution for chicken red blood cells. This medium is used to carry the sample and also the sheath stream. Substitution for the aqueous medium is facilitated by first running 95% ethanol through the entire system, to remove water, followed by the refractive index medium. The entire process of substitution may take over an hour, the test being disappearance of visible convection currents in the flow cell.

B. Effect of Refractility in Red Blood Cells

An interactive effect of refractility, shape, and orientation is seen on measuring acriflavin-Feulgen-stained mature red blood cells of chickens. These effects are superimposed upon a condensation effect (see Section IV, C) and all depend upon the maturity of the developing cell (73a). A WI-96 hybrid strain whose blood contains some circulating reticulocytes gives a fluorescence profile consisting of a sharp peak and a shoulder or small resolvable peak on the right. The value of the main peak is lower than that of reticulocytes or lymphocytes, is thought to reflect the effects of condensation, and can be at least partially countered by analyzing the sample in 1% SDS, which swells the cells to about 8 times their volume without changing their shape. The shoulder, as is the case with sperm cells, is due to cytoplasmic refractility, cell shape, and orientation (see Section V, Fig. 9).

The shoulder can be eliminated by increasing the refractive index of the medium with nonaqueous solutions (Fig. 6). Since the flattened reticulocytes and the SDS-swelled cells, also flat, show no or decreased shoulder, respectively, shape and orientation seem to be of consequence only where the influence of refractility passes some threshold level. The shoulder is also missing on analysis of isolated nuclei or lysed ghosts.

Sorting on the basis of fluorescence and reanalysis of cells gated on the shoulder and main peak showed that the two populations had similar ranges, but enrichment and depletion of the higher values in the cells from the shoulder and main peak, respectively (73a). Thus the shoulder is thought to be due to the presence in the population of two groups of cells. The smaller group, reticulocytes, is of lower refractility, and exhibits fluorescence values near the shoulder of the main population. The main group, mature erythrocytes, shows an overall decreased fluorescence due to condensation and the shoulder due to refractility and edge-to-detector orientation of some of the cells (73a).

Reported differences in DNA content attributed to differences in fluorescence, in the case of sperm cells (88) and chicken red blood cells (86), may have been the result of these effects.

C. Effects of Chromatin Condensation

Condensation can cause an underestimate of the amounts of stainable material. Highly condensed sperm cells show 60% of the fluorescence of early spermatids (29). SDS and DTT treatment of the cells between hydrolysis and staining causes increase in fluorescence staining by the sperm (29). The condensed nuclei of chicken erythrocyte ghosts show about 40% of the acriflavine–Feulgen fluorescence of their *in situ* counterpart (Fig. 7). Table

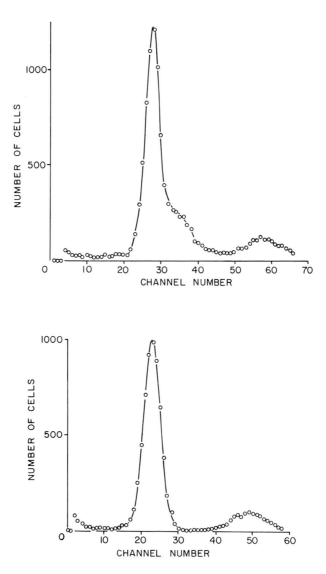

FIG. 6. Effect of refractive index on fluorescence profile of acriflavine–Feulgen-stained chicken red blood cells. The conditions are similar to those for Fig. 5. The upper curve is for cells run in an aqueous medium. The right-hand shoulder belongs to cells whose edge-to-detector orientation gives a higher signal than that of cells oriented face to detector, and also a minority population of circulating reticulocytes, not so subject to the effects of nuclear condensation on depressing fluorescence. The lower curve was obtained by running the cells in a mixture of aniline-ethylene glycol, whose refractive index more closely matched that of the dehydrated fixed cells. From Bloch *et al.* (*73a*).

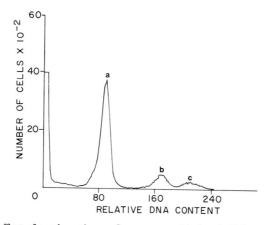

FIG. 7. The effect of condensation on fluorescence of isolated chicken erythrocyte nuclei. Acriflavine–Feulgen stain. The three peaks a, b, and c represent nuclei (or ghosts) (a), doublets or two adherent nuclei (b), and intact erythrocytes (c). The nuclei removed from the cell are very small and compact (73a).

IV shows the effects of a variety of treatments on the fluorescence of chicken erythrocytes and their nuclei (89–91).

The detergent "loosening" treatment also resulted in increased acriflavine–Feulgen fluorescence, but the values did not reach those of the intact cells. The effects of extreme condensation on staining of chromatin are well known in microspectrophotometry (92), and it is not surprising that they play a role in FCM as well.

TABLE IV

EFFECT OF VARIOUS TREATMENTS PRIOR TO FIXATION, ON FLUORESCENCE OF FORMALIN-FIXED, ACRIFLAVINE-FEULGEN STAINED CHICKEN ERYTHROCYTES, GHOSTS, AND NUCLEI

Treatment	State of stained cell	Fluorescence
None	Intact and undeformed	46.0 ± 0.04
SMTOG (89)[a]	Intact and undeformed	34.3 ± 0.07
SMTOG + Triton X-100	No cytoplasm, collapsed membrane, condensed nucleus	26.8 ± 0.06
Hanks' solution	Intact cell, irregular membrane	37.3 ± 0.04
NaCl–EDTA (89)	Intact cell, irregular membrane	36.6 ± 0.04
0.1 × SSC	Ghosts, dispersed chromatin	33.0 ± 0.08
0.01% NP-40 (90)	Intact polygonal cells	44.4 ± 0.05
0.1% NP-40 (90)	No cytoplasm, slightly collapsed ghosts, nuclei condensed and refractile	22.0 ± 0.04
0.1% Triton X-100 (91)	Collapsed ghosts, small nuclei	43.1 ± 0.04

[a] SMTOG: sucrose, Mg^{2+}, Tris, octanol, gum arabic.

The contribution of these effects toward the observed variation among cells in a population, and among different cell populations, is unknown. It is interesting in this connection that the C.V. observed using PI staining is often lower than that using acriflavine–Feulgen, in spite of the higher potentiality of the latter for stoichiometry and quantitation. The 20-minute hydrolysis routinely used by acriflavine–Feulgen staining may not be optimum. Rasch and Rasch (*93*) have described conditions of hydrolysis for minimizing the differences in response of different types of chromatin to staining conditions. Duijndam and Van Duijn (*54*) indicated that unassociated and protein-associated DNA are hydrolyzed similarly at much longer times than 20 minutes. Much of the variations among cells is due to differences in staining that are presently beyond our understanding. Present attempts to extract S-phase curves from DNA distribution profiles of growing cells, by subtracting artificially regenerated normal G_1 and G_2 distributions from the profile (see Section V) are instructive, but their validity rests on the elimination or assessment of errors that may affect G_1, G_0, S, G_2, and M cells differently.

D.   Instrumental Instability

Most FMF analyses have been for the purpose of determining the shapes of fluorescence distributions, and amounts are expressed in relative terms. Comparison of different samples run in succession requires that the stability of the light source and machine sensitivity be monitored. This can be done by including an internal standard to be used as a basis for comparison during different runs. A standard whose DNA content is known, stained in parallel with the samples being studied, can also provide a basis for expressing results in absolute terms.

The fluorescent microspheres mentioned above (Section III,A), or stored stained cells, fulfill the function of monitoring machine stability. Freshly stained and fixed standards, such as chicken erythrocytes, can provide the second function as well.

V.   Treatment of Data

The accumulation of large amounts of data with little effort permits analysis at a level that was unapproachable with older methods, where a typical experiment dealt with numbers of cells three or four orders of magnitude lower.

A. Analysis of Residuals

Obtaining the best fit for an observed curve and a theoretical normal distribution whose mean and spread are equal to that of the observed, results in the generation of "residuals" or departures from the theoretical. If these are randomly dispersed along the curve, the points can be assumed to have a normal distribution. If they are clustered, there probably exist two or more subpopulations with overlapping characteristics, and the fit can be improved by generating two normal overlapping curves (Moore *17*; Gray *et al., 21*). This method has been used effectively in the karyotyping of cells by FCM analysis of isolated chromosomes (*20*).

B. Cell Cycle Analysis

An interesting application of FCM to the study of rates of synthesis of DNA during the S period was done by Dean and Jett (*14*), in which the distribution of DNA amounts among S-period cells was determined by extracting artificially generated G_1 and G_2 distributions from the total profile, leaving the S period profiles. The G_1 and G_2 curves were approximated by fitting data, using least mean squares, to the sum of a series of normal curves of similar C.V.'s, representing cells in different intervals of the S period. The treatment rested on these assumptions: (a) normal distributions of fluorescence within the groups; (b) similar C.V.'s among the groups; (c) random disposition of cells among equal temporal intervals of the cell cycle; (d) similar characteristics of G_1 and G_0 phases, and G_2 and M stages. The results show a larger number of cells in the early and late part of the S period than in the middle. The tarrying of cells here indicates a lower synthetic rate during these subperiods of the S phase (*16*). The results accord with those using other approaches and suggest a gradual increase in the number of initiations during the first part of the S period, initiations equaling terminations during the middle part, and decreasing initiation during the later period. Figure 8 shows this treatment applied to Ehrlich ascites cells stained with PI and acriflavine–Feulgen for DNA and eosin Y for histone. Note the upward convex profile of the eosin-stained histone, indicating slowest histone accumulation in the nucleus during the S midperiod. The three analyses were carried out on different preparations, and the results are shown only for the purpose of demonstrating the value of the approach. The difference in the shapes of the DNA and histone S-period profiles is thought to be significant. However, a more precise correlation of the two events, increase in DNA and histone fluorescent staining, will have to await the improvement of double staining methods that would permit detection of both substances simultaneously in the same individual cells.

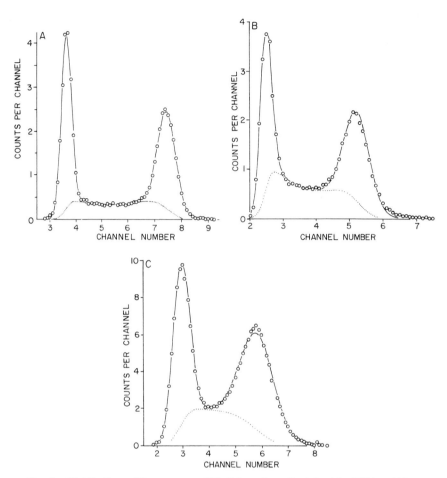

FIG. 8.   Distributions of fluorescence of Ehrlich ascites cells stained for DNA and histone, and their derived S-period profiles. The data are from Fu and Bloch (*72*). The S-period profiles, shown by the dotted lines, were determined by P. N. Dean. (A) PI stain; (B) acriflavine–Feulgen stain; (C) eosin Y stain.

C.   Multiparameter Analysis

Independent analysis of two parameters can be used to generate a two-dimensional plot, permitting correlation of the two, or a three-dimensional plot, using the x and y axes for the variables and the z axis for the number of cells, in a two-dimensional histogram. Figure 9 shows different perspectives of histograms obtained by simultaneous measurement of propidium iodide and eosin Y fluorescence of DNA and histone of doubly stained rat testis cells. These examples were referred to in Sections II and III.

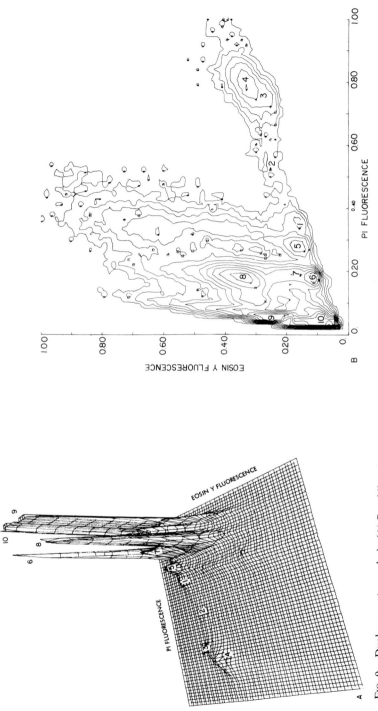

FIG. 9. Dual parameter analysis. (A) Propidium iodide fluorescence of DNA and eosin Y fluorescence of histone in doubly stained rat testes cells. The two fluorescence signals are separated by appropriate filters. This method resolves premeiotic G-1, G-2, and S stages (groups 1, 2, 3, 4) and five different spermatid—sperm stages (6, 7, 8, 9, 10). Notable are the apparent lag in premeiotic histone synthesis (group 3, 4), the histone transitions in the intermediate spermatid (group 8), and the typical lowering of the signals during sperm maturation (9, 10). (B) Contour view of (A). From Fu and Bloch, unpublished.

The program, PERSPEX (*87*), permits the perspective viewing of the plot from varying distances, angles of elevation above the x, y plane, and angles or rotation about a central vertical axis. It also incorporates "hidden line elimination" (renders the contorted plane opaque) and the drawing of contour lines at desired elevations above the x, y plane.

VI.   Future Directions

"Tomorrow is already here"—Robert Jungk

The main virtues of the flow-cytometer are speed and numbers. Its accuracy is probably not very different than that of the older two-wavelength method of microspectrophotometry (*94,95*) or of scanning methods (*96*). One drawback is lack of identification of the dispersed cells, but such identification means relating the measurement to other parameters, such as size, shape, location, amount of radioactivity, and this consideration is becoming less important with improved effectiveness of multiparameter analysis and cell sorting.

Multiparameter analysis, which can at present relate two-color fluorescence, low-angle and wide-angle scattering, Coulter volume, and fluorescence anisotropy, in any combination, can give information on, say, DNA, protein, size, shape, and orientation of each cell. Sorting permits recovery of cells with desired parameters, making possible more direct observation and identification on the basis of radioactivity or fine structure. Prior selection of cells for separate FCM analysis using other means, such as centrifugal elutriation (*97,98*) or unit sedimentation velocity (*99*), can accomplish a similar purpose.

Sorting by itself, at the present state of the art, will not give sufficient material for analysis at levels grosser than the cytochemical or perhaps the immunochemical. A regimen separating out 10% of a cell population, at the rate of 3×10^6 total cells per hour, will yield 3×10^6 cells per hour $\times 6 \times 10^{-6}$ μg of DNA per cell $= 18$ μg of DNA per hour, a range that lies within the limits of many of the more sensitive methods of biochemical analysis and provides many more cells than needed for routine cytological and cytochemical tests.

At present, chromosomal idiograms based on size (DNA content) are possible in organisms such as the muntjak and Chinese hamster (*21*), whose relatively few chromosomes have distinct differences in size. Human chromosomes are resolved into groups, consisting of chromosomes of similar sizes. Future karyotyping should take advantage of improvements in performance of the instruments and in refinements, such as orienting chromosomes in very thin streams, centromere marking, taking advantage of the affinity of centromeric regions for spindle fiber proteins (*100*), slit-scanning the lengths of the chromosomes as they flow by the detector (*101,102*) resulting in profiles of individual chromosomes, including positions of their centromeres.

The flow-cell and sorting process is compatible with cell life, and the development of harmless vital staining regimes should permit separation and cloning of cells with selected characteristics (*13,18,103*).

The practical use of the instrument as a diagnostic tool will depend on improvements in methods of cell dissociation and the development of a greater battery of discriminating cytochemical and immunochemical tests.

The flow-cytometer, cell sorter, and computerized systems for data analysis make an impressive array that leaves one with the feeling that "tomorrow is already here," or perhaps is already behind us. Several prognostications for future use have been vividly described (see Table I), and the reader is referred to these for further insights.

ACKNOWLEDGMENTS

The authors gratefully acknowledge the advice and help of Mr. Reuben Mitchell in preparing the computer programs used in the three-dimensional analysis. We thank Dr. Philip N. Dean, of the Lawrence Livermore Laboratories, for generously providing the analysis of the stage frequencies of the DNA and histone fluorescence profiles. We are grateful to Dr. Harry Crissman of the Los Alamos Scientific Laboratories for his help and tutelage in the experiment on the isolated nuclei shown in Fig. 1.

Much of this work was supported by Grant No. GM 09654 from the National Institutes of Health and from a Bio-Medical Research Support Grant from the N.I.H. to the University of Texas.

REFERENCES

1. Kamentsky, L. A., Melamed, H. R., and Derman, H. *Science* **150**, 630. (1965).
2. Mullaney, P. F., Van Dilla, M. A., Coulter, J. R., and Dean, P. N. *Rev. Sci. Instrum.* **40**, 1029. (1969).
3. Dittrich, W., and Göhde, W. *Z. Naturforsch., Teil B* **24**, 360. (1969).
4. Van Dilla, M. A., Trujillo, T. T., Mullaney, P. F., and Coulter, J. R. *Science* **163**, 1213. (1969).
5. Göhde, W., and Dittrich, W. *Z. anal. Chem.* **252**, 328. (1970).
6. Sprenger, E., Böhm, N., and Sandritter, W. *Histochemie* **26**, 238. (1971).
7. Mullaney, P. F., Steinkamp, J. A., Crissman, H. A., Cram, L. S., and Holm, D. M. *In* "Laser Applications in Medicine and Biology" (M. L. Wolbarsht, ed.), Vol. 2, p. 151. Plenum, New York. (1974).

8. Kraemer, P. M., Deaven, L. L., Crissman, H. A., and Van Dilla, M. A. *Adv. Cell Mol. Biol.* **2**, 47. (1972).
9. Crissman, H. A., Mullaney, P. F., and Steinkamp, J. A. *In* "Methods in Cell Biology" (D. M. Prescott, ed.), Vol. 9, p. 179. Academic Press, New York. (1975).
10. Melamed, M., Mullaney, P., and Mendelsohn, M. "Flow Cytometry and Sorting." (In preparation) (1976).
11. Crissman, H. A., and Steinkamp, J. A. *J. Cell Biol.* **59**, 766. (1973).
11a. Stöhr, M., and Petrova, L. Histochemie **45**, 95 (1975).
12. Steinkamp, J. A., Hansen, K. M., and Crissman, H. A. *J. Histochem. Cytochem.* **24**, 292 (1976).
13. Arndt-Jovin, D. J., Ostertag, W., Eisen, H., Klimek, F., and Jovin, T. M. *J. Histochem. Cytochem.* **24**, 332. (1976).
14. Dean, P. N., and Jett, J. H. *J. Cell Biol.* **60**, 523. (1974).
15. Gray, J. W. *J. Histochem. Cytochem.* **22**, 642 (1974).
15a. Freed, J. *Comput. Biomed. Res.* **9**, 263 (1976).
16. Dean, P. N., and Anderson, E. C. *In* "Pulse-Cytophotometry," pp. 77–86. Medikon, Ghent (1975).
17. Moore, D. H., II "Use of Residuals in Fitting Normal (Gaussian) Distributions," Publ. UCRL 76507. University of California, Berkeley. (1975).
18. Julius, M. H., Masuda, T., and Herzenberg, L. A. *Proc. Natl. Acad. Sci. U.S.A.* **69**, 1934 (1972).
19. Steinkamp, J. A., Fulwyler, K. M., Coulter, J. R., Hiebert, R. D., Horney, J. L., and Mullaney, P. F. *Rev. Sci. Instrum.* **44**, 1301. (1973).
20. Carrano, A. V., Gray, J. W., Moore, D. H., II, Minkler, J. L., Mayall, B. H., Van Dilla, M. A., and Mendelsohn, M. L. *J. Histochem. Cytochem.* **24**, 348. (1976).
21. Gray, J. W., Carrano, A. V., Moore, D. H., II, Steinmetz, L. L., Minkler, J., Mayall, B. H., Mendelsohn, M. L., and Van Dilla, M. A. *Clin. Chem.* **21**, 1258. (1975).
22. Gray, J. W., Carrano, A. V., Steinmetz, L. L., Moore, D. H., II, Mayall, B. H., and Mendelsohn, M. L. *Proc. Natl. Acad. Sci. U.S.A.* **72**, 1231. (1975).
23. Van Dilla, M. A., Gray, J. W., Carrano, A. V., Minkler, J. L., and Steinmetz, L. L. *Proc. Symp. Pulse-Cytophotometry, 1975* (in press).
24. Mendelsohn, M. L. *Proc. 3rd Life Sci. Symp.* (E. Z. Anderson and E. M. Sullivan, eds.), p 95. N.T.I.S., Springfield, Va. (1976).
25. Earle, W. R., and Schilling, E. L. *J. Natl. Cancer Inst.* **4**, 165 (1943).
26. Beutler, E. "Red Cell Metabolism." Grune & Stratton, New York. (1971).
27. Hanks, J. H. *J. Cell. Comp. Physiol.* **31**, 235. (1948).
28. Cameron, G. "Tissue Culture Technique." Academic Press, New York. (1950).
29. Gledhill, B. L., Lake, S., Steinmetz, L. L., Gray, J. W., Crawford, J. R., Dean, P. N., and Van Dilla, M. A. *J. Cell. Physiol.* **87**, 367. (1976).
30. Parry, J. S., Cleary, B. K., Williams, A. R., and Evans, D. M. *Acta Cytol.* **15**, 163. (1971).
31. Husain, O. A. N., Allen, R. W. B., Hawkins, E. J., and Taylor, J. *J. Histochem. Cytochem.* **22**, 678 (1974).
32. Wheless, L. L., and Onderdonk, M. A. *J. Histochem. Cytochem.* **22**, 522. (1974).
33. Fowlkes, B. J., Herman, C. J., and Cassidy, M. *J. Histochem. Cytochem.* **24**, 322. (1976).
34. Youngner, J. S. *Proc. Soc. Exp. Biol. Med.* **85**, 202 (1954).
35. Cavanaugh, D. J., Berndt, W. O., and Smith, T. E. *Nature (London)* **200**, 261. (1963).
36. Rodbell, M. *J. Biol. Chem.* **239**, 375. (1964).
37. Mallette, J. M., and Anthony, A. *Exp. Cell. Res.* **41**, 642. (1966).
38. Rappaport, C., and Howze, G. B. *Proc. Soc. Exp. Biol. Med.* **121**, 1016. (1966).
39. Barnard, P. J., Weiss, L., and Ratcliffe, T. *Exp. Cell Res.* **54**, 293. (1969).

40. Pitelka, D. R., Kerkof, P. R., Gagne, H. T., Smith, S., and Abraham, S. *Exp. Cell Res.* **59**, 43 (1969).

41. Liu, T. C., and Gorski, J. *Endocrinology* **88**, 419. (1971).

42. Sayers, G., Swallow, R. I., and Giordano, N. D. *Endocrinology* **88**, 1063 (1971).

43. Steinkamp, J. A., and Crissman, H. A. *J. Histochem. Cytochem.* **22**, 616. (1974).

44. Drochmans, P., Wanson, J. C., and Mosselmans, R. *J. Cell Biol.* **66**, 1. (1975).

45. Frazier, M. E., Hadley, J. G., Andrews, T. K., and Drucker, H. *Lab. Invest.* **33**, 231. (1975).

46. Koss, L. G., Dembitzer, H. M., Herz, F., Herzig, N., Schreiber, K., and Wolley, R. B. *In* "Automation of Uterine Cancer Cytology" (G. L. Wied, G. Bahr, and P. Bartels, eds.), pp. 54–60. Univ. of Chicago Press, Chicago, Illinois. (1976).

47. Meistrich, M. L., and Trostle, P. K. *Exp Cell Res.* **92**, 231. (1975).

47a. Bellvé, A. R., Millette, C. F., Bhatnagar, Y. M. O'Brien, D. A. *J. Histochem. Cytochem* **25**, 480 (1977).

48. Romrell, L. J., Coppe, M. R., Munro, D. R., and Ito, S. *J. Cell Biol.* **65**, 429. (1975).

49. Hirsch, M. A. Thesis, University Microfilms, Ann Arbor, Michigan. (1972).

50. Mayall, B. H. *In* "Automation of Uterine Cancer Cytology" (G. L. Wied, G. Bahr, and P. Bartels eds.), pp. 61–68. Univ. of Chicago Press, Chicago, Illinois (1976).

51. Hossainy, E., Zweidler, A., and Bloch, D. P. *J. Mol. Biol.* **74**, 283. (1973).

52. Kasten, F. *Stain Technol.* **33**, 39. (1958).

53. Gill, J. E., and Jotz, M. M. *J. Histochem. Cytochem.* **22**, 470. (1974).

54. Duijndam, W. A. L., and Van Duijn, P. *J. Histochem. Cytochem.* **23**, 891. (1975).

55. Rigler, R., Jr. *Acta Physiol. Scand.* **67**, Suppl. 267, 1. (1966).

56. Böhm, N., Sprenger, E., and Sandritter, W. *In* "Fluorescence Techniques in cell Biology" (A. A. Thaer and M. Seruetz, eds.), p. 67. Springer-Verlag, Berlin and New York. (1973).

57. Porro, T. J., and Morse, H. T. *Stain Technol.* **40**, 173. (1965).

58. LePecq, J. B. *Methods Biochem. Anal.* **20**, 41. (1971).

59. Angerer, L. M., and Moudrianakis, E. N. *J. Mol. Biol.* **63**, 505. (1972).

60. LePecq, J. B., and Paoletti, A. *J. Mol. Biol.* **27**, 87. (1967).

61. Weisblum, B., and Haenssler, E. *Chromosoma* **46**, 255. (1974).

62. Latt, S. A. *Proc. Natl. Acad. Sci. U.S.A.* **70**, 3395. (1973).

63. Latt, S. A. and Stettin, G. *J. Histochem. Cytochem.* **24**, 24. (1976).

64. Crissman, H. A., and Tobey, R. A. *Science* **184**, 1297. (1974).

65. Crissman, H. A., Oka, M. S., and Steinkamp, J. A. *J. Histochem. Cytochem.* **24**, 64 (1976).

65a. Barlogie, B., Spitzer, G., Hart, J. S. Johnson, D. A., Büchner, T., Schumann, J., and Drewinko, B. *Blood* **48**, 245 (1976).

66. Hiraoka, T., and Glick, D. *Anal. Biochem.* **5**, 497. (1963).

67. Krishan, A. *J. Cell Biol.* **66**, 188. (1975).

68. Alvarez, M. R. *Cancer Res.* **33**, 786. (1973).

69. Campbell, G. LeM., and Gledhill, B. L. *Chromosoma* **42**, 385. (1973).

70. Darzynkiewicz, A., Traganos, F., Sharpless, T., and Melamed, M. R. *Exp. Cell. Res.* **90**, 411. (1975).

71. Braunstein, J. D., Good, R. A., Hansen, J. A., Sharpless, T. K., and Melamed, M. R. *J. Histochem. Cytochem* **24**, 378. (1976).

72. Fu, C. T., and Bloch, D. P. *Exp. Cell Res.* **93**, 363. (1975).

73. Bloch, D. P., and Teng, C. *J. Cell Sci.* **5**, 321. (1969).

73a. Bloch, D. P., Beaty, N., Fu, C. T., Chin, E., Smith, J., and Pipkin, J. L. Jr. Submitted (1977).

74. Erlanger, B. F., and Beiser, S. M. *Proc. Natl. Acad. Sci. U.S.A.* **52**, 68. (1964).

75. Dev, J. G., Warburton, D., Miller, O. J., Miller, D. A., Erlanger, B. F., and Beiser, S. M. *Exp. Cell Res.* **74**, 288. (1972).

76. Schreck, R. R., Warburton, D. M., Miller, O. J., Beiser, S. N., and Erlanger, B. F. *Proc. Natl. Acad. Sci. U.S.A.* **70**, 804. (1973).

77. Stoller, B. D. *In* "The Antigens" (M. Sela, ed.), Vol. 1, p. 1. Academic Press, New York. (1973).

78. Graetzner, H. G., Pollack, A., Ingram, D. J., and Leif, R. C. *J. Histochem. Cytochem.* **24**, 34 (1976).

79. Desai, L. S., Pothier, L., and Foley, G. E. *Exp. Cell Res.* **70**, 468. (1972).

80. Pothier, L., Gallagher, J. F., Wright, C. E., and Libby, P. R. *Nature (London)* **255**, 350. (1975).

81. Stenman, S., Rosenquist, M., and Ringertz, N. R. *Exp. Cell Res.* **90**, 87 (1975).

82. Silver, L. M., and Elgin, S. R. C. *Proc. Natl. Acad. Sci. U.S.A.* **73**, 423. (1976).

83. Mackay, R. R., and Burnett, F. M. "Autoimmune Diseases." Thomas, Springfield, Illinois. (1963).

84. Cram, L. S., Forslund, J. C., Horan, P. K., and Steinkamp, J. A. *In* "Automation in Microbiology and Immunology" (C. G. Heden and T. Illium, eds.), p. 47, Wiley, New York. (1975).

85. Wray, W., and Stubblefield, E. *Exp. Cell. Res.* **59**, 469. (1970).

86. Beaty, N. B., and Bloch, D. P. *J. Cell. Biol.* **63**, 18a. (1974).

87. Phillips, D. M. "Perspective Representation of Function of Two Variables, with Overlaid Contours," Publ. UTJ5-O5-CC086 University Texas, Austin (revised May 1972). University of Texas Computation Center. Identification: J5 UTEX PERSPEC). (1972).

88. Sarkar, S., Jones, O. W., and Shioura, N. *Proc. Natl. Acad. Sci. U.S.A.* **71**, 3512. (1974).

89. Walmsley, M. E., and Davies, H. G. *J. Cell. Sci.* **17**, 113. (1975).

90. Ringertz, N. R., and Bolund, L. *Exp. Cell. Res.* **55**, 205. (1969).

91. Moss, B. A., Joyce, W. G., and Ingram, V. M. *J. Biol. Chem.* **248**, 1025 (1973).

92. Kernell, A. M., Bolund, L., and Ringertz, N. R. *Exp. Cell Res.* **65**, 1. (1971).

93. Rasch, R. W., and Rasch, E. M. *J. Histochem. Cytochem.* **12**, 1053 (1973).

94. Ornstein, L. *Lab. Invest.* **1**, 250. (1952).

95. Patau, K. *Chromosoma* **5**, 341. (1952).

96. Deeley, E. M., Richards, B. M., Walker, P. M. P., and Davies, H. G. *Exp. Cell Res.* **6**, 569. (1955).

97. Glick, D. E., von Redlich, E., Juhos, E. T., and McEwen, C. R. *Exp. Cell Res.* **65**, 23. (1971).

98. Grabske, R. J., Lake, L., Gledhill, B. L., and Meistrich, M. L. *J. Cell. Physiol.* **86**, 177. (1975).

99. Lam, D. M. K., Furrer, R., and Bruce, W. R. *Proc. Natl. Acad. Sci. U.S.A.* **65**, 192. (1970).

100. McGill, M., and Brinkley, B. R. *J. Cell Biol.* **67**, 189. (1975).

101. Wheless, L. L., and Patten, S. F. *Acta Cytol.* **17**, 333. (1973).

102. Wheless, L. L., and Patten, S. F. *Acta Cytol.* **17**, 391. (1973).

103. Kwan, D., Epstein, M. B., and Norman, A. *J. Histochem. Cytochem.* **24**, 355 (1976).

Chapter 15

The Laser Microbeam as a Probe for Chromatin Structure and Function

MICHAEL W. BERNS

Developmental and Cell Biology,
University of California, Irvine,
Irvine, California

I. Introduction

One of the most intense areas of research over the past decade has centered around the structural and functional organization of chromatin. As in any major area of scientific research, the quest for understanding leads to the development of new approaches. Some of these approaches turn out to be useful and provide valuable information; others do not. The usefulness of the laser for studying basic problems in chromatin research has yet to be determined. However, there are several areas in which laser light is being used to study the control and organization of the genetic material.

The laser is a device that can provide electromagnetic radiation from the ultraviolet to the infrared region of the spectrum. The radiation produced by this instrument is monochromatic, extremely intense, and coherent. Furthermore, a laser beam can be in the form of a continuous beam over time, or it can be in the form of short "pulses" down to a few picoseconds (10^{-12} second). These unique features have led to the use of the laser in several areas of genetic research: (a) stimulated Raman spectroscopy (*1,2*), (b) diffraction-limited fluorescent microscopy (*3*), (c) selective laser microbeam irradiation (*4*).

In the first area, the intense monochromatic light is used to stimulate the emission of Raman spectra, which in turn reveal precise information about the atomic and molecular structure of the compounds of interest. Previous studies using classical radiation sources have been difficult because the Raman spectra are very weak and therefore difficult to resolve from background noise. The intensity and monochromaticity of laser light have overcome these problems. Several Raman studies have been conducted on DNA and RNA solutions (5–7), and it should not be long now before studies are conducted on living cells, extracted chromatin, and condensed chromosomes.

The second area of chromatin research where the laser is being applied is in fluorescent analysis of chromosomes (3). Currently, there are only a few laboratories working in this area, and the studies are still at a very "developmental" stage. The general approach is to focus a laser beam down to a near-diffraction limited spot (a micrometer or less) on individual chromosomes that have been treated with different fluorochromes. For example, it has been possible to stimulate fluorescence in the "quinacrine bands" of chromosomes. Preliminary measurements were made in the band and interband regions of quinacrine mustard-stained metaphase chromosomes of *Vicia faba* (3). Using this technique, it was possible to determine that within the bands the adenine–thymine (A-T) concentration was 60% of the total DNA. Furthermore, it was determined that the fluorescent waveforms contained a rapid and a slow decay component. It was suggested that the rapid decay component was due to the intercalation of the quinacrine binding to guanine residues, and the slower-decay component was due to the intercalation of the quinacrine between the A-T base pairs. Presently, the apparatus cannot give accurate measurements for out-of-band (interband) regions, and the authors are in the process of improving the resolution of the instrumentation to permit these kinds of determinations.

The third area of laser application to chromatin research is laser microbeam irradiation. Since 1969 (8), studies have been conducted in which the major approach has been to focus a laser beam down to an effective spot size of 0.25–1 μm on selected individual chromosomes of mitotic tissue culture cells. Over the past 7 years, this technique has been used to study problems on (a) the organization of chromosomes, (b) gene location and function, (c) chromosome stability, and (d) mitotic mechanisms. In this chapter, these studies will be reviewed very briefly. Emphasis will be placed on "how to do" the kinds of experiments and the special ancillary techniques that have been developed for subsequent analysis of the laser microirradiated cells.

II. Instrumentation

The laser is the critical part of the system. There are literally hundreds of laser instruments on the market ranging in price from a couple of hundred dollars to $100,000. Unfortunately, if an investigator is going to build his/her own microbeam system, one of the more expensive instruments should be considered. Because lasers are by nature monochromatic, the number of available wavelengths will be limited unless one of the more sophisticated instruments is chosen. A problem encountered by earlier investigators was the selection of the wrong laser—thus resulting in much wasted time, effort, and expense. Our work on chromosome microirradiation (9, 10) has shown that visible laser light up to about 560 nm can be used to affect chromosomes either in combination with a vital dye or alone. In addition, ultraviolet (UV) laser light down to 240 nm is quite effective for making precise chromosome lesions (Berns, unpublished). The most versatile lasers are the tunable organic-dye lasers (11). There are several types of dye lasers available. Some of these lasers are stimulated by a flashlamp, and others employ another laser as the stimulating light source. The advantage of dye lasers is the availabilty of wavelengths throughout the spectrum. A major disadvantage of the dye laser is the 10–20% variability from pulse to pulse.

For many years we have employed a pulsed argon-ion gas laser in our microbeam systems (12). These lasers provide a power output of 35 watts in a 50-μsec pulse duration. The beam can be used with the mixed argon wavelengths of 514 nm and 488 nm, or each wavelength can be used separately. In addition, it also has laser lines at 351.1 nm and 363.8 nm. Even though this laser does not provide a wide range of wavelengths, the wavelengths available are in the right region of the spectrum, and the cost is below $10,000. If an investigator does choose to build a microbeam system, the argon-ion laser may be the most inexpensive way to go. However, the limitation in available wavelengths will restrict the kinds of experiments possible.

Recently, a state-of-the-art laser microbeam system has been constructed that provides a wide range of visible and ultraviolet wavelengths (Fig. 1). The laser system employs two lasers: (a) a tunable dye laser, (b) a neodymium YAG laser. The neodymium laser emits 13 infrared laser lines which are converted by a lithium iodate crystal to the corresponding visible second harmonic wavelengths. By employing different crystal angles and different laser cavity mirrors, it is possible to obtain high-power visible output at any one of the 13-second harmonic lines (Fig. 1). It is possible to frequency-shift the 532 nm line using a KDP crystal (potassium dihydrogen phosphate) to get high laser output at 265 nm. In addition to the above manipulations, several of the visible second harmonic wavelengths are used to stimulate the dye laser. By using any of four different dyes (11) in combination with the YAG laser wavelengths, it is possible to generate a wide spectrum of visible laser wavelengths $\pm$ Å. Insertion of different KDP crystals into the path of

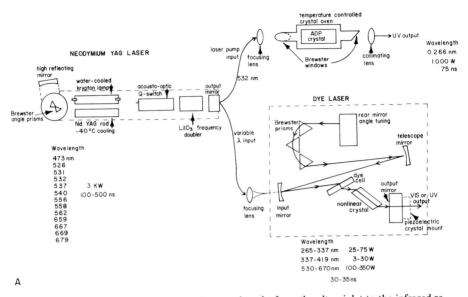

FIG. 1A. Laser system that provides wavelengths from the ultraviolet to the infrared regions of the spectrum. The three major components in this system are (a) the YAG laser, (b) the dye laser, (c) the external fixed-frequency second harmonic doubling unit. Discrete wavelengths in the range of 473–679 nm can be obtained directly from the YAG laser. High power in the ultraviolet at 266 nm can be obtained by passing the YAG 532 nm wavelength into the external fixed-frequency ADP doubling unit. A wide spectrum of ultraviolet and visible wavelengths (265–670 nm) can be obtained by using the wavelengths of the YAG laser to stimulate the dye laser.

the visible-dye laser wavelengths results in the production of the ultraviolet second harmonic wavelengths. It is therefore possible by using the YAG laser and dye laser system in combination with second harmonic generation to produce laser wavelengths from 240 nm to 700 nm. Using this system, we have already plotted the precise action spectrum of one of the chromosome responses (13,14).

The combination of the laser with the microscope is a critical step. Virtually any good microscope will be adequate. However, we have found that the microscopes with "straight-through" accessory ports are the best. This enables the beam to be directed down into the primary optical system (i.e., the objective lens) without passing through a prism or reflecting system. This minimizes internal reflection, loss of energy due to scatter and reflection, and damage produced to the reflecting surfaces. The objective selected for focusing the laser light can be varied depending upon the desired focused

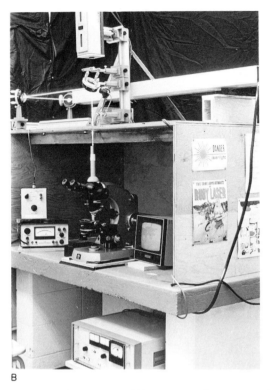

B

Fig. 1B. Actual laser microbeam system. This system employs an argon-ion gas laser. The laser beam is reflected by three mirrors; the mirror mounted above the microscope is a dichroic filter that reflects the blue-green beam into the microscope and transmits the longer wavelengths that carry the microscope image to the television camera. In the tunable laser system, a set of different dichroic filters would be employed depending upon the wavelengths used.

spot diameter. For lesion sizes below 1 μm, an $\times 100$ oil objective is used. If uv laser light is to be used, either a quartz-ultrafluar objective or a reflecting objective must be employed.

Another important feature of the microbeam system is the dichroic filter that reflects the laser beam into the microscope. If a tunable system is to be used, then a set of these filters with reflectivity of 80–90% in the desired ranges must be employed (see Table I for a typical set of filters). The filter with the appropriate reflectivity is placed above the microscope for the particular laser wavelengths employed. The filters are coated so that the microscope image at the wavelengths longer than the laser wavelength are transmitted to a television camera mounted above the dichroic filter. This permits

TABLE I

A Typical Set of Dichroic Filters for Tunable
Laser Microbeams

No.	Reflectance (nm)		Transmittance (nm)	
1	95%	240–300	80%	350–800
2	98%	300–450	60%	560–700
3	90%	430–500	60%	650–700
4	90%	530–580	70%	600–800
5	90%	560–650	70%	450–550

the experimenter to view the target structure on a television monitor and move it under a cross hair that denotes the focal point of the laser. However, it should be pointed out that the television system is not absolutely necessary. It is also possible to place a calibrated reticule in one of the microscope oculars and align the laser beam underneath it. The only disadvantage is that the specimen cannot be easily viewed during the course of irradiation.

A final component of the microbeam system is a system for recording the laser output. One of the most ignored yet most important features of microbeam experimentation is dosimetry (15). The laser must be continually monitored, either with a vacuum photodiode or a thermopile. The former device when connected to an oscilloscope permits monitoring of pulse duration, pulse shape, and peak power. The latter instrument permits direct measurement of energy. It is important to be able to make both kinds of measurements at the output end of the laser and at the focal plane of the microscope objective. Because of the cumbersome configuration of most standard thermopiles, it is virtually impossible to focus onto the surface through an $\times 100$ oil objective. However, it is possible to focus onto the surface of most photodiodes. Problems with dosimetry are great, and they are among the major drawbacks of the microirradiation approach.

III. Applications

A. Chromosome Organization

In a rather extensive series of experiments using the argon laser microbeam, we have shown that it is possible to selectively alter DNA and/or chro-

TABLE II

LASER IRRADIATION EFFECTS ON CHROMOSOMES

Laser	Pretreatment acridine orange at 0.1 mg/ml, 5 min	Wavelength (nm)	Energy ($\mu J/\mu m^2$)	Light microscope morphology	Electron microscope morphology	Alkaline fast green	Feulgen reaction
Argon laser	–	488 514	1000	Large pale region	Central electron dense mass + peripheral aggregates	–	–
	–	488 514	500	Small pale region	Electron dense aggregates, 0.08–0.19 μm	–	+
	+	488 514	500	Large pale region	Numerous electron dense aggregates, 0.05–0.15 μm	+	–
	+	488 514	50	Small pale region	Numerous electron dense aggregates, 0.05–0.15 μm	+	–
Dye laser	–	460	Na [a]	Large pale region	Central electron dense mass + peripheral aggregates	NA	NA
	–	460	NA	Small pale region	Electron dense aggregates, 0.08–0.19 μm	NA	NA
	+	460	NA	Large pale region	Numerous electron dense aggregates, 0.05–0.15 μm	NA	NA
	+	460	NA	Small pale region	Numerous electron dense aggregates, 0.05–0.15 μm	NA	NA

[a] NA, not available.

mosomal basic proteins in a chromosome region of less than 1 μm. These experiments were all done in mitotic tissue culture cells. By employing various energy levels of combined 514 and 488 nm laser light, it was possible to damage a selected chromosome region (Table II). The immediate effect on the irradiated chromosome was a change in refractive index that was evident as a "phase paling" (Fig. 2). Otherwise, the chromosome maintained its integrity and underwent normal mitotic movements.

When the chromosomes were subsequently analyzed by cytochemical procedures, a selective effect on either the DNA or the basic chromosome protein was demonstrated (10). Selective damage to the DNA was produced by pretreating the cells with acridine orange (AO) in concentrations of 1–0.01 μg/ml for 5 minutes prior to laser microirradiation. The laser energy was classified as "low" (50 μJ/μm^2). The basic proteins were damaged by exposure to "moderate" laser energy levels (500 μJ/μm^2) and no vital dye, such as AO, was employed. It was possible to destroy both molecular components by moderate laser energies with AO or "high" laser energies (1000 μJ/μm^2) without AO. The cytochemical procedures used following irradiation were the Feulgen–Schiff reaction for DNA (16) and the alkaline fast-green procedure (17) for basic protein. In addition, a more general protein stain, naphthol yellow S, was also used (17). This staining reaction was normal in the chromosome regions that had been damaged for the basic protein component. This result further substantiated the selective disruption of the basic protein component. With respect to the fundamental organization of chromosomes, these studies demonstrate that the selective disruption of two of the major chromosome components does not result in any major breakdown in chromosome organization or behavior during mitosis. Furthermore, the results indicate that alteration of either DNA or basic protein is possible without seriously affecting the other component. The significance of this result will be discussed in a later section on gene function.

In addition to the previous cytochemical analysis of irradiated chromosomes, extensive electron microscope studies have been undertaken. These studies involve making electron microscope observations on the irradiated chromosome (see subsequent section for discussion of the methodology). Observations on serial thin sections of laser microirradiated chromosomes (18) demonstrate two different kinds of ultrastructural damage. In regions of DNA laser damage, the lesion material is electron dense, spherical, and about 0.05 μm in diameter (Fig. 3). The lesion material is not connected, and there is considerable chromosomal material between the spherical lesion bodies. The protein lesion material is electron dense, varies in diameter from 0.08 to 0.19 μm, and is often interconnected (Fig. 4). Both the DNA and protein lesions are localized very precisely to the region detected as a phase paling with the light microscope. The electron microscope lends further credence to the cytochemical studies with respect to selective molecular effects.

B. Gene Deletion

One of the primary goals of the earlier studies was to use the laser microbeam to delete specific genetic sites on selected chromosomes. With the ability to produce localized damage as small as 0.25 μm, it was felt that the laser microbeam could be used for studies on gene mapping and gene function. Extensive studies were conducted on the ribosomal (nucleolar) genes of salamander (*Taricha*) and rat kangaroo (*Potorous*) cells in culture. The secondary constriction regions of the chromosomes in these cells were irradiated and subsequent nucleolus formation was assayed (*19,20*). These studies were able to specifically locate the ribosomal genes and in addition provided an experimental method to study the relationship between the ribosomal genes and the control of postmitotic nucleolus organization.

It was possible to demonstrate that the deletion of nucleolar genes resulted in an increased activity of the remaining nucleoli (*19*). Deletion of all of the nucleolar organizers resulted in the production of numerous micronucleoli (*20*). These results suggested the possibility of either accessory nucleolar organizers or the aggregation of premitotic nucleolar components. Another aspect of these experiments was the demonstration that irradiation of the nucleolar organizer under the DNA-damaging conditions resulted in gene inactivation, and irradiation under the protein-damaging conditions did not alter the ability to synthesize a functional nucleolus. These results indicate that the integrity of the irradiated chromosomal protein was not necessary for gene function. In addition, it confirmed the earlier cytochemical results and further suggested that the basic protein and DNA in the irradiated chromosome region were not so intimately associated that one could not be affected without altering the functionality of the other.

One of the major problems of this approach is the fact that in the cells studies (*Taricha* and *Potorous*), virtually no genetics has been performed. These cells were chosen because they remain flat during mitosis, thus making all the chromosomes easily identifiable during mitosis. For this approach to be more generally useful, human cells should be used. The fact that most human cells in culture round up during division and they have a large number of chromosomes (compared to 12 and 22, respectively, for the rat kangaroo and salamander) poses severe limitations to the applicability of the laser microbeam to selective gene mapping. However, the development of artificial flattening procedures and/or the use of cell fusion in combination with the laser technique may prove to be a viable solution to the problem. The

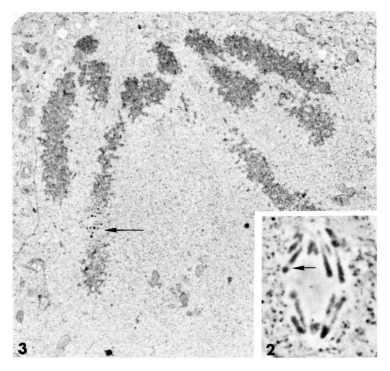

FIG. 2. Light micrograph of anaphase chromosomes from live cell demonstrating the "paling" response following laser microirradiation (arrow). The lesion diameter is 1–2 μm. This cell was treated with acridine orange, 0.1 μg/ml for 5 minutes, prior to laser exposure.

FIG. 3. Electron micrograph of cell depicted in Fig. 2. The small spherical lesion material is indicated by the arrow and represents damaged DNA.

ability to selectively mutate a desired chromosome at a desired point could be an extremely valuable tool.

C. Chromosome Stability

In addition to the previous studies on selective gene deletion, the laser microbeam has also been used to remove entire chromosomes from mitotic cells (21). This is accomplished by irradiating the centromere region of metaphase chromosomes. As a result of this kind of irradiation, the chromosomes fall off the mitotic spindle and are either excluded from both daughter nuclei or both chromatids are included in the same nucleus. It was felt that this kind of selective manipulation could be used to produce cell lines that were deficient in specific chromosomes. However, unexpectedly, the results with PTK_2 cells indicated that these cells may have a mechanism to replace the lost chromosome. This could come about by (a) selective nondisjunction of

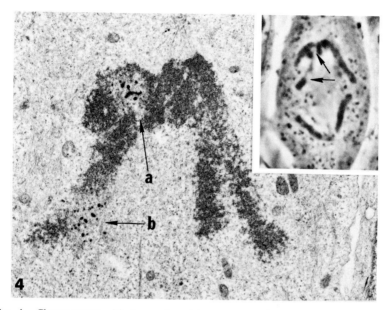

FIG. 4.   Chromosomes with damaged protein component (arrows) following high-power laser microirradiation without acridine orange treatment. The inset is the light microscope picture immediately after irradiation, and the electron micrograph demonstrates the nature of the lesion material.

the remaining homolog at a subsequent mitosis, (b) selective endoreduplication of one chromosome, (c) the gradual selection process *in vitro* over several months. At this time we are not sure which, if any, of these mechanisms are operational. However, the studies have been repeated six times in two different *Potorous* cell lines with the same result. In all the studies conducted, one of the larger chromosomes was removed. It is possible that if there is a selective chromosome replacement mechanism, it may be for only certain chromosomes. Present investigations are designed to see whether this mechanism exists for the other chromosomes in the karyotype.

D.   Mitosis

One of the areas in which the laser microbeam is enjoying considerable success is in studies on mitosis. In these studies, the laser beam is focused onto kinetochores and centrioles. The purpose of these studies is to elucidate the role of these two organelles in the structure and function of the mitotic spindle (22,23).

The basic experimental approach is to irradiate the kinetochore or centriole with laser light either alone or in combination with a vital dye. In this way, selective "molecular disruption" of either the protein or nucleic acid can be produced. The cells are then analyzed (a) for chromosome behavior, (b) ultrastructurally, and (c) for specific alterations in spindle organization.

IV.   Methodologies

In performing a laser microbeam experiment, there are several specific methodological sequences that we usually follow. These are (a) alignment of the laser microbeam system, (b) preparation of the cell cultures, (c) appropriate monitoring and recording of the actual experiment, (d) subsequent analysis of single cells by either isolation and cloning of the cells or single-cell electron microscopy.

Since part of the purpose of this volume is to provide "cook book" type recipes, it seems appropriate to provide detailed information on each of the four areas listed above. However, it should be pointed out that each of these procedures can, and undoubtedly would, be modified to suit the specific requirements of individual investigators.

A.   Alignment and Calibration of the Laser Microbeam

Since the entire system combines reflective surfaces, lenses, microscope optics, filters, and television components, it is not unusual for one or more of the components to move with respect to each other. Because the laser is ultimately focused to a spot less than 1 μm, any movement further up in the optical system could result in gross change in the focused spot location. It is therefore necessary to check the alignment at the beginning of an experimental series and periodically during the course of experimentation.

There are two major optical alignment procedures: (a) alignment of the laser beam through the optical center of the microscope, (b) alignment of the focused spot over the target cross hair on the video screen. In the first procedure, the laser beam is brought through the top monocular or triocular port of the microscope and down through the microscope nose piece that has *no* objective attached. The laser beam should be superimposed over the substage condenser spot that is stopped down to as small as spot as possible on a microscope slide (about 1 mm in diameter). If the optical image-producing system of the microscope is properly adjusted, then the stopped-down condenser spot will be in the center of the microscope optical field. The laser beam is carefully centered over the condenser spot by adjusting the micrometer controls on the dichroic filter that reflects the laser beam down into the microscope. Aligning the laser beam to the substage condenser spot ensures that the beam will pass into the center of the microscope objective when it is rotated into place. This alignment procedure may take from 2 to 10 minutes depending on the degree of misalignment.

In the second alignment procedure, a test specimen is placed under the microscope objective, and the laser is fired. The test specimen should have absorption characteristics such that enough laser light is absorbed to result in a small hole or lesion. For visible lasers, a dried smear of red blood cells is excellent, and for ultraviolet wavelengths, any living cells are adequate. When the laser is fired through an $\times 100$ objective, a lesion of between 0.25 and 3 μm is produced (Fig. 5). The cell with the lesion can be viewed directly on the television monitor. When the system is properly aligned, a cross hair on the television screen will be directly over the lesion. If the cross hair is not directly over the lesion, the television camera is moved by two micrometer controls until the lesion is directly under the cross hair. When this is done, alignment is complete, and the investigator need only move the "real" target

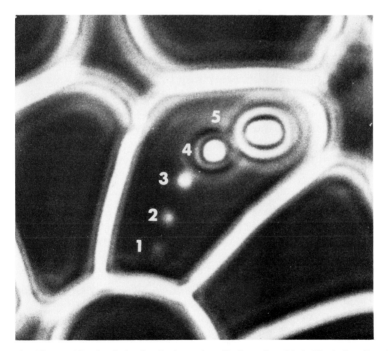

FIG. 5.   Human blood cell that has had a series of lesions placed in it. The lesion diameters range from 0.25 μm (No. 1) to 3.0 μm (No. 5) in diameter.

structure under the cross hair. This alignment procedure should not take more than 2–3 minutes.

B.   Cell Cultures

The cells can be from either short-term primary tissue cultures or long-term established cell lines. The choice of cell culture type really depends upon the nature of the particular experiments. If the purpose is to study chromosome structure, or assay immediate cell function or behavior, then primary culture may be the system of choice. However, if cell cloning is desired or long-term cell functions are to be assayed, then the established cell lines may be more desirable.

A primary culture system that we have found very useful for studies on nucleolar organizer function and chromosome structure is the salamander lung culture system described by Seto and Rounds (24). In this procedure, lung tissue is removed from the animal, minced into 1–2 mm³ pieces, and placed under a dialysis membrane in a Rose tissue culture chamber. Culture medium is a standard Eagle's MEM fortified with 10% fetal calf serum, antibiotics, and phenol red. After 1 week to 10 days in culture, numerous cells have migrated out from the primary explant and formed a monolayer sheet with abundant mitotic cells. The culture can last up to several weeks. The culture medium is changed once a week. Prior to irradiation of a traget chromosome, the investigator may wish to treat the cells with a sensitizing agent (25). For example, laser microbeam damage to chromosomal DNA is facilitated by AO intercalation into the DNA molecule. The solution of dye can be injected directly into the Rose chamber without exposing the cells to the outside environment. A typical sequence of pre-irradiation treatments is: removal of culture medium with syringe → injection of AO solution (0.1 µg/ml culture medium) → leave AO solution in chamber for 5 minutes → removal of culture medium with syringe → injection of AO solution (0.1 by syringe injection and removal → replace fresh culture medium → place chamber under microscope, locate target chromosome, and irradiate with laser microbeam. The cells may then be either fixed for subsequent cytochemical and electron microscope analysis or be analyzed behaviorally.

For established cells, the cell culture procedures are somewhat different. Stock cultures are maintained in plastic T25 or T60 flasks. The cells are enzymically removed from the flasks, gently centrifuged to a pellet, and resuspended in fresh culture medium. The cells can then be injected into Rose chambers at desired densities. The established lines we work with are from the rat kangaroo (*Potorous tridactylis*). They are ideal for chromosome studies because the cells remain flat throughout cell division, thus making all the chromosomes readily identifiable. In addition, *Potorous* cells have only 12–14 chromosomes as compared to 20 or more for most mammalian cell lines (*20,26*). The procedures for treatment and irradiation are basically the same as in the primary cultures.

C. Monitoring and Photography

It is critically important to have an accurate record of the experiment. This is routinely accomplished by two methods: (a) making photographs with a 35 mm camera that is part of the microscope system, (b) videotaping the entire experiment; 35 mm photography is important because the television video image is not of high enough quality to provide the fine-detail pictures necessary for publication. Routine 35 mm pictures of the target chromosome are taken before irradiation, within 10 seconds of irradiation, and at 1–5 minutes after irradiation. In addition, since the experiment is performed while viewing the cell on the television monitor, it is possible to store the entire experiment by sending the video signal into a videotape machine. This procedure is particularly valuable because of (a) the immediate playback capability, (b) the ability to erase and reuse the videotape, (c) the capacity for inexpensive storage of large quantities of information. In addition, the availability of time-lapse videotape systems now permits long-term monitoring of a cell with the capability of immediate playback.

D. Single-Cell Analysis

It has been necessary to devise a special set of procedures for analyzing single cells. For genetic studies, the irradiated cell must be isolated and cloned into a viable population (*21,27,28*). The procedure devised is outlined in Fig. 6. The essential steps are: (a) irradiation of a specific chromosome in mitosis, (b) removal of unirradiated cells from near the target cell using micromanipulators, (c) laser destruction of cells migrating into the area of the cell being followed, (d) monitoring of proliferation of the target cell using either the videotape or simply observing the cells every 4 hours, (e) enzymic removal of descendant "clonal" cells and transfer into culture flask. Once the clonal cells are safely growing in a T flask, they can be subjected to karyotypic and biochemical analysis.

The other major single-cell procedure is the electron microscope analysis of the irradiated cell (*18*). Immediately after irradiation, the cell is photographed under both high ($\times 100$) and low ($\times 16$) magnification of the microscope. Using an ink-marking objective, a small circle (about 1 mm in diameter) is drawn on the cover glass around the region of the irradiated

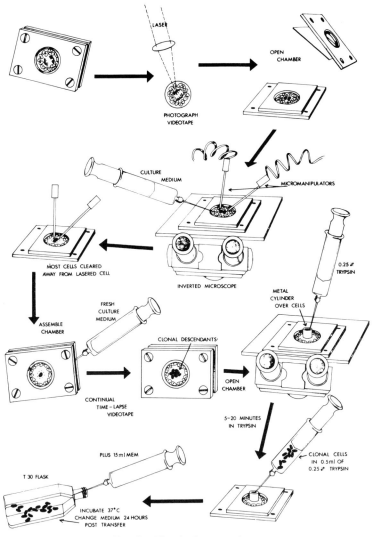

FIG. 6.   The cloning procedure.

cells. The chamber is next removed from the microscope, the medium is
removed, and 3% glutaraldehyde solution is injected directly into the
chamber. The chamber is placed back on the microscope, the cell is relocated
inside the ink circle, and another circle is drawn with a wax pencil. (The
ink subsequently will come off during chemical treatment.) After 30 minutes
to 24 hours in the glutaraldehyde, the chamber is opened, and cells are
washed with Millonig's phosphate buffer with sucrose (pH 7.4) and post-

fixed for 1 hour in 1% osmium tetroxide. They are then washed with distilled water and dehydrated in a cold ethanol series including 2% uranyl acetate in 70% ethanol for 2 hours. After two changes in 100% ethanol, they are passed into hydroxypropyl methacrylate and embedded in Epon (29). The polymerized cell sheet is separated from the coverglass by submersion in liquid nitrogen. Then the irradiated cell is relocated from light micrographs, circled with a Zeiss diamond-tipped object marker, cut out, and mounted for serial sectioning.

V. Summary

In this manuscript, I have attempted to introduce a technique that could be of significant use to investigators who are studying chromatin structure and function. The introduction provides some basic background on the various potential applications of lasers to genetic research. In Section III, Applications, a brief review is provided of the four major areas in chromosome research where the laser microbeam approach is being actively employed. Section IV, Methodologies, is an attempt to provide some of the basic practical information concerning laser instrumentation, cell culture procedures, and the special single-cell accessory procedures that are necessary for microbeam work.

In general, the laser microbeam provides a highly specialized and unique approach to studying the genetic material. However, it must be pointed out that the expense of the equipment, the tedious nature of single-cell work, and the difficulty in performing large-scale biochemical analysis are limiting factors that should be carefully considered by every investigator contemplating using the technique.

ACKNOWLEDGMENTS

Some of the research reported in this manuscript were supported by grants NSF GB 43527, NIH HL 15740, and GM 22754.

REFERENCES

1. Behringer, J. In "Raman Spectroscopy" (H. A. Szymanski, ed.), Vol. 1, p. 168. Plenum, New York (1967).
2. Heyde, M. E., Rimai, L., Kilponen, R. G., and Gill, D. J. Am. Chem. Soc. 94, 5222 (1972).
3. Andreoni, A., Sacchi, C. A., Cova, S., Bottiroli, G., and Prenna, G. In "Lasers in Physical Chemistry and Biophysics" (J. Joussot-Dubien, ed.), p. 413. Elsevier, Amsterdam (1975).

4. Berns, M. W., and Salet, C. *Int. Rev. Cytol.* **33**, 131 (1972).
5. Thomas, G. J., Jr. *Biochim. Biophys. Acta* **213** 417 (1970).
6. Tsuboi, M., Takahashi, S., Muraishi, S., Kajiura, T., and Nishimura, S. *Science* **174**, 1142 (1971).
7. Erfurth, S., and Peticulas, W. L. *Biopolymers* **14**, 247 (1975).
8. Berns, M. W., Olson, R. S., and Rounds, D. E. *Nature (London)* **221**, 74 (1969).
9. Berns, M. W., Cheng, W. K., Floyd, A. D., and Ohnuki, Y. *Science* **171**, 903 (1971).
10. Berns, M. W., and Floyd, A. D. *Exp. Cell Res.* **67**, 305 (1971).
11. Berns, M. W. *Nature (London)* **240**, 483 (1972).
12. Berns, M. W. *Exp. Cell Res.* **65**, 470 (1971).
13. Berns, M. W. *Biophys. J.* **16**, 973 (1976).
14. Berns, M. W. *In* "Photochemical and Photobiological Reviews" (K. C. Smith, ed.), Vol. 2. Plenum, New York (in press).
15. Berns, M. W. "Biological Microirradiation." Prentice-Hall, Englewood Cliffs, New Jersey (1974).
16. Leuchtenberger, C., *Gen Cytochem. Methods* **1**, 219 (1958).
17. Deitch, A. D. *In* "Introduction to Cytochemistry" (G. L. Wied, ed.), p. 327, Academic Press, New York (1966).
18. Rattner, J. B., and Berns, M. W. *J. Cell Biol.* **62**, 526 (1974).
19. Berns, M. W., Ohnuki, Y., Rounds, D. E., and Olson, R. S. *Exp. Cell Res.* **60**, 133 (1970).
20. Berns, M. W., Floyd, A. D., Adkisson, K., Cheng, W. K., Moore, L., Hoover, G., Ustick, K., Burgott, S., and Osial, T. *Exp. Cell Res.* **75**, 424 (1972).
21. Berns, M. W. *Science* **186**, 700 (1974).
22. Berns, M. W., and Rattner J. B. *J. Cell Biol.* **67**, 30a (1975).
23. Berns, M. W., Rattner, J. B., Meredith, S., and Witter, M. *Ann N. Y. Acad. Sci.* **267**, 160 (1976).
24. Seto, T., and Rounds, D. E. *Methods Cell Physiol.* **3**, 75 (1968).
25. Berns, M. W., and Rounds, D. E. *Sci. Am.* **222**, 98 (1970).
26. Branch, A., and Berns, M. W. *Chromosoma* **56**, 33 (1976).
27. Basehoar, G., and Berns, M. W. *Science* **179**, 1333 (1973).
28. Berns, M. W. *Cold Spring Harbor Symp. Quant. Biol.* **38**, 165 (1974).
29. Luft, J. H. *J. Biophys. Biochem. Cytol.* **9**, 409 (1961).

Chapter 16

Utilization of Neutron Scattering for Analysis of Chromatin and Nucleoprotein Structure

R. P. HJELM JR., J. P. BALDWIN, E. M. BRADBURY

Biophysics Laboratories,
Portsmouth Polytechnic,
Portsmouth, United Kingdom

I. Introduction

The structure of the chromosome of higher eukaryotes has been the object of electron microscopic and X-ray scatter investigation for a number of years. However, the basic limitations of these techniques combined with the lack of a conceptual framework within which to interpret the results has limited the interpretation of the data obtained. The development of neutron small-angle scattering as a viable method for probing biological organization has removed some of the constraints imposed by the more established techniques. At the same time the conceptual resolution of chromosome organization into a repeating subunit structure has improved experimental design and has given a basis upon which to interpret the results.

The potential power of the application of neutrons to biologically related problems has been tapped in the past few years by the development of neutron scatter apparatus which can be used for experimentation with biological materials (*1–3*). This in turn has stimulated renewed interest in analysis of small-angle data (*4–7*). As a result of these advances it is now possible with the use of neutron scatter techniques to reduce the scatter of a solution of identical particles to contributions from the particle shape and internal structure, and gain information about particle structure and shape by application of the scattering theory.

The chromosome of higher plants and animals is a complex of DNA and well-defined basic proteins, the histones. In addition, they contain variable amounts of RNA and proteins collectively known as nonhistones, which are generally not well defined (*8*). Recent experiments have suggested that the organization of the chromosome is based on a repeating subunit structure (*9–12*). Each subunit is supposed to contain 200 base pairs of DNA wrapped around a core of eight histones. The core contains two copies each of histones H2A, H2B, H3, and H4 (the first two are the slightly lysine-rich histones; the last two are the arginine-rich histones) (*13–16*). The lysine-rich histone H1 (also H5 in avian erythrocytes) is not included in this basic structure and may be involved in higher order organization (*12,18*). The strings of subunits make up the higher orders of chromatin structure by further coiling and/or folding.

It has been demonstrated that the chromosome subunit can be produced by digestion with micrococcal nuclease (*14,19*). In this form the basic structure of the chromosome can be investigated by rigorous application of neutron small-angle scattering theory. Here we review the methods as they apply to the study of the structure of chromosomes. Of great importance in this respect is the application of these techniques to the determination of the structure of the chromosome subunit particle (*19a*). We also give an account of the earlier applications of neutron techniques to chromatin and interpret the results in terms of the arrangement of the subunits within the chromatin fiber.

II.   Principles of Neutron Scatter

A.   Interaction of Neutrons with Matter

A neutron of velocity *v* and mass *m*, like any other particle, has kinetic energy $\frac{1}{2}mv^2$, a momentum, *mv*, and a wavelength, λ, given by the de Broglie equation,

$$\lambda = h/mv \tag{1}$$

where *h* is Planck's constant. Because of this wave character of neutrons, their scatter or diffraction can be described in the same forms usually associated with X-rays [Eqs. (2–7)].

It is in the interaction with the scattering object that neutrons have unique properties. Whereas X-rays interact with, and are scattered by, the electron distributions in atoms, neutrons interact with nuclei.[1] This has two important consequences: First, because nuclear forces are very short ranged, the nuclei are point scatterers. As a result neutron scatter from an isolated atom is isotropic and can be described by a single parameter, scattering length, *b*,

---

[1] Neutrons do have a magnetic moment which can interact with the magnetic moments of electrons and their orbitals, but in most cases of biological interest scattering due to these interactions can be neglected.

TABLE I

SCATTERING LENGTHS AND CROSS SECTIONS OF ATOMS OF BIOLOGICAL IMPORTANCE

| Element | Neutrons[a] | | | | | X-rays[b] | |
	b^c	$\sigma_{coh}{}^d$	σ_{abs}	σ_t	$\sigma_{inc}{}^g$	f_x^e $s = 0$	$s = 5\ nm^{-1}$
^1H	−0.374	1.76	0.19	81.5	80.3	0.28	0.02
^2D	+0.667	5.59	5×10^{-4}	7.6	2.0	0.28	0.02
C^f	+0.665	5.56	3×10^{-3}	5.6	—	1.69	0.48
N^f	+0.940	11.10	1.1	11.14	—	1.97	0.53
O^f	+0.580	4.23	1×10^{-4}	4.24	—	2.25	0.62
P^f	+0.51	3.27	9×10^{-2}	3.6	—	4.23	1.83
S^f	+0.285	1.02	0.28	1.2	—	4.5	1.9

[a] Values for 0.1 nm neutrons recompiled from Engelman and Moore (2).
[b] X-ray scattering lengths shown for comparison. Recompiled from Bacon (20).
[c] Neutron scattering lengths expressed in Fermis, 1 F = 10^{-12} cm.
[d] Neutron scattering cross sections expressed as barns, 1 b = 10^{-24} cm².
[e] X-ray scattering lengths expressed in units of 10^{-12} cm.
[f] Values are for samples with a natural abundance of their different isotopes.
[g] Neutron incoherent scattering lengths calculated from the other cross sections. Reported for hydrogen and deuterium only.

the measure of the probability of neutron scatter. This contrasts with X-rays and electrons, where even in the case of scatter from an isolated atom (Table I), the scattering centers have spatial distribution, resulting in an angular dependence of scatter. Second, unlike X-ray scattering factors (Table I), the neutron scattering by an isolated atom is not a monotonic function of its atomic number (Table I) (2,20). In particular, hydrogen, which essentially gives no contribution to the scatter of an object by X-rays, contributes a large proportion of the neutron scatter. In addition, the isotope of hydrogen, deuterium has a neutron scattering length vastly different from that of hydrogen. These two considerations make neutron scattering experiments useful in that (a) the positions of hydrogens in a molecule can be determined and, (b) the substitution of hydrogen with deuterium can be used to change the total scattering and scattering distribution of a molecule.

Another important property of neutrons as they are used in biological experiments is that they have kinetic energies comparable with the dynamic energies of macromolecules. This allows neutrons to be used in studying these motions. Also, at these energies there is very little absorption of neutrons by biological molecules. This means that there is negligible damage to the sample from the irradiating neutron beam.

B. The Basic Scattering Equations

When more than one scattering center is present in an object, interference occurs between the scattered waves such that the intensity of the scattered radiation has a distinct relationship to the distances between the scattering centers in the sample. As a result, with neutrons the scatter is no longer isotropic.

The basis for this is shown in Fig. 1. A plane wave with direction of propagation given by S_0 is incident on two scatterers at positions 0 and 0' separated by a distance r. Sampling the scatter in one direction, and denoting its direction by S_1, the path difference between the waves scattered by the two centers is $\Delta = r \cdot S_1 - r \cdot S_0$. If we define the vector describing the change in direction between the incident and scattered beams as $s = (S_1 - S_0)/\lambda$, then it is seen from Fig. 1 that

$$s = 2 \sin \theta / \lambda \qquad (2)$$

where 2θ is the scattering angle. The phase difference between the two scattered waves is just $(2\pi/\lambda)\Delta$; thus, $\phi = r \cdot s$. Taking the scattering center at 0 to be at the origin, and assuming any number of scattering centers r_i from 0, the amplitude of the scattering radiation at any given

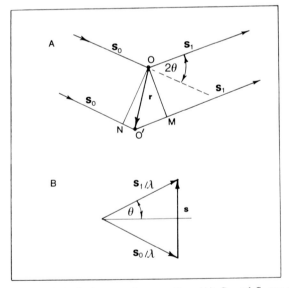

FIG. 1.   Derivation of the basic scattering equation. (A). S_0 and S_1 are the unit vectors giving the directions of the incident and scattered waves, respectively. The path difference between the waves scattered through 2θ by the points at 0 and 0' is $\overline{NO'} + \overline{O'M}$. The phase difference between the two waves is then $\phi = 2\pi(\overline{NO'} + \overline{O'M})/\lambda = 2\pi r \cdot (S_1 - S_0)/\lambda$. (B) The definition of the scattering vector s as $(S_1 - S_0)/\lambda = 2 \sin \theta / \lambda$.

s [Eq. (2)] becomes

$$A(\mathbf{s}) = \Sigma b_i \exp{(2\pi \ i\mathbf{s} \cdot \mathbf{r}_i)}. \qquad (3)$$

In this equation, b_i is the scattering length of the ith scattering center, (Table I). If the scattering centers are presumed to be distributed continuously through the object, then Eq. (3) is (integrating over volume in real space, v_r)

$$A(\mathbf{s}) = \int \rho(r) \exp{(2\pi \ i\mathbf{s} \cdot \mathbf{r})} \ dv_r \qquad (4)$$

where $\rho(r)$ is now the density distribution of scattering length [for X-rays $\rho(r)$ is proportional to electron density]. Equation (4) is the Fourier transform of the scattering-length density distribution of the object.

If the scattering amplitude is known, then the structure of an object can be determined by inversion of Eq. (4)

$$\rho(\mathbf{r}) = \int A(s) \exp{(-2\pi \ i\mathbf{s} \cdot \mathbf{r})} \ dv_s \qquad (5)$$

(the integration is over volume in s-space; v_s). However, it is not the amplitude of the scatter that is observed; rather it is the intensity

$$I(\mathbf{s}) = \Phi |A(\mathbf{s})|^2 = \Phi A(\mathbf{s}) \ A^*(\mathbf{s}) = \Phi \int P(\mathbf{u}) \exp{(2\pi \ i\mathbf{s} \cdot \mathbf{u})} dv_u, \qquad (6)$$

(* indicates complex conjugate) where Φ is the flux (neutrons per second per square centimeter). Therefore inversion of the scattered intensity yields the Patterson of the object. The Patterson is defined as

$$P(\mathbf{u}) = \int \rho(\mathbf{r}) \ \rho(\mathbf{u} - \mathbf{r}) \ dv_r \qquad (7)$$

where u is the distance between any two scattering centers.

C.   Coherent and Incoherent Scatter

Equations (2) through (7) describing the relationship between the structure of a sample and the form of its scattered radiation assume that the scattering event is coherent and that the wavelength of the incident beam is not altered. When this condition is met, the scatter is elastic and coherent. Any contribution to the scatter intensity from inelastic and incoherent scatter must therefore be subtracted. In the case of neutrons there are many ways that incoherent and/or inelastic scatter can occur, and the contribution to the total scatter can be very large. This is especially true with samples containing large amounts of hydrogen with its large incoherent cross section (Table I).

D. Scattering Cross Sections

The scattering cross section is defined as the ratio of the total number of neutrons per second scattered to the incident neutron flux (neutrons/sec/ cm²). Cross sections may be defined for coherent σ_{coh}, and incoherent σ_{inc} neutron scatter, and σ_{abs} for absorption of neutrons by the nucleus. The sum of these scattering cross sections is then proportional to the total number of neutrons removed by a nucleus from the incident beam per second and is called the total cross section σ_t. Each cross section represents the effective area presented by the nucleus to the incident beam. For an isolated atom $\sigma_{coh} = 4\pi b_i^2$. The various cross sections are tabulated in Table I along with the coherent scattering lengths of some common elements found in biological materials. Included for comparison are some X-ray scattering lengths.

Using the definition of coherent scattering cross section for a macromolecule, we get from Eq. (6)

$$\sigma_{coh} = \frac{\int I(s)\, d\Omega}{\phi} = \int |A(s)|^2\, d\Omega$$

where Ω is solid angle. Therefore the squared scattering amplitude of Eq. (6) is the differential coherent scattering cross section

$$d\sigma/d\Omega = |A(s)|^2 \tag{8}$$

It is the ultimate aim of small-angle scattering and diffraction experiments to determine $A(s)$ from its measured modulus $|A(s)|$, and thence to solve Eq. (5) for the distribution of scattering density. This can be done, however, only for samples consisting of highly ordered—that is, crystallized—molecules. Many systems of biological interest have not been crystallized; thus methods have been developed for solving the analogous equations describing scatter from disordered systems. The procedure involves techniques to which neutron scattering experiments are best adapted. We describe these below.

III. Structure Information from Small-Angle Scatter

From the measured scatter curve of X-rays or neutrons, information on the structure of macromolecules can be obtained. We have outlined the basic relationships between structure and scattering intensity in the first sections [Eqs. (2–7)]; we now turn to the implications of these relationships and their use in obtaining information about the system of interest. The theory of small-angle scatter has been developed for systems of monodis-persed, noninteracting particles, rods and lamellar structures.

Applications of the theory to particles are particularly important in that neutron small-angle scattering measurements have been made on mono-dispersed solutions of chromosome subunit particles (19a, 21–24). The sub-units, which are believed to be the basic structural unit of the chromosome (9–13, 19a, 25), can be produced by judicious use of micrococcal nuclease (14, 19a) and conform to the requirement that the particles do not interact if the proper solvent conditions are used. Each contains 200 base pairs of DNA and all but the lysine-rich histones. The methods of small-angle scat-tering are described in the context of their application to the structure of the subunits, and the results and interpretation for the chromosome beads are given. Applications to other, similar systems are also given to indicate the scope of future experimentation in chromosome particles.

It should be borne in mind that the theories presented can be, and have been, applied to a lesser degree to systems not meeting the criteria of homo-geneity and noninteraction, such as films of monomer beads and chromatin fibers, but that interpretation of the data is greatly complicated in these cases.

A.   The Method of Contrast Variation

The basic physical principle expressed in the scattering equations [Eqs. (3 and 4)] is that scatter of radiation through a finite angle is due to the difference in scattering power of the particle compared to the solvent and to the variation of scattering power within the particle. With X-rays the variation is proportional to fluctuations of electron density; with neutrons it is dependent upon the changes in nuclear densities and composition (scat-tering-length density; Table I and II).

Fluctuations in the sample are conveniently expressed in terms of scat-tering-length density, $\rho(\mathbf{r}) = \Sigma b_i/V$, where the summation of scattering lengths is over the scattering centers in the sampled region of volume V (for X-rays this is proportional to the electron density).

Assume, now, that the scattering-length density, ρ_{sol}, of the solvent can be varied continuously from a value far below the average scattering-length density of the particle, $\bar{\rho}_p$, to a value far higher than the particle average ($\bar{\rho}_p$, $= \Sigma b_i/V_f$ for neutrons, and is proportional to $\Sigma e/V_f$ for X-rays; V_f is the volume of the particle); then the fluctuations between the particle and the solvent can be expressed as the difference between the average scattering of the region occupied by the particle and that occupied by the solvent (Fig. 2). Mathematically this is $\bar{\rho} \, \Omega_f(\mathbf{r})$; where the contrast is, $\bar{\rho} = \bar{\rho}_p - \rho_{\text{sol}}$,

TABLE II

AVERAGE SCATTERING LENGTH DENSITIES OF SOME IMPORTANT
BIOLOGICAL MACROMOLECULES

Substance	$\bar{p}_p{}^a$	%D$_2$O at contrast[b] match	Reference
H$_2$O	−0.55	0	26
Hydrocarbon chains	−0.3	3.7 (calc.)	26
Light density, lipoprotein	+0.48	15	26
Hemoglobin	+2.24	40	27
Myoglobin	+2.24	40	6
Apoferritin	+2.36	42	28
Bovine fibrogen	+2.4	43	29
Histone	+2.51	44.3 (calc.)	
Egg white lysozyme	+2.55	45	30
Ferritin-2[c]	+2.63	46	28
Chromatin subunits			
Calf thymus	+2.85	49	21
Chicken	+2.85	49	24
erythrocyte	+2.95	51	19a
Ferritin-3[c]	+3.17	53	28
Full ferritin[c]	+3.33	56	28
30 S Ribosome	+3.37	56.8	31
50 S Ribosome	+3.53	59	31
DNA	+3.94	64.5 (calc.)	
16 S RNA	+4.11	67.5 (calc.)	31
23 S RNA	+4.17	68.5 (calc.)	31
D$_2$O	+6.36	100	

[a] Units of scattering-length density are 10^{10} cm^{-2}.

[b] Percent D$_2$O at which the scatter at zero angle goes to zero. Unless otherwise noted, values reported have been measured by the procedures given in the text. Calculated (calc.) values take into account the proportion of exchangeable hydrogens expected to be deuterated.

[c] Ferritins with different amounts of bound iron.

and $\Omega_f(\mathbf{r})$ is the structure function. The latter has the property that it equals one in the space occupied by the particle and zero elsewhere.

If the solvent scatter is made to match the average scattering-length density of the particle, the particle shape is no longer observable; only fluctuations of scattering power within the particle are seen (Fig. 2). The internal structure observed when this condition is met—zero contrast—is termed the internal structure function, $\rho_s(\mathbf{r})$. When this function is added to the term $\rho\,\Omega_f(\mathbf{r})$ a complete description is made of the fluctuation of scattering length

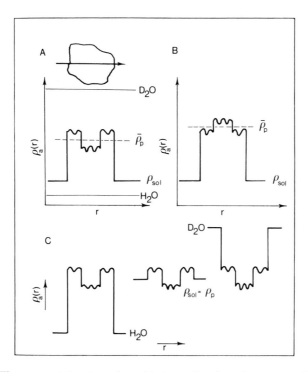

FIG. 2. The apparent structure of an object as a function of contrast and deuteration. (A) A hypothetical particle with cross section demonstrating a central portion of lower scattering-length density than the periphery. The particle is shown contrasted against a solvent scattering length, ρ_{sol}, which is lower than the average scatter of the particle, $\bar{\rho}_p$. The neutron scattering-length densities of H_2O and D_2O are shown. The ordinate is labeled $\rho_a(r)$ to indicate that the scattering values are absolute. (B) The same particle, but with the components making up the central part selectively deuterated, raising the average neutron scattering-length density of the particle and changing its apparent internal structure. (C) The apparent changes in the particle as seen against different solvent densities. From left to right the figures represent the form of the particle seen when in H_2O, as seen when in a mixture of D_2O and H_2O such that the solvent scattering-length density, ρ_{sol}, equals that of the average of the particle, $\bar{\rho}_p$, and as observed in D_2O. The equation for the excess scattering-length density of the particle in any solution is $\rho(r) = \bar{\rho}\Omega_f(r) + \rho_s(r)$, as described in the text.

density in the system; thus (5–7),

$$\rho(\mathbf{r}) = \bar{\rho}\Omega_f(\mathbf{r}) + \rho_s(\mathbf{r}). \tag{9}$$

When defined in this manner the two functions have the properties that

$$\int \Omega_f(\mathbf{r})dv_r = V_f; \qquad \int \rho_s(\mathbf{r})dv_r = 0 \tag{10}$$

where V_f is the volume of the particle.

From the above discussion and from Fig. 2, it is obvious that the appearance of an object as seen by X-rays or neutrons is dependent not only on intrinsic scattering properties, but also on the scattering strength of the background upon which it is observed. With alteration of contrast, the contribution of each region of the molecule to the total scatter changes; thus varying the scattering-length density of the solvent—the method of contrast variation—is roughly equivalent to the method of isomorphous replacement in X-ray or neutron diffraction.

The manner in which observed structure of the particle changes with contrast reflects the shape, $\Omega_f(\mathbf{r})$, and internal structure, $\rho_s(\mathbf{r})$, of the particle, and their spatial correlation. At vanishing contrast, the internal structure function is seen. Further, if the contrast is extrapolated to infinity, then the internal structure is no longer important, and only the shape, $\Omega_f(\mathbf{r})$, is observed. We shall see in later sections how these two considerations allow the separation of the contribution to the scatter intensity of the internal structure of the particle and the particle shape.

The method of contrast variation can be used both by X-rays and neutrons. With X-rays the electron density of the solvent may be altered by the addition of small organic molecules or salts, such as NaBr, NH_4SO_4, or Na_2SO_4 (5.6,32–34). In neutron scattering, reference to Table I indicates that the substitution of deuterium for hydrogen in the solvent gives a large increase in scattering power density; thus by mixing appropriate amounts of D_2O and H_2O the solvent may be made to have any scattering power between the limits shown in Table II. This range spans all the scattering-length densities observed in hydrogen-substituted biological macromolecules. The relative ease with which the contrast may be changed for neutrons makes neutrons ideal for small-angle scattering studies and is one justification for their use. Small organic molecules have also been used to vary contrast in neutron scattering experiments (28).

It should be pointed out that the quantities calculated for a particle, $\Omega_f(r)$, V_f, $\bar{\rho}_p$, etc., depend upon the method used to vary the contrast. Differences result from the extent of penetration of the molecule and its surrounding water of hydration by atoms being used to alter the contrast of the solvent (we define here the water of hydration as H_2O bound to the particle which excludes small solvent molecules). Therefore, if salts or small organic molecules are used, V_f, $\Omega_f r$, etc., will include the solvating water. If D_2O is used to vary the contrast, then the quantities found for the macromolecule will include a factor which accounts for H/D exchange. In this case of subscript $c(V_c\Omega_c(\mathbf{r})$, etc.) is used for all quantities dependent on the degree of H/D exchange (6). Inasmuch as D_2O is used most often in neutron scatter, we will use this convention.

B. Scattering Power and Particle Composition

The average scattering-length densities, $\bar{\rho}_p$, of macromolecules are dependent on their elemental density and composition, particularly with hydrogen with its negative scattering length (Table I). It is easily seen from Table II that molecules rich in the polar groups, oxygen, nitrogen, and phosphorus, and low in nonexchangeable hydrogen, will scatter both X-rays and neutrons more strongly than molecules rich in nonexchangeable hydrogen and depleted of polar elements. Table II illustrates neutron scattering-length densities for several important classes of biological macromolecules.

If one can determine the form of the internal structure function, $\rho_s(r)$, of a particle, then it is possible to gain some idea of the distribution of the macromolecule's component parts. Examples of determinations of the spherical averages, $<\rho_s(r)>$, are shown in Fig. 3. In many proteins (5, 6, 30, 35) the negative part of $<\rho_s(r)>$ is found toward the center of the molecule, reflecting the well-known fact that in globular proteins the hydrophobic—thus weaker scattering—amino acid residues are grouped in the center of the molecule whereas the periphery of the protein is rich in the polar, hydophilic residues. In the RNA and protein-containing tomato bushy stunt virus a positive region of $<\rho_s(r)>$ is seen between two negative regions, giving the conclusion that the strongly scattering RNA is sandwiched between, a protein core and outer cover (32). In contrast is the form of $<\rho(r)>$ of ferritin, which indicates that the iron-containing moiety is at the center (28,36). Figure 3 also illustrates the distributions determined for light-density human serum lipoproteins (LDL), (26,33,34), where the form of the internal structure function is similar to that the above proteins owing to the concentration of hydrocarbon chains in the center of the molecule. A similar pattern has been observed for chromatin subunit particles (19a, 21–23,37, also see below), and this has been interpreted to mean that DNA must be folded or coiled about a protein core.

C. Scattering by a Particle and the Basic Scatter Functions

The square of the Fourier transform of Eqs. (6) and (8) may be used to describe the differential cross section of a sample containing N identical particles if the particles are independent. In this case the scattering intensities add, and the scattering of the sample is simply N times that of a single particle. Scattering from the solutions of independent particles usually implies that the particles are randomly oriented; thus the observed intensity curve is the N times spherical average of the squared Fourier of Eq. (9); thus

$$I \propto N < |\int [\bar{\rho}\Omega_c(\mathbf{r}) + \rho_s(\mathbf{r})] \exp(2\pi i\mathbf{s} \cdot \mathbf{r}) \, dv_r|^2 > \qquad (11)$$

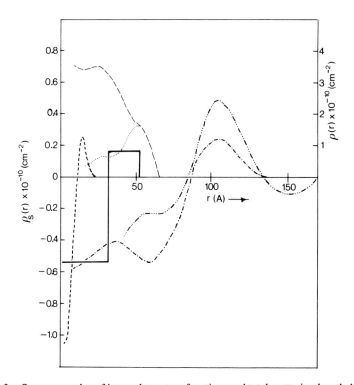

FIG. 3. Some examples of internal structure functions and total scattering-length densities. Spherically averaged internal structure functions, $\rho_s(r)$: ———, the model for chromatin subunit particles (*19a*), $\alpha = 4.55 - 5.1 \times 10^{-4}$; ------, calculated for lysozyme from X-ray diffraction data (*30*), $\alpha = 3.5 \times 10^{-5}$; -.-.-., low-density lipoprotein from neutron small-angle scattering data (*26*), $\alpha = 2.8 \times 10^{-3}$; -...-...-, tomato bushy stunt virus determined from X-ray small-angle data (*32*), Fourier inversion of the spectrum taken with $\rho_{sol} = 0.408$ electron/Å^3 very near the contrast match at 0.412 e/Å^3. Total particle scattering-length densities $\rho(r) = \bar{\rho}\Omega_c(r) + \rho_s(r)$ for apoferritin (.....), $\alpha = 3 \times 10^{-4}$; and ferritin (------), $\alpha = -1.4 \times 10^{-3}$ (*28*). The negative value of α for ferritin reflects the fact that for this molecule $\rho_s(r)$ is positive near its center, and negative at its periphery.

The term between the brackets is the scattering amplitude of the particle, $A(s) = \bar{\rho}A_c(s) + A_s(s)$, where each term in Eq. (9) has its corresponding scattering amplitude; squaring this and taking the spherical average, the terms maintain their identity with the creation of a cross term. Equation (11) then becomes

$$I(s) = <|A(s)|^2> = \bar{\rho}^2 I_c(s) + \bar{\rho} I_{cs}(s) + I_s(s) > \tag{12}$$

where $I_{cs}(s) = < A_c(s)A_s^*(s) + A_c^*(s)A_s(s) >$.

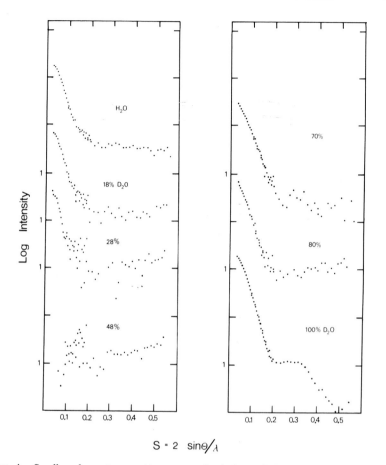

FIG. 4. Small-angle neutron scatter curves of solutions of chromatin subunits. Neutron scattering intensities of chromatin subunits from chicken erythrocytes in solutions of different contrasts. The contrast is varied by the proportion of D_2O in each solution. Data from Hjelm et al. (19a).

Equation (12) is quadratic with contrast; thus if one plots intensities, $I(s)$ (as in Fig. 4; the intensities are clearly a function of contrast) versus contrast, a series of parabolas should result (Fig. 6). The regression coefficients from a least-squares analysis will then give the three basic scattering functions $I_c(s)$, $I_{cs}(s)$, and $I_s(s)$. They are shown for chromatin subunit particles (19a) in Fig. 7–9. As discussed above, $I_s(s)$ results from interpolation to zero contrast and $I_c(s)$ comes from values determined from extrapolation to infinite contrast.

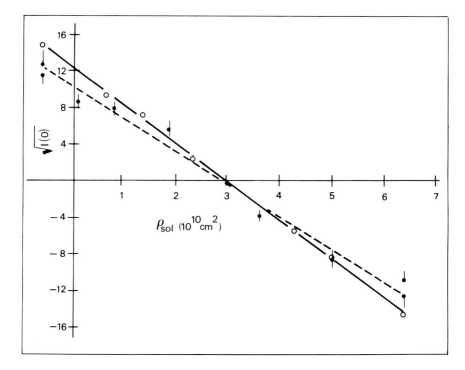

FIG. 5. Neutron zero-angle scatter from solutions of chromatin subunits. Scattering-intensity curves like those shown in Fig. 4 are extrapolated to zero-angle with the aid of the Guinier approximation (see text). The square root of the zero-angle intensity $I(0)$ is then plotted against the solvent scattering-length density ρ_{sol}. The intercept at $I(0) = 0$ gives the average scattering-length density of the particle, $\bar{\rho}_p$. O, Chicken erythrocyte chromatin subunits (*19a*); ●, calf thymus subunits (*21–23,37*).

D. Zero-Angle Scatter

The equation for scattering intensity [Eqs. (10) and (11)] becomes at zero angle ($s = 0$)

$$I(0) = \bar{\rho}^2 V^2_c$$

$$\sqrt{I(0)} = (\bar{\rho}_p - \rho_{sol})\, V_c \tag{13}$$

Therefore, if the measured scattered curves are extrapolated to zero-angle and the square root of the resulting forward scattering intensities are plotted versus the solvent density, a straight line will result. By interpolating to the solvent scattering-length density where the forward-angle scattering vanishes (zero contrast), then $\bar{\rho}_p = \rho_{sol}$, and the average scattering-length density

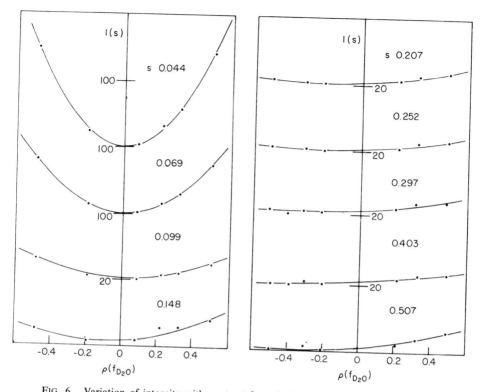

FIG. 6. Variation of intensity with contrast for solutions of chromatin subunits. Several representative points from the $I(s)$ of Fig. 1 are plotted versus the contrast. The value of each s sampled for this example is given in small numbers to the upper right of the curve. The scales for the intensities at $s = 0.044$, 0.069, and 0.099 nm^{-1} are in units of 100. Intensities at all larger s are in units of 20. Contrast is expressed as fraction D_2O in solution at zero contrast—in this case 0.51 (Fig. 5)—minus the fraction of D_2O in the solution at which the intensity was measured. The parabolic form of these curves is in agreement with Eq. (12). Data from Hjelm et al. (19a).

of the scattering particles can be determined. Furthermore, the slope of the line gives V_c if the scattering curve is scaled to intensity per particle, $d\sigma/d\Omega$. V_f can be calculated from the measured $\bar{\rho}_p$ from the definition of scattering-length density $\rho_p = \Sigma b_i/V_f$, if the elemental composition is known. The application of this procedure to chromosome subunit particles from chicken erythrocyte (19a) and calf thymus (21–23) is shown in Fig. 5. The volume calculated for subunits is given in Table III. Determinations of $\bar{\rho}_p$ from other biological systems are tabulated in Table II. Once $\bar{\rho}_p$ is measured for a macromolecule, the contrast $\bar{\rho} = \bar{\rho}_p - \rho_{sol}$ for any given solvent can be calculated.

TABLE III

SUMMARY OF PARAMETERS DERIVED FROM NEUTRON SMALL–ANGLE SCATTER DATA OF CHROMATIN SUBUNITS[a]

The model of the spherically averaged shape and structure functions $\langle\Omega_c(r)\rangle$ and $\langle\rho_s(r)\rangle$ determined from the derived $I_c(s)$, $I_s(s)$, and $I_{cs}(s)$ (Figs. 7–9)	Radius of $\langle\Omega_c(r)\rangle$, 52 Å	Radii of $\langle\rho_s(r)\rangle$		
		Protein core, 32 Å	DNA-rich shell, 52 Å (20 Å thick)	
Parameters calculated from the model	Total volume 590,000 Å³	Volume of protein core, 137,000 Å³	Volume of DNA-rich shell, 450,000 Å³	α, 4.55 × 10⁴
Compared with the measured parameters	V_f^b, 249,000 Å³	Volume of histone present 139,000 Å³	Volume of DNA present 110,200 Å³	Rc, α, 40.5 Å 4.55 × 10⁻⁴
Large discrepancy in volume, possibly due to the structure being very accessible to water			Volume of water in DNA-rich shell, 340,000 Å³	
Limits on the deviation from spherical symmetry of the shape function $\langle\Omega_c(r)\rangle$ in terms of limiting ellipsoids	Oblate ellipsoid axial ratio, 0,5 Radial dimensions 30 × 60 × 60 A Volumes 452,000 Å³ Rc 40.5 Å		Prolate ellipsoid axial ratio, 1,4 45 × 45 × 64 Å 542,000 Å³ 40.5 Å³	

[a] Compiled from Hjelm et al. (*19a*).
[b] Calculated as $V_f = \bar{p}_p \Sigma b_i$, where $\bar{p}_p$ is 2.95 × 10¹⁰ cm⁻² (Table II) taken from Fig. 6.

E. Radius of Gyration

The determination of the radius of gyration, Rg, of the scattering-length density is a natural result of analysis of the small-angle intensity curve of a sample of independent particles. If intensities are measured at sufficiently small values of s, then the approximation holds, $I(s) = \rho^2 \exp(4/3\ \pi^2 Rg^2 s^2)$ (*38*). Therefore, a plot of the log of intensity versus s^2 will give a line with slope proportional to Rg² at the limit of $s = 0$.

From Fig. 2 it is obvious that the radius of gyration of a macromolecule—defined as the second moment of the scattering-length density divided by the total scattering-length density, $\int \rho(r) r^2\, dv_r / \int \rho(r)\, dv_r$—must change with contrast in a manner related to the internal structure. Stuhr-

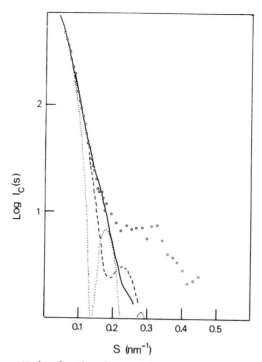

FIG. 7. Basic scattering function for particle shape of chromatin subunits, $\circ$, $I_c(s)$ calculated by G. G. Kneale of this laboratory for chicken, erythrocyte chromatin subunits; ..., $I_c(s)$ calculated for a solid sphere of radius 52 Å ------, $I_c(s)$ calculated for a solid prolate ellipsoid of axial ratio 1.4; ———, $I_c(s)$ calculated for a solid oblate spheroid of axial ratio 0.5. The ellipsoids were chosen to have radius of gyration equal to Rc. Data from Hjelm *et al.* (*19a*).

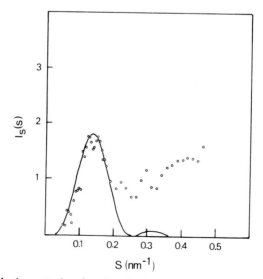

FIG. 8. The basic scattering function for internal structure of chromatin subunits. $\circ$, $I_s(s)$ calculated by G. G. Kneale for chicken erythrocyte subunits; ———, $I_s(s)$ calculated from the model for chromatin subunits (see Fig. 3 and Table III). Data from Hjelm *et al.* (*19a*).

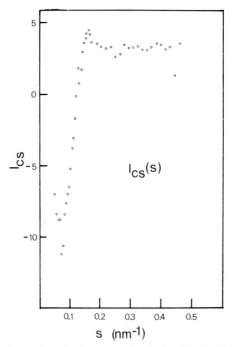

FIG. 9. $I_{cs}(s)$ of chromatin subunits. O, $I_{cs}(s)$ calculated by G. G. Kneale for chromatin subunits from chicken erythrocytes. This function gives the correlations between particle shape and internal structure. Data from Hjelm *et al.* (*19a*).

mann (*5*) and Ibel and Stuhrann (*6*) have shown that Rg^2 varies with contrast as

$$Rg^2 = Rc^2 + \alpha/\rho - \beta/\rho^2 \tag{14}$$

Rc is the radius of gyration of the particle at infinite contrast; and thus corresponds to the radius of gyration of the shape of the particle. The coefficients α and β give information on the internal structure function. They are defined as:

$$\alpha = \int \rho_s(r) r^2 \, dv_r / V_c; \quad \beta = (\int \rho_s(r) \, dv_r / V_c)^2 \tag{15}$$

Therefore, by plotting the square of the radius of gyration versus the reciprocal of the contrast, information can be obtained on the shape of the particle and the first and second moments of the internal structure function, $\rho_s(r)$.

From Fig. 3 it is clear that, if the higher scattering-length density regions are to the outside of the particle, then the fact that in this case $\rho_s(r)$ is negative in the center and positive toward the periphery dictates that α should be

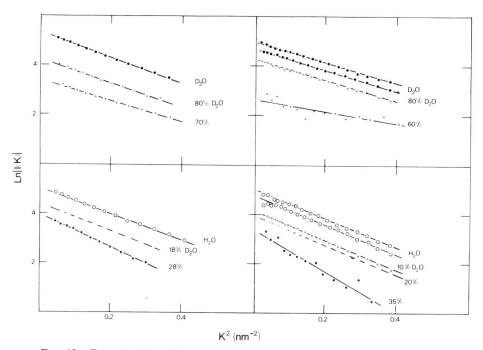

FIG. 10. Determination of the radius of gyration of chromatin subunits in solutions of different contrasts. Shown are Guinier plots from solutions of subunits from (left) chicken erythrocyte, and (right) calf thymus. The contrast is varied with D_2O, the percentage of which is shown to the right of each curve. Data are plotted versus $K^2 = (2\pi s)^2$. Data from Bradbury *et al.* (*21–23,37*) and Hjelm *et al.* (*19a*).

positive. The converse is also true. Further, the moment of the internal scattering function will be zero if the structure function has center of mass which coincides with the geometric center of the molecule—that is, if the center of scattering-length density does not change with contrast.

The validity of Eqs. (14) and (15) has been demonstrated for myoglobin and lysozyme, systems for which the structure is known from X-ray diffraction (*5,6,30*). The method has been used to determine the general form of the internal structure function—thus the distribution of components—for proteins (*5,6,26,30,36*), human serum lipoproteins (*26,33,34*), and chromatin subunit particles (*19a,21–24,37*).

In Fig. 10 is shown the Guinier plots for subunit particles, and the values of Rg^2 taken from these are plotted according to Eq. (14) in Fig. 11. The positive α is clear indication that $\rho_s(r)$ has the general form sketched in Fig. 3, with the regions of higher scattering-length density concentrated at the periphery of the subunit. The figures given in Table II for the scattering-length densities of protein and DNA make this consistent with the statement that

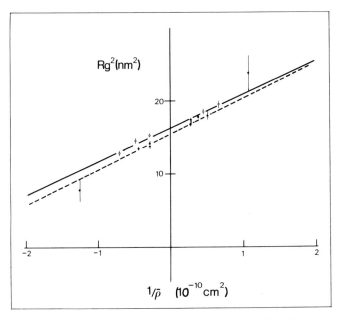

FIG. 11. Contrast variation of radius of gyration of chromatin subunits. O, Subunits from chicken erythrocytes (*19a*); ●, subunits from calf thymus (*21–23,37*).

the DNA is concenentrated on the outside of the particle, the protein being concentrated in the core. The points in Fig. II give no evidence that the lines describing the relationship between Rg^2 and $1/\rho$ are curved; thus $\beta \cong 0$, probably meaning that the particles have a center of symmetry.

F. Structure Determination from the Basic Scattering Functions

The basic scattering functions can be Fourier-inverted to give the Pattersons of the shape and internal structure functions [i.e., the distribution of interatomic distances, Eq. (7)]. Much more information can be obtained if $I_c(s)$, $I_s(s)$, and $I_{cs}(s)$ can be converted into correctly phased amplitudes. Theoretically, this can be done if each basic scattering function is broken down into two classes of terms: one, $\sigma^2(s)$ corresponds to the square of the Fourier transform of the spherical average of the structure; the other, $\delta^2(s)$, corresponds to a series of terms related to the square of the Fourier of the deviations of the structure from the spherical average (*4,7,32,35*). Using the notation of Luzzati *et al.* (*7*), which groups all the nonspherical terms together, the decomposition becomes

$$I_c(s) = \sigma_c^2(s) + \delta_c^2(s)$$

$$I_s(s) = \sigma_s^2(s) + \delta_s^2(s) \tag{16}$$

$$I_{cs}(s) = \sigma_c(s)\,\sigma_s(s) + <\delta_c(s)\,\delta_s(s)>$$

where

$$\sigma_c(s) = \int <\Omega_c(r)> \exp(2\pi\,is\,r)dv_r$$
$$\delta_c(s) = \int (\Omega_c(r) - <\Omega_c(r)>) \exp(2\pi\,is\,r)dv_r$$
$$\delta_c^2(s) = <\delta_c(r)\,\delta_c^*(s)>$$

and likewise for $\sigma_s(r)$ and $\delta_s(r)$.

This type of analysis has been used successfully to determine the spherically averaged structures $[<\Omega_c(r)>$ and $<\rho_s(r)>]$ of several systems including proteins (28–30), tomato bushy stunt virus (32), and serum lipoproteins (26,34).

More information can be obtained by analysis of the residual, $\delta(s)$. Although the requirements for doing this are very stringent, the analysis has been shown to be feasible in well-defined systems such as fibrogen (29), lysozyme (30), and ferritin (28). These analyses give a much better idea of the internal structure and shape; they cannot, however, give a unique solution. The random orientations of the particles result in the loss of information concerning the relative orientation of the structural components determined; thus the ultimate solution of small-angle scattering curves can lead only to an envelope of possible structures.

Unfortunately, the requirements are too strict for the straightforward application of this theory to the measured basic scattering curves of chromatin subunit particles. However, using the spirit of this theory we can still decompose the scattering data into a spherically symmetric part and a residual. To do this, we use the fact that $I_{cs}(s)$ is zero at 0.12 nm^{-1} (Fig. 9) and that at small s only the first term of $I_{cs}(s)$ in Eqs. (16)—that is, the spherically averaged terms—are significant. From this we conclude that $\sigma_c(r)$ is zero at 0.12 nm^{-1}, and that $\sigma_s(r)$ is negative in the first maximum at 0.13 nm^{-1} (Fig. 8). The resulting derived form of $\sigma_c^2(r)$ is consistent with a solid sphere, 52 Å in radius. This fits the dimensions observed in the subunits by electron microscopy (9,11,39,40), and with our own measurements of the radius of gyration (Table III; 19a,21–23,37). However, the discrepancies between V_c and the volume of the sphere indicate that the subunit cannot be a solid sphere. The structure must be highly convoluted, with water filling the crevices. We obtain $\sigma_s^2(r)$ by assuming that only the spherical terms contribute to the first maximum of $I_s(s)$. That $\sigma_s(s)$ is negative in this region dictates that $<\rho_s(r)>$ must be negative nearest its center, reflecting the results of measurements on the radius of gyration (Fig. 11). Fitting the form of the derived $\sigma_s^2(r)$ to a model of $<\rho_s(s)>$, we obtain dimensions for the low (protein) and high (DNA) scattering-length density region of the molecule

(Table III). The data are consistent with a spherically averaged structure of the particle having a protein core about 64 Å in diameter surrounded by a shell of DNA 20 Å thick, resulting in a particle of diameter 104 Å (*19a*). This model gives a calculated α in good agreement with that measured in Fig. 11 (Table III).

G. Substitution of Deuterated Components

Information about the internal structure of a particle—thus the distribution of its components—provided by the method of contrast variation results from changes in the relative contribution to the total scatter curve of each component as the contrast changes. Clearly, this can be accomplished also by incorporating selectively, into one or more components, atoms which change the total scatter of the components (Fig. 2). This approach, which is analogous to the method of X-ray isomorphous replacement, can be done with neutrons, by the specific inclusion of sufficient amounts of deuterium into nonlabile hydrogen sites. Methods have been developed for the incorporation of deuterium into algal proteins and DNA, supplying a source of material which can be incorporated into the chromosomes of higher organisms (*41*). Deuterated amino acids have successfully been incorporated into the chromatin of mouse tissue culture fibroblasts (*21*), and work is under way to obtain chromatins with deuterated proteins and/or DNA by reconstitution or biological methods (*41a*). A comparison of the internal structure functions derived from materials with labeled components with those obtained from unsubstituted material should improve our concepts of the distribution of protein and DNA within the subunits.

If deuterated components can be incorporated into components in a highly specific manner, then other methods may be used to determine the relative positions of each component. It has been demonstrated (*42–44*) that if a complex can be reconstituted with two different components appropriately labeled, then by taking a difference spectrum between a 50–50 mixture of doubly labeled and nonlabeled complex and 50–50 mixture of the two possible monolabeled complexes a spectrum results which gives the distance between the two subunits. By repeating the procedure for appropriate pairs, the arrangement of all the subunits in the particle can be completely determined, except for handedness. That the technique can work has been demonstrated for ribosomes (*45*) in which the distances were determined for the S2 and S5 and S8 proteins of the 30 S ribosomal subunit. The power of this technique makes certain that its application to chromosome structure will at some time be tested.

H.   Other Applications of the Method of Contrast Variation

In addition to the rather lengthy procedures outlined above for determining the distribution of components in a complex, less extensive small-angle measurements may be performed using the method of contrast variation. The protocol—which works best in two-component complexes—is to contrast-match with the solvent the average scattering-length density of one component. In this way information may be obtained about the distribution of the other components. For example, approximate values of the radii of gyration of the distribution of protein and DNA of chromatin subunits have been measured with neutrons as 30 and 50 A, respectively (21, 24). In a similar manner—aided by selective deuteration of the components—the distance has been determined between the centers of scattering mass of protein and RNA in 30 and 50 S ribosomes (31, 46). In both cases, however, the results must be interpreted with care, in that contrast-matching applies only to scattering intensity at zero angle. The fluctuations of the contrast-matched component are not removed, and they may contribute significantly to the scatter at finite angles, and thus contribute to the measured radius of gyration. The radius of gyration cannot, therefore, be measured for, one component of a particle without some contribution from the other components.

IV.   Application of Neutron Diffraction and Scatter to Chromosome Structure

Having reviewed the theory of neutron small-angle scatter, we turn to its application in determination of structure of the chromosome.

The theory of neutron small-angle scatter, as we have seen, can be used in its rigorous application only in very specific cases. These include the isolated chromosome subunits or any other chromosome structure which can be treated as an ensemble of independent particles, rods, or lamellar structures. Chromatin and whole chromosomes do not meet this criterion in that interactions are present which result in large low-angle scatter, interfering with the analysis at small angles. Regardless, there are still important applications of neutrons and the method of contrast variation to the larger, intact structures. For example, neutron detection techniques are more advanced than those of X-rays, and allow relative intensities as a function of scattering angle to be measured to high accuracy. In addition, the wavelengths of neutrons can be varied continuously over a wide range. Important in this respect is the availability of long-wavelength neutrons (greater than 8 Å), which when combined with the current design features of neutron small-angle cameras (see Appendix) results in the ability to sample very long structural repeats in chromatin and chromosomes.

A. Summary of Neutron Scatter Results of Chromatin Subunits

The application of neutron small-angle scatter to chromosome subunits has already been reviewed above as part of an introduction to the theory of neutron scatter. From these experiments (*19a*) information was obtained on the spherically averaged shape and internal structure functions $[< \Omega_c(r) >$ and $< \rho_s(r) >]$. The results (Table III) indicate that the particle consists of a protein core about 64 Å in diameter surrounded by a DNA-rich shell approximately 20 Å thick (Fig. 3), resulting in a particle of about 104 Å average diameter. The outer dimension is consistent with other available evidence (*9–11,39,40*), and the model fits very well the parameters derived from radius of gyration measurements. By fitting the scatter curve of the shape function $I_c(s)$, to the scatter curves of oblate and prolate ellipsoids (Figs. 7 and 8), an envelope of possible shapes is found. It is concluded that under no circumstances can the shape of the particle be such that the ratio of the shortest to longest distance of the equivalent ellipsoid is smaller than 0.5. In addition, the form of the derived $< \Omega_f(r) >$ and the discrepancy between the actual calculated particle volume and the volumes of the limiting sphere and ellipsoids (Table III) is consistent with the DNA layer being highly accessible to the solvent, while the protein core is relatively inaccessible. The highly convoluted nature of the particle implied by the large amount of free solvent within it means that $< \Omega_c(r) >$ cannot be described as a solid object; rather, the shape function, $< \Omega_c(r) >$, varies within the particle and has larger values toward the center. This "soft" appearance will account for the form of $I_c(s)$, even if the particle is spherical. Finally, the form of the basic scattering functions, $I_c(r)$, $I_s(r)$, and $I_{cs}(r)$ (Figs. 7–9) indicate that there are large contributions to the scatter from deviations of the structure from the derived spherical average; thus there is yet a large amount of information to be gleaned from these experiments.

B. Studies on Chromatin

X-Ray diffraction experiments of chromatin show a series of rings at equivalent spacings of about 10, 5.5, 3.7, 2.7, and 2.2 nm (*47–49*). Assuming that this series results from a single structural component of chromatin, a model was proposed with the DNA arranged in a superhelix of pitch 110 Å and diameter 130 Å (*50*). The first experiments using neutron small-angle scattering with the method of contrast variation were designed to test the underlying assumption of this model (16); for, if all the peaks in the X-ray series are from a single structural repeat, then all the diffraction rings should show similar behavior as the contrast is changed. The observed changes in the series of rings with contrast were not correlated; rather it was found that the ca. 10 and 3.7 nm peaks changed approximately together, but that their behavior was distinctly different from that of the 5.5 and 2.7 nm rings. It was concluded that there must be at least two, if not more, structural components responsible for the diffraction pattern of chromatin. Most likely one was due in large part to protein, the other to DNA.

Further studies on deuterated chromatins—one with the histones 60% deuterated, the other with DNA completely deuterated (see above)—have supported this conclusion (21,22).

Neutron scatter curves for chromatin (see Fig. 2 of Baldwin et al., 16) and films of chromatin subunit particles are similar in appearance and contrast behavior, (21,50a). However, if the films are dissolved and the scatter is measured at progressively lower concentrations, the 5.5 nm peak attenuates and the 10 nm peak moves to smaller angles (21). Eventually, in solutions of moderate ionic strength at concentrations less than 20%, no Bragg peak is observed in the region of s less than 0.1 nm $^{-1}$ (10 nm equivalent spacing). In this case all that remains of the scatter pattern observed in sub-unit films and chromatin in a broad peak about 3.7 nm (Fig. 4; 19a,22,23, 37). The inference is clear: The peak at ca. 3.7 nm in chromatin is part of the structure of the subunit. The ca. 10 nm peak is not; rather it arises from the arrangement of the particles in the chromatin fiber. The contrast behavior of this diffraction peak, along with its intensity variation with concentration in subunit films and in chromatins, deuterated and nondeuterated, supports this conclusion; but it also indicates that at this resolution the repeat cannot be thought of in terms of an array of discrete subunits. The repeat appears to be due to variation in scattering-length density along the chromatin fiber due to a regular spacing of protein in a continuous matrix of solvated DNA (21).

Studies on the contrast and concentration behavior of the ca. 3.7 nm peak demonstrates that it is part of the scatter from the protein core of the sub-unit (16,21).

The arrangement of the protein and DNA within the fiber may be further explored by experiments with oriented chromatin fibers (16,51). These studies have shown that the diffraction of the oriented fibers the ca. 10 nm peak has maxima at approximately 8° off the meridian. Translated into structural terms, this is consistent with the chromatin fiber consisting of a flat helix of average diameter 200 Å and pitch 100 Å.

According to this interpretation, there may be relatively little variation of scattering-length density along the chromatin strand or "string of beads" making up the helix, but there must be large fluctuations of scattering length density along the axis of the helix. This arrangement can be achieved if it is assumed that the strand which is coiled into the helix consists of subunit particles with histone cores in contact and DNA wrapped in a continuous coil—two turns per subunit—about the inner protein core.

The strand is then 100 Å thick and when coiled results in a helix 300 Å in maximal diameter with a 100 Å hole in the center. With these dimensions, each turn of the helix contains approximately 6 subunits. At the resolution of the data on which this model is based, the subunit particles lose their identity in the chromatin fiber. The model when refined at higher resolution must include the obvious discontinuities seen between the subunits by micrococcal nuclease.

Appendix

Neutron Cameras and Diffractometers

Neutron small-angle scattering cameras and diffractometers, like their X-ray counterparts, require a source, monochromator, collimator, sample holder, and a detector.

1. NEUTRON SOURCE

The source of neutrons for instruments routinely used in biological experiments is the pile of a steady-state reactor. The neutrons produced have very high velocities and correspondingly small wavelengths [Eq. (1)], which are not suitable for experiments of biological interest; thus the source is viewed through a moderator which slows the neutrons. The neutrons approximately equilibrate with the moderator and emerge with a Maxwell energy distribution of about the same temperature as the moderator. For neutrons the relationship between the temperature of the beam and the average speed of the neutrons is $\bar{v} = (2kT/m)^{1/2}$. The most probable wavelength is given by inserting $\bar{v}$ into the de Broglie equation [Eq. (1)]. The moderator temperature is chosen so that the beam contains an adequate flux of neutrons at a wavelength suitable for the desired measurement. For example, values of s [Eq. (2)] appropriate for neutron diffractometry of crystals (*52*) might require neutron wavelengths of 0.1–0.2 nm. A D_2O or H_2O moderator at room temperatore (300°K) would be best in this case. On the other hand, low-angle neutron scattering would best be performed at longer wavelengths, say 0.4–1.0 nm, and a cold source of liquid deuterium at 25°K would be used.

2. GEOMETRIES OF NEUTRON APPARATUS

The two major considerations in the design of neutron cameras and diffractometers is to provide an instrument of good resolution in s, while allowing sufficient neutron flux at the detector to give reasonable flexibility in measuring times and sample concentrations. From the equation for the scattering vector, $s(s = 2 \sin \theta/\lambda)$, it is seen that good monochromatization and angluar resolution are essential for good resolution of s. The former depends upon the monochromator (see below), and the latter depends upon the collimation, solid angle subtended by the detector elements, $\Delta \Omega$, and the geometry of the instrument.

Small wavelength bandwidth and detector angular resolution invariably imply low intensities. This problem is especially difficult with neutrons: even the most intense neutron sources are orders of magnitude less intense than modern X-ray generators (1). The designs of neutron cameras and diffractometers are a compromise between these two requirements. The low flux of neutron sources is compensated partly by their large area. If large source areas are combined with a large sample area, considerable gain is made in the number of neutrons irradiating the samples. Large source and sample areas will decrease resolution considerably if special features are not designed into the camera or diffractometer. Figure 12 illustrates two designs used for neutron apparatus.

The design in Fig. 12A is for small-angle neutron scatter measurements. Its striking feature is its great length (D11 at Institut Laue-Langevin, Grenoble, France is 80 meters long), which is necessary to maintain suitable resolution given the large source and sample areas. In this design, the source-sample (l) and sample-detector (L) distances may be altered to comply with resolution, s range, and intensity requirements. The largest sample-detector distances allow very small values of s to be explored. The detector used with this instrument allows simultaneous measurement of scattering intensity in a plane (3).

The scheme in Fig. 12B is not unlike that of some X-ray diffractometers, and is principally used for high-angle scatter measurements. The large source and sample areas are compensated in this design by the use of Söller slits. One set, in front of the sample, allows only a beam that is collimated at a chosen angle to the sample fact to irradiate the sample. A second set, after the sample, select a specific angle of scatter to be monitored by the detector. In this design only one value of s is sampled at a time. Other designs, which are hybrids of those shown in Fig. 12, are now coming into use.

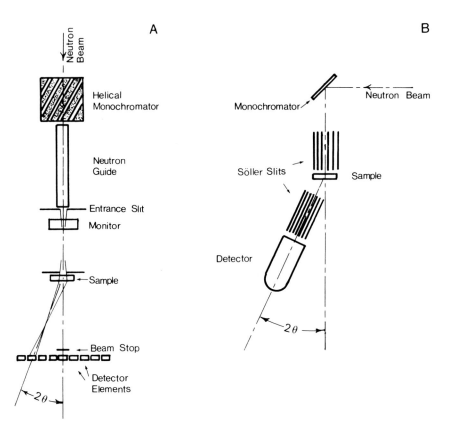

Fig. 12. Neutron small-angle camera and diffractometer. (A) Schematic of the basic layout for the neutron small-angle scattering camera, D11, at the Institut Laue-Langevin, Grenoble, France. The entrance slit to sample distance (*l*) and the sample to detector distance (*L*) can be changed. Drawing is after Schmatz *et al.* (*1*). Further details can be found in the description by Ibel (*3*). (B) Schematic of a neutron diffractometer. In this design, angles may be varied.

3. MONOCHROMATERS

Like X-rays, monochromatization may be done by crystals or filters (*53*). Unlike X-rays, neutrons may be monochromatized by velocity selectors. These devices select from the beam, neutrons with a relatively narrow speed distribution. A velocity selector, which is a drum with helical channels cut near its surface, is used in the small-angle detector in Fig. 12A. The desired wavelength is achieved by setting the angular velocity of the drum. The large wavelength bandwidth of this type of monochromator results in a large flux of neutrons incident on the sample. There is, however, a concomitant loss in resolution. In instruments where small wavelength bandwidth is important, a graphite (crystal) monochromator is employed. Such a device is used in the diffractometer of Fig. 12B. Here the monochromator is placed in the beam before the sample. It can also be used as an analyzer positioned between the Söller slits after the sample and the detector (2).

4. DETECTORS

Reactions which occur in nuclei after absorption of a neutron are used to detect neutrons in a scattering experiment.

Reactions (54) commonly used are:

$$^1n + {}^{10}B \nearrow {}^7Li^* + {}^4He \to {}^7Li + {}^4He \qquad (93\%)$$
$$\searrow {}^7Li + {}^4He \qquad (7\%)$$

$$^1n + {}^3He \to {}^1H + {}^3T$$

$$^1n + {}^6Li \to {}^3T + {}^4He$$

The charged reaction recoil products create secondary ionizations, which are in turn detected by their interaction with a charged electrode or a scintillating phosphor. Solid lithium or boron counters utilize scintillation counting, whereas gas counters of ^3He or BF_3 count ionizations by their charged pulses.

In our experience, the most common detectors are those using $^{10}BF_3$ or ^3He gas. Counters filled with these gases are of two types: the first, a one-dimensional counter, is associated with neutron diffraction designs such as that shown in Fig. 12B. These detectors are usually long gas-filled tubes. Because only one position need be counted at a time, the tube may be placed with the long axis parallel to the scattered neutron beam. This allows for a long flight path of neutrons in the tube, increasing the probability that the neutrons will be counted. A detector of high efficiency is the result.

The second basic detector design is that found in neutron small-angle scattering cameras, as illustrated in Fig. 12a, which count events in a two-dimensional field. The simplest of these position-sensitive detectors is an array of single-dimension counter tubes (1). More advanced designs, such as Dll, utilize a single flat, gas-filled envelope (3) containing a matrix or position-sensitive electrodes. Electrons produced by the secondary ionizations are attracted to a parallel array of cathode strips. The position of the strip receiving the maximum pulse gives one coordinate of the incident neutron. The pulse activates an array of anode wires set at right angles to the cathodes. The location of the anode receiving the positive ions gives the other coordinate of the neutron event. If no pulse is received by the anode within a time specified by a coincidence circuit in the detector, the cathode count is forfeited. This arrangement results in a detector with 50% efficiency. In the detector in Dll, 64 cathodes and 64 anodes, each 1 cm apart, are used, allowing events to be counted at 4096 discrete locations. The graininess implicit in this arrangement along with the sample-detector distance defines the solid angle of each element, $\Delta \Omega$, and thus the ultimate resolution of the machine.

ACKNOWLEDGMENTS

We wish to acknowledge the contributions of Mr. K. Simpson and Drs. B. G. Carpenter, P. Suau, and R. Hancock to the neutron small-angle scattering work presented here. Also, credit is due to Dr. G. G. Kneale for many of the calculations. The work was supported by the Science Research Council (U.K.). Dr. R. P. Hjelm was supported by a NATO fellowship (1974–1975), the Jane Coffin Childs Memorial Fund for Medical Research (1975–1976), and the Science Research Council (U.K.) (1976–1977).

REFERENCES

1. Schmatz, W., Springer, T., Schelten, J., and Ibel, K. *J. Appl. Crystallogr.* 7, 96. (1974).
2. Engelman, D. M., and Moore, P. B. *Annu. Rev. Biophys. Bioeng.* 4, 219. (1975).
3. Ibel, K. *J. Appl. Crystallogr.*, 9, 630. (1976).
4. Stuhrmann, H. B. *Acta Crystallogr., Sect. A* 26, 297. (1970).
5. Stuhrmann, H. B. *J. Mol. Biol.* 77, 363. (1973).
6. Ibel, K., and Stuhrmann, H. B. *J. Mol. Biol.* 93, 255. (1975).
7. Luzzati, V., Tardieu, A., Mateu, L., and Stuhrmann, H. B. *J. Mol. Biol.* 101, 115. (1976).
8. Huang, R. C. C., and Hjelm, R. P. *In* "Handbook of Genetics" (R. C. King, ed.), Vol. 5, pp. 55–116, Plenum, New York (1976).
9. Woodcock, C. L. F. *J. Cell Biol.* 59, 358a. (1973).
10. Olins, A. L., and Olins, D. E. *J. Cell Biol.* 59, 252a. (1973).
11. Olins, A. L., and Olins, D. E. *Science* 183, 330. (1974).
12. Hewish, D. R., and Burgoyne, L. A. *Biochem. Biophys. Res. Commun,* 52, 504. (1973).
13. Kornberg, R. D. *Science* 184, 868. (1974).
14. Noll, M. *Nature (London)* 251, 249. (1974).
15. Noll, M. *Nucleic Acids Res.* 1, 1573 (1974).
16. Baldwin, J. P., Boseley, P. G., Bradbury, E. M., and Ibel, K. *Nature (London)* 253, 245. (1975).
17. Littau, V. C., Burdick, C. J., Allfrey, V. G., and Mirsky, A. E. *Proc. Natl. Acad. Sci. U.S.A.* 54, 1204. (1965).
18. Bradbury, E. M., Carpenter, B. G., and Rattle, H. W. E. *Nature (London)* 241, 123. (1973).
19. Van Holde, K. E., and Isenberg, I. *Acc. Chem. Res.* 8, 327. (1975).
19a. Hjelm, R. P., Kneale, G. G., Baldwin, J. P., Suau, P., Bradbury, E. M., and Ibel, K. *Cell* 10, 139–151. (1977).
20. Bacon, G. E. "Neutron Diffraction," 3rd ed. Oxford Univ. Press, London and New York. (1975).
21. Bradbury, E. M., Baldwin, J. P., Carpenter, B. G., Hjelm, R. P., Hancock, R., and Ibel, K. *Brookhaven Symp. Biol.* 27, IV 97. (1975).
22. Bradbury, E. M., Hjelm, R. P., Carpenter, B. G. Baldwin, J. P., and Hancock, R. *In*

"CalTech Symposium of the Structure of the Genetic Apparatus" (P. O. P. T'so, ed.) p. 53. Elsevier, Amsterdam (1977).

23. Bradbury, E. M., Hjelm, R. P., Carpenter, B. G., Baldwin, J. P., and Kneale, G. G. *In* "Organization and Expression of Chromosomes" (V. G. Allfrey *et al.*, eds.) p. 223. Dahlem Konferenzen, Berlin (1977).

24. Pardon, J. F., Worcester, D. L., Wooley, J. C., Tatchell, K., Van Holde, K. E., and Richards, B. M. *Nucleic Acids Res.* **2**, 2163. (1975).

25. Burgoyne, L. A., Hewish, D. R., and Mobs, J. *Biochem. J.* **143**, 67. (1974).

26. Stuhrmann, H. B., Tardieu, A., Mateu, L., Sardet, C., Luzzati, V., Aggerbeck, L., and Scanu, A. M. *Proc. Natl. Acad. Sci. U.S.A.* **72**, 2270. (1975).

27. Schelten, J., Schlect, P., Schmatz, W., and Mayer, A. *J. Biol. Chem.* **247**, 5436. (1972).

28. Stuhrmann, H. B., Haas, J., Ibel, K., Koch, M. H. J., and Crichton, R. R. *J. Mol. Biol.* **100**, 399. (1976).

29. Marguerie, G., and Stuhrmann, H. B. *J. Mol. Biol.* **102**, 143. (1976).

30. Stuhrmann, H. B., and Fuess, H. *Acta Crystallogr., Sect. A* **32**, 67. (1976).

31. Moore, P. B., Engelman, D. M., and Schoenborn, B. P. *J. Mol. Biol.* **91**, 101. (1975).

32. Harrison, S. C. *J. Mol. Biol.* **42** 457. (1969).

33. Mateu, L., Tardieu, A., Luzzati, V., Aggerbeck, L., and Scanu, A. M. *J. Mol. Biol.* **70**, 105. (1972).

34. Tardieu, A., Mateu, L., Sardet, C., Weiss, B., Luzzati, V., Aggerbeck, L., and Scanu, A. M. *J. Mol. Biol.* **101**, 129. (1976).

35. Stuhrmann, H. B. *J. Appl. Crystallogr.* **7**, 173. (1974).

36. Stuhrmann, H. B., and Duee, E. D. *J. Appl. Crystallogr.* **8**, 538. (1975).

37. Bradbury, E. M., Hjelm, R. P., and Baldwin, J. P. *In* "Nucleic Acid-Protein Recognition" (H. J. Vogel, ed.) p. 117. Academic Press, New York. (1977).

38. Guinier, A. "X-ray Diffraction." Freeman, San Francisco, California. (1963).

39. Finch, J. T., Noll, M., and Kornberg, R. *Proc. Natl. Acad. Sci. U.S.A.* **72**, 3320 (1975).

40. Oudet, P., Gross-Bellard, M., and Chambon, P. *Cell* **4**, 281. (1975).

41. Katz, J. J., and Crespi, H. L. *ACS* **167**, 286 (1970).

41a. Simpson, K., *et al.* In preparation. (1977).

42. Hoppe, W. *Isr. J. Chem.* **10**, 321. (1972).

43. Hoppe, W. *J. Mol. Biol.* **78**, 581. (1973).

44. Engelman, D. M., and Moore, P. B. *Proc. Natl. Acad. Sci. U.S.A.* **69**, 1997. (1972).

45. Engelman, D. M., Moore, P. B., and Schoenborn, B. P. *Proc. Natl. Acad. Sci. U.S.A.* **72**, 3888. (1975).

46. Moore, P. B., Engelman, D. M., and Schoenborn, B. P. *Proc. Natl. Acad. Sci. U.S.A.* **71**, 172. (1974).

47. Wilkins, M. H. F., Zubay, G., and Wilson, H. R. *J. Mol. Biol.* **1**, 179. (1959).

48. Luzzati, V., and Nicolaieff, A. *J. Mol. Biol.* **7**, 142. (1963).

49. Subirana, J. A., and Puigjaner, L. C. *Proc. Natl. Acad. Sci. U.S.A.* **71**, 1672. (1974).

50. Pardon, J. F., and Wilkins, M. H. F. *J. Mol. Biol.* **68**, 115. (1972).

50a. Baldwin J. P. *et al.* In preparation. (1977).

51. Carpenter, B. G., Baldwin, J. P., Bradbury, E. M., and Ibel, K. *Nucleic Acids Res.* **3**, 1739–1796. (1976).

52. Schoenborn, B. P. *Cold Spring Harbor Symp. Quant. Biol.* **36**, 569. (1971).

53. Iyengar, P. K. *In* "Thermal Neutron Scattering" (P. A. Egelstaff, ed.), p. 98. Academic Press, New York. (1965).

54. Cocking, S. J., and Webb, F. J. *In* "Thermal Neutron Scattering" (P. A. Egelstaff, ed.), p. 142. Academic Press, New York. (1965).

Chapter 17

Circular Dichroism Analysis of Chromatin and DNA–Nuclear Protein Complexes[1]

GERALD D. FASMAN

Graduate Department of Biochemistry,
Brandeis University,
Waltham, Massachusetts

I. Introduction

The conformation of biological molecules plays an important role in determining their function and activity. Through conformational transitions (e.g., allosteric) the activity of many biological entities determines the control and regulation of the genome in cellular metabolism, replication, and development. One of the most studied biological molecules, DNA, exemplifies the importance of conformation in the cell cycle.

Of the many physical–chemical probes available to study conformation and conformational transitions, circular dichroism (CD) is one of the most sensitive and most widely used techniques.

The optical activity of nucleic acids is made up of several components. The basic asymmetry of the individual components, e.g., deoxyribose, the nature of the anomeric configuration at the C-1′ of the pentose, etc., contribute to the overall CD of nucleic acids, but it is the conformation of the nucleic acid chain, i.e., the asymmetry of the double helix, which makes the largest contribution to the CD spectra. It is the interaction between the chromophores, the purine and pyrimidine bases, as the chain is built through dimer, trimer, etc., to the final nucleic acid structure, which plays the most significant role in the CD spectra, and which is the main concern in this article.

[1] Contribution No. 1091 of the Graduate Department of Biochemistry, Brandeis University, Waltham, Massachusetts. The writing of this article was generously supported by grants from the U.S. Public Health Service (GM 17533), National Science Foundation (GB 29204X), and the American Cancer Society (P-577).

327

II. The Phenomena of Circular Dichroism and Optical Rotatory Dispersion

The main elemental aspects of the theory of CD will be reviewed briefly. For more detailed discussions the following articles are recommended: Moscowitz (*1*), Mommaerts (*2*), Yang and Samejima (*3*), Brahms and Brahms (*4*), Gratzer (*5*), Bush and Brahms (*6*), and Bush (*7*).

A beam of linearly polarized light of wavelength λ can be considered as the sum of two components: beams of right- and left-handed circularly polarized light, with electric vectors E_R and E_L, respectively. When such light interacts with an asymmetric molecule (such as most biological macromolecules), two phenomena, CD and optical rotatory dispersion (ORD), are observed, and the molecule is said to be optically active. These phenomena arise from the following events:

1. E_R and E_L travel at different speeds through the molecule. This difference in refractive index leads to optical rotation, the rotation of the plane of polarization, measured in degrees of rotation, α_λ. ORD is the dependence of this rotation upon wavelength. In a region where the molecule does not absorb light, the rotation plotted against wavelength yields a plain curve. In the region of light absorption, however, the dispersion is anomalous. The rotation first increases sharply in one direction, falls to zero at the absorption maximum, and then rises sharply in the opposite direction. This anomalous dispersion is called a Cotton effect.

2. In the region of its Cotton effect, an asymmetric molecule which exhibits ORD will also show unequal absorption of left- and right-handed circularly polarized light; this difference in extinction coefficient ($\epsilon_L - \epsilon_R$) is known as circular dichroism and can be measured directly in some instruments as a differential absorbance. When CD occurs the emerging light beam is no longer linearly polarized, but instead is elliptically polarized. Thus, the ellipticity of the resulting light, θ_λ, is another measure of CD, and is proportional to ($\epsilon_L - \epsilon_R$). A typical absorption band with its associated ORD and CD cotton effects is shown in Fig. 1. Both dispersive and absorption phenomena are caused by the same charge displacements in a molecule; therefore ORD and CD are closely related to each other. By means of the Kronig–Kramers transform equations developed by Moscowitz (*1*), ORD curves can theoretically be computed from CD data and vice versa. This calculation is sometimes useful for evaluation of CD bands at very low wavelength.

The two requirements necessary for optical activity are absorption and asymmetry. The four main bases found in DNA are composed of two pyrimidines, thymine and cytosine, and the two purines, adenine and guanine.

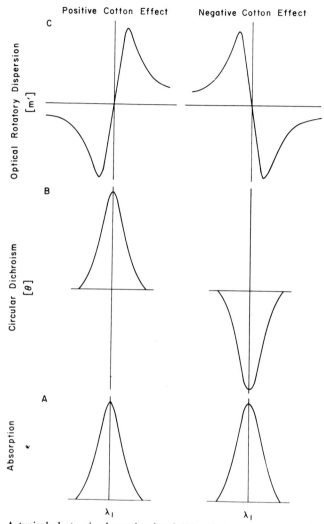

FIG. 1. A typical electronic absorption band (A) with its associated circular dichroism (B) and optical rotatory dispersion (C) curves.

These are the basic chromophores responsible for light absorption in the ultraviolet region (300–185 nm). The main electronic transitions responsible for this absorption are the $\pi \rightarrow \pi^*$ and $n \rightarrow \pi^*$ transitions. The asymmetry present has been mentioned above.

Circular dichroism has frequently been employed to detect the asymmetry of DNA, i.e., a conformational probe. Although the exact conformation of DNA cannot always be obtained from the CD spectra, despite the fact that theoretical studies along these lines are well advanced (see Brahms and Brahms, 4), it is an excellent probe to follow or detect changes in asymmetry, i.e., conformational changes. This is the main application to be utilized in this discussion of the application of CD spectra for the study of chromatin structure.

When two or more chromophores, which contain asymmetric centers are stacked, e.g., the two DNA strands of the double helix, and their

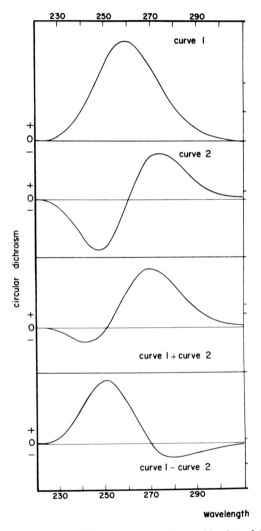

FIG. 2. The shape of circular dichroism curves. The combination of the top two curves gives all others. From Tinoco (8). Reprinted with permission of *Journal de Chimie Physique*.

transition moments interact, several CD spectra are theoretically possible, as seen in Fig. 2 (*8*). The B form of DNA has a spectrum similar to curve 2, which has been called a conservative spectrum. Once the double strand has been formed, it is still possible to alter its dimensions (asymmetry) without destroying the base-pairing, by tilting the bases relative either to the vertical axis or to the horizontal axis. This is shown in Fig. 3, where the tilt and twist of the planar bases are shown. Changes of asymmetry of this type cause different transition moment interactions and consequently changes in the CD spectra. Thus if binding a protein to DNA causes a tightening of the DNA helix, or causes a twisting of the bases, i.e., a conformational change, a different CD spectrum will be produced.

X-ray analysis of DNA has demonstrated three canonical forms termed A, B, C (*9–11*). These forms differ in the number of residues per turn, pitch, translation per residue, angle between perpendicular to helix axis and bases, etc. (see Table 1 of Davies, *12*). These various forms can be produced in oriented fibers of DNA under specific conditions of relative humidity and ionic strength. One of the most significant differences in these forms is the angle between the perpendicular to the helix axis and the bases (tilt). The A form has a 20° tilt and the B form has a 2° tilt. Utilizing the same set of conditions, Tunis-Schneider and Maestre (*13*) obtained films of DNA and examined their CD spectra, they found that each of the A, B, and C forms of DNA had characteristic spectra, shown in Fig. 4 (*14*).

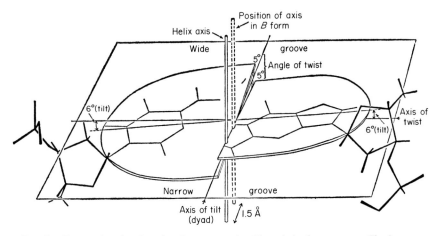

FIG. 3. Perspective drawing showing one nucleotide pair in the structure. The bases are shown as if drawn on solid plates which are rotated around the axes of tilt and twist by 6° and 5°, respectively. The corresponding angles in the model for the B-form DNA are 2° tilt, and 5° twist. The base pair is moved 1.5 Å away from the helix axis relative to its position in B-form DNA. From Marvin *et al.* (*9*). Reprinted with permission of Academic Press.

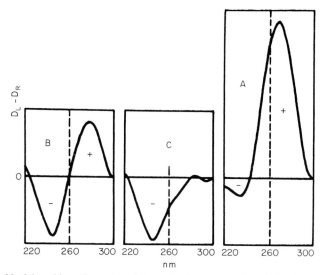

Fɪɢ. 4. Nucleic acid conformation. Schematized representation of the circular dichroism spectra for the A, B, and C forms of DNA. The dotted lines are drawn through the absorption maximum. From Ivanov *et al.* (*14*). Reprinted with permission of John Wiley & Sons, Inc.

thus CD spectra can easily identify which of three forms of DNA exists under a particular set of conditions. The B-form spectrum is also that found in aqueous solution.

Recently, Lerman *et al.* (*15,16*) have shown that DNA undergoes a cooperative structural change which results in a compact molecular conformation, a condensed state, by the addition of a neutral polymer, such as polyethylene oxide (PEO), in the presence of salt. This spontaneous rearrangement to an ordered tertiary structure is characterized by a CD spectrum differing greatly from that found in aqueous solution. As a critical concentration of polymer (PEO) and s salt are required for this induced transition, they have termed this PSI or Ψ-DNA. The CD spectrum of Ψ-DNA is seen in Fig. 5, produced by various concentrations of PEO, after various times of mixing, which causes an extremely large negative ellipticity band, totally eliminating the positive peak. Lerman *et al.* (*15–17*) believe that this spectrum is caused by a superfolding of the DNA (i.e., ordered spaghetti) and may be comparable to the condensed form of DNA in the chromosome. This new spectrum is similar to that obtained by other means (Fig. 6), such as by complexing with histone Hl or poly-ʟ lysine (*16*).

The CD spectrum of DNA changes drastically upon addition of alcohol, presumably indicating conformational changes. An example of such studies is seen in Fig. 7 (*14*). At concentrations less than 65% ETOH, the

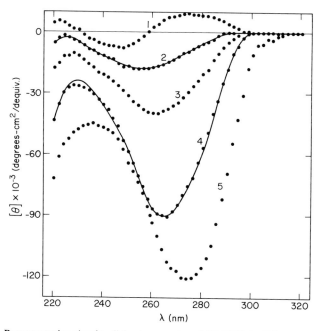

FIG. 5. Representative circular dichroism spectra of T7 DNA in different concentrations of polyethylene oxide (PEO) at different times after mixing. The circles represent measured spectra in PEO concentrations at the time after mixing as follows: curve 1, 80.9 mg/ml, 0.7 hour; 2, 98.1 mg/ml, 0.7 hour; 3, 126.5 mg/ml, 0.7 hour; 4, 90.1 mg/ml, 0.7 hour; 5, 126.5 mg/ml, 48 hours. DNA = $22 \times 10^{-6} N$ in 0.26 M Na$^+$. From Jordon *et al.* (*16*). Reprinted with permission of Macmillian Journals, Ltd.

CD spectrum is that of B-DNA. As additional alcohol is added, the positive ellipticity band increases to a maximum at 78% ETOH; the spectrum is now similar to that of A-type DNA.

The other main component to be discussed herein is the group of proteins which are associated with DNA in chromatin. Proteins and polypeptides are capable of assuming three main conformations, the α-helix, the β-pleated sheet, and the random coil (or irregular conformation). The CD spectra of these three conformations are seen in Fig. 8 (*19*). Curve 1, that observed for the α-helix, has two negative troughs at 222 and 208 nm and a positive peak at 191 nm. Curve 2, that found for the β-pleated sheet, has a negative band at 217 nm and a positive peak at 195 nm; and curve 3, that observed for the random conformation, has a small positive peak at 217 nm and a large negative band at 197 nm. Therefore, the three basic conformations of proteins are easily distinguished by their CD spectra. Thus the basic CD spectrum of DNA in its various forms, plus the basic protein CD spectrum of the known structures, are available to interpret the spec-

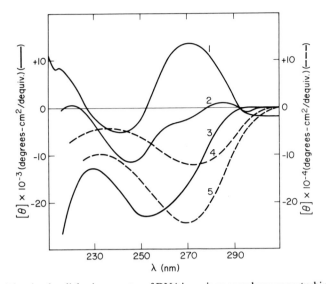

FIG. 6. The circular dichroism spectra of DNA in various complexes or perturbing solvents. Curves are redrawn from published data. 1, DNA:H4 complex, $r = 1.5$; 2, DNA in ethylene glycol; 3, DNA:H1 complex, $r = 1.0$; 4, T7 DNA in 0.2 M NaCl and 126.5 mg/ml PEO, 48 hours after mixing; 5, DNA:poly-L-lysine complex, $r = 1.1$. All data except for curve 4 are with calf thymus DNA. The ordinate calibration on left applies to solid curves, and the calibration on right, corresponding to 10-fold larger values, applies to the dashed curves. From Jordon *et al.* (*16*). Reprinted with permission of Macmillian Journals.

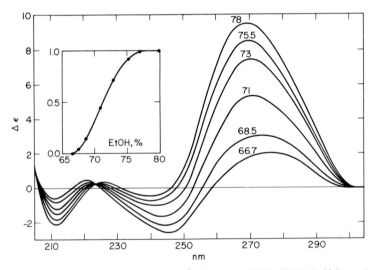

FIG. 7. Influence of ethanol upon the circular dichroism (CD) of DNA (calf thymus). DNA concentration = 5 × 10^{-5} M (PO$_4$) containing 5 × 10^{-4} M NaCl. Curves are labeled with the percent ethanol. The dependence of the relative CD change at 270 nm upon ethanol concentration is shown in the inset. From Ivanov *et al.* (*14*). Reprinted with permission of John Wiley & Sons.

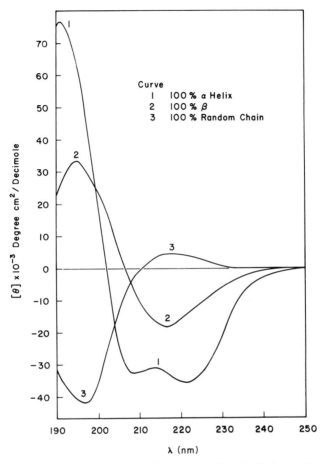

FIG. 8. Circular dichroism spectra of poly-L-lysine in the α-helical, β, and random conformation. Reprinted with permission from Greenfield, N., and Fasman, G. D. *Biochemistry*, **8**, 4108 (1969). Copyright by the American Chemical Society.

trum of chromatin. However, the complex spectrum of chromatin is not so easily unraveled into its component parts.

The CD spectrum of chromatin (Fig. 9) shows that DNA in the presence of chromosomal proteins is considerably altered relative to DNA (also shown) (*18*). The positive band of DNA at 277 nm (8500°) is diminished to ≅ 4000°. The chromatin CD spectrum below 240 nm is attributable to proteins in a partial α-helical (≅ 40%) and random conformation. Upon removal of some of the chromosomal proteins by the addition of sodium dodecyl sulfate, the DNA CD spectrum returns to that of native DNA, indicating that it is the proteins which are responsible for the altered CD

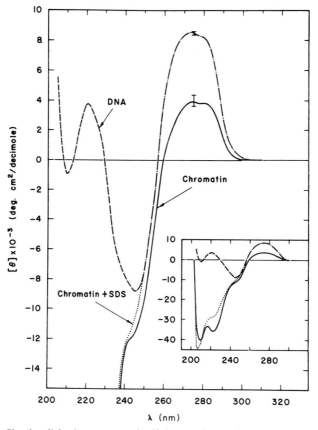

FIG. 9. Circular dichroism spectra of calf thymus chromatin, chromatin in 0.1% sodium dodecyl sulfate (SDS), and DNA. —, Calf thymus chromatin; . . . . . , chromatin in the presence of 0.1% (w/w) SDS; ---, pure calf thymus DNA. Solutions in 0.14 M NaF–10 mM Tris-HCl (pH 8.0). Concentrated DNA = 1.6 × 10^{-4} M (PO$_4$). Mean residue ellipticity is based on DNA residue concentration. From Shih and Fasman (18). Reprinted with permission of Academic Press.

spectrum. The proteins which are dissociated contain less α-helical content than when bound to DNA. Thus the conformations of the DNA and proteins are interdependent. This evidence would indicate that DNA in chromatin has been altered from the B form, or the overall asymmetry of the DNA in chromatin is different from that found in native DNA alone. The supercoil model of Wilkins *et al.* (see review by Pardon and Richards, 20) would also introduce a new larger asymmetric unit capable of producing different CD spectra. Recent studies (21) also indicate that assemblies of nucleosomes are arranged in an asymmetric supercoil.

Studies of the binding of histones to DNA (22; see also review by Fasman,

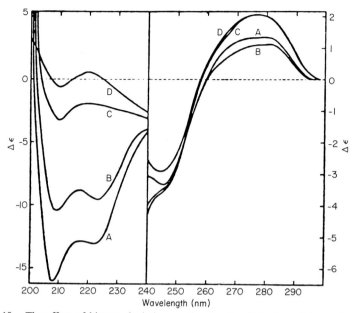

FIG. 10. The effect of histone depletion on the circular dichroism of deoxyribonucleo-histone (DNH) calf thymus. Curve A,. Whole DNH; B, DNH treated with 0.7 M NaCl; C, DNH treated with 1.0 M NaCl; D, DNA. Solvent 0.7 mM sodium phosphate (pH 6.8). From Hensen and Walker (24). Reprinted with permission of Springer-Verlag, New York, Inc.

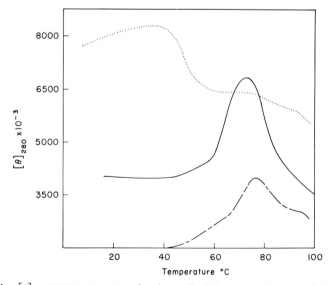

FIG. 11. $[\theta]_{280}$, versus temperature for chromatin (—), mononucleosomes (-·-), and DNA (....) in 2.5×10^{-4} M EDTA (pH 7.0). From Mandel and Fasman (26).

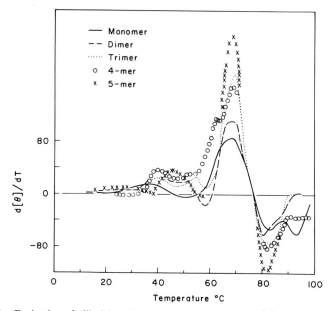

FIG. 12. Derivative of ellipticity with respect to temperature, $d[\theta]_{280}/dT$, versus temperature. The curves have been artificially displaced for clearer display. From Mandel and Fasman (*26*).

23) have shown changes in the CD spectrum of DNA, each histone yields a characteristic DNA CD spectrum.

Another approach to the study of chromatin is to remove the histones stepwise and examine the resulting nucleoprotein complex to evaluate the effect of the deleted histone. The CD spectra of Fig. 10 are illustrative of this approach (*24*).

A technique widely used in DNA stability studies is that of following the denaturation by hyperchromacity change as a function of temperature. A similar approach has been developed to follow DNA denaturation by observing the CD changes during such transitions (*25*). The CD behavior, $[\theta]_{280}$, as a function of temperature is shown in Fig. 11 (*26*) for DNA, chromatin, and mononucleosomes (*21*), and differences are clearly discernible. The derivative plot $d[\theta]/dT$ gives a clearer indication of the various melting temperatures, T_{m}s, as well as the relative amounts of nucleoprotein melting at each temperature. Such a derivative plot is shown in Fig. 12, for the CD meltout, of chromatin nuclease digests, for monomers, dimers, trimers, tetramers, and pentanucleosomes (*21*).

The studies briefly discussed have demonstrated the utility of using CD as a very sensitive probe of conformational changes associated with chromatin. Both the secondary structure of DNA (double helix) as well as the unfolding of the tertiary structure (e.g., supercoil) can be monitored by this technique.

III. Experimental Methods

A. General Considerations

The measurement of the optical activity (ORD or CD) is dependent upon the light which traverses the sample cell and strikes the photomultiplier tube. In order to obtain a meaningful reading, sufficient light must pass through the cell to ensure a suitably high signal-to-noise ratio. This requirement becomes important to consider below 225 nm, where light sources begin to lose intensity, and when light is absorbed more strongly by the chromophores measured, by the optics and by the solvents and buffers employed. In such cases the parameters of protein concentration, cell path length, and solvent must be chosen with care. Under conditions of high absorbance and low CD signal, it is desirable to reduce path length and/or concentration.

To ensure that a rotational band (or spectral) is fully resolved and has reached its true amplitude, the natural bandwidth[2] of the peak should be greater than ten times the spectral bandwidth[2] of the light beam; i.e., the light must be monochromatic.

At high absorbance another problem is stray light, since it can give rise to rotational artifacts. Solutions of high absorbance eliminate light for which the monochromator is set, while permitting stray light outside the pass band to reach the analyzer and gives rise to rotations characteristic for those wavelengths. Rotational artifacts occur at optical densities greater than 1.5–2.5, but in practice an optical density greater than 1.5 should be avoided. To investigate whether a CD band is genuine, one employs the Beer's law test. One measures the observed anomaly at two different concentrations of the chromophore. The difference in concentration should cause no change in amplitude, shape, or wavelength. Most recording spectropolarimeters have a variable pen period (τ), which is usually critically damped. Therefore, a brief discussion of how the time constant is relevant to CD data may

---

[2] The natural bandwidth (nbw) of a sample is the width at half height of the absorption peak. It is independent of instrument bandwidth, being an intrinsic sample characteristic. The spectral bandwidth (sbw) of an instrument is defined as that band of wavelengths from which 75% of the energy in the passed beam originates. The sbw is numerically equal to the physical slit width in millimeters multiplied by the resolution (or reciprocal dispersion function) of the monochromator.

be in order. The noise level often becomes too high to allow precise data to be gathered when CD measurements are light-limited (due to high absorbance or low lamp intensity). Under these conditions the experimenter should switch to a longer pen period, if possible, to help average out the noise, since the root-mean-square noise is inversely proportional to the square root of τ of the recording system.[3] Furthermore, a large τ is essential for well resolved narrow Cotton effects, especially when the slit width must be small, but this is not usually a problem in nucleic acid work.

To ensure proper resolution, CD spectra should not be scanned at a rate greater than the spectral bandwidth divided by τ. This consideration determines the maximum allowable scan speed. For example, if the sbw is 1.5 nm and τ is 3 seconds, then the scan speed should be slower than 0.5 nm per second. In practice, a compromise must often be struck between a tolerable noise level and a scan speed sufficient for data collection.

Optical rotation is very sensitive to temperature, and this is particularly true when heating induces changes in macromolecular conformation. Oligonucleotides, RNA, and polyribonucleic acids are especially subject to gradual transitions, spread out over a wide temperature range. Thus, temperature should be carefully controlled and recorded. When practical, solutions should be measured in waterjacketed, thermostatted cells. At the very least, "room temperature" measurements should be made in constant-temperature rooms. The reproducibility of the data is aided by an instrument with thermostatted optical and electrical components.

Most modern spectropolarimeters are well suited for nucleic acid measurements. Data are reproducible to within 1 m°, optical densities of up to 2 can usually be tolerated without artifacts, the noise is easily averaged by eye, the solvent base line can be made to read zero at all wavelengths (through use of the multipot mechanism), and the entire optical system can be thermostatted as well as the sample. Significant data can be taken, on nucleic acids, over the range 187–600 nm, provided transparent solvents are used (see below), and provided the polarimeter compartment is flushed with pure nitrogen at $\lambda \leq 200$ nm. Mononucleotides, because of their small rotations even in the far ultraviolet, begin to have low signal-to-noise ratios at 210 or 220 nm and thus cannot be accurately measured at lower wavelengths. A note on the use of nitrogen: it has been found (a) that a flow rate of 4 cubic feet per hour through each compartment (monochromator, lamp housing, and polarimeter when required) is sufficient, (b) that half this flow rate is quite enough for a constant purge when the equipment is not in use, which is recommended to maintain the optical system in optimum condition,

---

[3] Cary Instruments, "Optimum Spectrophotometer Parameters," Report A R 14-2 (1964).

TABLE I

EXAMPLE OF SCANNING CONDITIONS[a]

Pen period (sec)	Maximum scan speed (nm/min)	Maximum dynode voltage (kV)
3	12	0.4
10	4	0.5
30	1	0.6

[a] Recommended values for the Cary 60 recording spectropolarimeter.

and (c) that nitrogen must be run through the polarimeter compartment to prevent water condensation when experiments are carried out at below room temperature.

The following may serve as an example of the relationships between scan speed, pen period, and the amount of light reaching the photomultiplier. With a constant spectral band width of 1.5 nm and a chart calibration of 5 nm/cm, the scanning conditions recommended are indicated in Table I. This table shows how the variables may be adjusted to help compensate for the increased noise level at high absorbance and at low wavelengths. Under these conditions, the noise level varies from 0.5 m° (for all samples down to about 250 nm, and for solvent blanks down to 210 nm) to 4 m° (at the 195 nm through for polycytidylic acid with an optical density of nearly 3 at that wavelength). The Cary instrument can be equipped with a differential cell holder for direct measurement of difference CD spectra.

B.   Solutions

1.   SOLVENTS

A solvent must be chosen which transmits throughout the wavelength region of interest. Water is the usual solvent for nucleic acid work. It should be degassed, if measurements in the far-ultraviolet region (≤ 200 nm) are required, to remove dissolved oxygen, which absorbs in this region. Tris-(hydroxymethyl)aminomethane and acetate buffers, in moderate concentration, may be used at $\lambda \geq 200$ nm, but in the far ultraviolet there is no buffer of suitably low absorbance. Fluoride and perchlorate salts may be used down to the spectral limit of CD instrumentation, but most common salts, especially hydroxides and phosphates, are to be avoided in the far ultraviolet. The absorbance of various ions is shown in Fig. 13 (27).

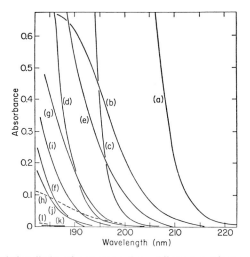

FIG. 13.   Transmission limits of common anions. All spectra refer to aqueous solution in a 2-mm path cell. Analytical grade reagents are used, without further purification: (a) sodium hydroxide (pH 11.86). (b) sodium hydroxide (pH 10.9); (c) 10 mM sodium chloride; (d) 1 mM sodium chloride; (e) 10 mM disodium phosphate; (f) 10 mM monosodium phosphate; (g) 10 mM sodium sulfate; (h) 1 mM disodium phosphate; (i) 10 mM boric acid + sodium hydroxide (pH 9.1); (j) 0.1 M perchloric acid; (k) 0.1 M sodium fluoride; (l) 10 mM boric acid. From Gratzer (27). By courtesy of Marcel Dekker, Inc.

2.   Preparation of Solutions

The solutions to be measured must be optically pure; solutions containing suspended material or gellike particles may be birefringent and cause optical anomalies. However, opalescence produced artificially, e.g., added Celite or colloidal sulfur, causes no rotational artifacts. These suspensions did increase the optical density and hence the noise level.

3.   Sample Concentration

It has already been implied that the choice of a suitable nucleic acid concentration is often a compromise between sufficient rotation and adequate light intensity. For measurements in the vicinity of the 260 nm Cotton effects, a nucleotide concentration (in terms of phosphate or sugar or base, but not of macromolecule concentration) of about 10^{-4} M is often suitable. Such solutions, in 1-cm cells, have maximum optical densities of 0.5–1 in this region, and CD measurements require only about 25 μg of material. Farther in the ultraviolet, lower concentrations are advisable. The absorption spectrum in the range of interest, or at least the maximum optical density should be recorded under carefully controlled conditions of pH, salt concentration, and temperature. Preferably the same cell should be used for CD and for absorption measurements.

The concentration of samples must be known accurately in order to calculate molar or residue rotations. Extinction coefficients, in terms of base or phosphate concentration, are known for many nucleosides and nucleotides[4,5] (28,29), for some polyribonucleotides (29–31), and for some poly-deoxyribonucleotides (29,32,33). The absorption spectra of natural RNAs and DNAs vary with the source. With oligonucleotides, hypochromicity is strongly dependent upon chain length (34). In cases where the extinction coefficients are unknown, the nucleotide concentration must be determined either by analytical phosphate determination (35) or by perchlorate digestion (36), by orcinol assay (37) (for ribose species only) or by hydrolysis to the component monomers (for homopolymers only).

The concentration of DNA in chromatin or complexes can be determined by ultraviolet absorption at 260 nm in the presence of 0.1% (w/v) SDS, using ϵ_p 6800 (36).

C. Cells

Only fused quartz, circular cells should be used for ORD and CD. Such cells are available[6] in a great variety of path lengths, and special adaptations (such as water jacketed), and should be tested for birefringence (38). For path lengths of 1 mm or less, face-filling cells and double-necked cells are recommended because they are relatively easy to fill and clean. In any case, syringes fitted with thin Teflon tubing will be needed for filling thin cells. The exact path length of thin cells (1 mm or less) may be measured by counting interference fringes in the near infrared (39); this procedure is recommended for cells thinner than 1 mm.

Cell holders should be designed so that cells can be positioned reproducibly and firmly; V-blocks are not usually suitable. A cell holder assembly for CD measurements in the Cary 60 has been described (38). A very similar assembly can be machined for use in the Cary CD compartment. However, because of space limitations, the brass block with the groove (part A, Fig. 3, of reference 38) must be removable; it can be equipped with two holes on its underside which fit onto pins on a small brass plate permanently fixed to the floor of the sample compartment. Similar assemblies can be designed for other instruments.

Many organic solvents in which DNA is soluble, such as dimethylforma-

[4] Pabst Laboratories Circular OR-10, Pabst Brewing Co., Milwaukee, Wisconsin, 1956.
[5] Schwarz BioResearch, Inc. Catalog, Orangeburg, New York, 1966.
[6] Manufactured by Opticell Co., 10792 Tucker St., Beltsville, Maryland 20705.

mide and dimethylsulfoxide are not suitable for rotatory studies at $\lambda >$ 250 nm unless cells of extremely thin path length are used (40,41).

D. CD Measurements

The choice of path length is an important consideration under conditions of high absorbance and low rotation. Shorter path lengths often are desirable in order to reduce the absorbance due to the sample, especially in regions displaying Cotton effects. The transmittance increases logarithmically with concentration and path length (Beer–Lambert law), while the rotation increases linearly with these variables (Biot law). Thus by a reduction of the path length, an overall gain is achieved by diminishing the absorbance, which more than compensates for the loss in rotation. For example, if the path length is shortened from 1 cm to 1 mm, a 10-fold loss in rotation is accompanied by a 60-fold increase in light intensity at an absorbance of 2.0.

Measurements should be made with the solutions, and then a cell blank with the identical solvent should be run immediately afterward, at the same sensitivity and over the same wavelength range. Frequent checks should be made on the air zero of the instrument to monitor whether the lamp arc position has jumped. The cell should fit snugly in the holder, but should not be forced or strained, and should be assembled in exactly the same position for each CD spectrum. Ideally, it is best not to remove the cell from the holder at all between measurements of the solution and the solvent blank. Cells may be filled and emptied by means of a syringe. The outside optical surfaces should not be touched even with pure solvent, between runs, so that the cell blank will not change.

Cells should be cleaned, at the end of a set of experiments, first with the solvent used, then with water, then with methanol, and should be blown dry with nitrogen. The outside surfaces may be washed with methanol and then dried gently, in a circular direction, with lens paper. More vigorous cleaning can be accomplished by brief contact with chromic acid, or by longer treatment with equal parts of 3 N HCl and 95% ethanol.

E. Reporting of Data

It is of value to include experimental details in publications, so that other workers can assess the precision of the data. Tables or plots of molar rotations should be accompanied by an indication of (a) the magnitude of measured rotations, (b) noise level, (c) concentration and maximum optical density, (d) path length, (e) pen period and scan speed, (f) spectral bandwidth or physical slit width, and (g) temperature control. It is advisable to check and note whether the rotation is proportional to concentration and to path length, especially at high absorbance.

F. Calibration

The absolute accuracy of any quantitative measurement depends upon the standardization of the method. It is easy to check the calibration of ORD instruments: specific rotation values (at several wavelengths from 250 to 589 nm) of a 0.25% solution of National Bureau of Standards sucrose can be compared to literature values (42). If necessary the polarimeter can be re-calibrated.

However, there is at present no CD standard available with the consistent purity of National Bureau of Standards sucrose. The compound commonly used for calibration of circular dichrometers is d-10-camphorsulfonic acid in 0.1% aqueous solution, which displays a large ellipticity band at 290 nm. But d-10-camphorsulfonic acid forms a hydrate containing about 7% water under normal laboratory conditions (43,44), so that weight may not be an accurate measure of concentration. Furthermore, yellow impurities were found[7] in some batches of reagent grade (Eastman Kodak) d-10-camphor-sulfonic acid. The acid may be purified by recrystallization from acetic acid (43), followed by vacuum sublimation, drying at 80° under vacuum, and storage in a desiccator.[7] It is then suitable as a CD standard by means of which the signal gain adjustment controlling the magnitude of the observed CD signal on an instrument can be manipulated.

The exact value of the peak molecular ellipticity, $[\theta]_{290}$, of d-10-camphor-sulfonic acid is not known with certainty, partly because of impurity problems. Fortunately, the $[\theta]_{290}$ value can be calculated (43–45) by means of the Kronig–Kramers transform, from accurate ORD data on the same d-10-camphorsulfonic acid sample obtained on a well calibrated polarimeter. For this calculation to be valid the sample must not contain optically active impurities, although small amounts of water are tolerable. A simple way to obtain an absolute CD value for a rotationally pure aqueous solution of 10-camphorsulfonic acid standard is to use Cassim and Yang's (45) calculated ratios of peak molecular ellipticity to peak and trough molecular rotations: $[\theta]_{290}/[M]_{306} = 1.76$ and $[\theta]_{290}/[M]_{270} = -1.37$. For example, one dry, purified sample of 10-camphorsulfonic acid[7] yielded measured rotations of $[M]_{306} = +4480$ and $[M]_{270} = -5700$, from which $[\theta]_{290}$ equals the average of $4480 \times 1.76 = 7880$ and $-5700 \times -1.37 = 7800$, or $[\theta]_{290} = 7840$ (corresponding to $\Delta\epsilon_{L-R} = 2.37$). The resulting $[\theta]_{290}$ magnitude can then be used to calibrate the circular dichrometer, even though the 10-

---

[7] G. D. Fasman and P. Lituri, unpublished data.

camphorsulfonic acid sample may contain some water. Note that the calibration setting recommended in the Cary 60 operating manual corresponds to $[\theta]_{290} = 7150$, and should not be used with 10-camphorsulfonic acid standards of unknown purity and dryness. The value of 7840 agrees well with that for another (44) pure sample of 10-camphorsulfonic acid, but not with a third (43) value, possibly because of different handling of the Kronig–Kramers relationships.

G. Calculations of Circular Dichroism

Data for CD are reported either as $[\theta]$, the molar or residue ellipticity, or as $(\epsilon_L - \epsilon_R)$, the differential molar circular dichroic extinction coefficient. The two measurements are proportional: $[\theta] = 3300(\epsilon_L - \epsilon_R)$. The rotational strength, R_K, of each optically active absorption band is sometimes calculated, provided that the experimental CD bands can be resolved; a du Pont 310 Curve Resolver[8] is useful for this purpose. There is no method of phenomenological CD analysis comparable to the Drude or Moffitt equations.

1. Molar or Residue Ellipticity, $[\theta]$

The molar ellipticity (for small molecules) or mean residue ellipticity (for nucleic acids) is defined as:

$$[\theta]\lambda = \frac{\theta_{obs} \times MW \text{ (or MRW)}}{10 \times d \times c''}$$

where λ = wavelength; θ_{obs} = observed ellipticity, in degrees; MW = molecular weight; MRW = mean residue weight of repeating unit; c'' = concentration in grams per milliliter; d = path length in centimeters. If the molar concentration of phosphate residues, c', is known directly, then $[\theta]$ may be calculated from:

$$[\theta] = \theta_{obs} \times \frac{10}{(l \times c')}$$

where l = path length in decimeters, c' = concentration in moles of residue/liter. The units for $[\theta]$ are degrees per square centimeter per decimole.

2. Reduced Molar or Residue Ellipticity, $[\theta]$

The Lorentz refractive index correction is not usually applied to CD data. However, this correction is occasionally useful for literature comparisons:

$$[\theta'] = [\theta] \times 3/(n^2 + 2)$$

[8]Manufactured by du Pont Instrument Products Division, Wilmington, Delaware 19898.

3. Differential Molar CD Extinction Coefficient

In some instruments the difference in absorbance between left- and right-handed circularly polarized light, $(A_L - A_R)$ is measured directly. In such cases, the molar circular dichroic extinction coefficient (also called the molar dichroic absorption), $\epsilon_L - \epsilon_R$, is obtained from:

$$(\epsilon_L - \epsilon_R) = (A_L - A_R)/(c' \times d)$$

where d = path length in centimeters; c' = concentration in moles of residue per liter; $(\epsilon_L - \epsilon_R)$ has units of liters per mole centimeter.

4. Relation between Ellipticity and Differential Absorption

The proportionality, $[\theta] = 3300(\epsilon_L - \epsilon_R)$, has already been given. Another useful relationship, for comparison of raw CD data, is $[\theta]_{obs} = 33 (A_L - A_R)$. Thus, an observed ellipticity of 0.001 degree corresponds to an observed differential dichroic absorbance of 3×10^{-5} absorbance unit.

5. Rotational Strength

The rotational strength R_K of the Kth optically active absorption band is defined by an integral which can be found elsewhere (Moscowitz, 1960). If the CD band is nearly Gaussian in shape, then:

$$R_K \approx 1.23 \times 10^{-42}[\theta_{max}]_K(\Delta/\lambda_{max})$$

where λ_{max} = wavelength of the Kth transition; $[\theta_{max}]_K = [\theta]$ at λ_{max}; Δ = half-width of the band.

IV. Artifacts

Two major artifacts of the optical activity of suspensions are recognized (46–49). First scattering of the light beam by particles might disturb the CD spectrum. Since CD is the measurement of the difference in the absorption between left- and right-handed circularly polarized light, only asymmetric but not symmetric scattering of these two polarized components contribute to the measured CD. The direct experimental measurement of the magnitude of asymmetric scattering in systems such as chromatin is not available, but a theoretical consideration concludes that for particles of polypeptides or polynucleotides with a radius less than 1 μm the scattering distortion of CD spectrum is quite small (46). The second possible artifact is the flattening effect of the absorption and optical activity spectra near the absorption region, arising from the shadowing effect of chromophores in the particles, which is equivalent to the reduction of the effective concentration (50).

An excellent review by Schneider (51) dealing with the analysis of optical activity spectra of turbid biological suspensions can be consulted for greater detail. The problem has been circumvented to a large degree by the development of fluorescent scattering cell (fluorescat). The cell design is such that *all* incident beam intensity which is either transmitted or scattered through the sample cell, except light scattered back out of the entrance path, is absorbed in a scintillator solution which surrounds the usual sample cell. The fluorescence emission is thus detected by the photodetector (52). In this manner the artifacts of scattering are largely eliminated. Studies utilizing the fluorescat cell on histone–DNA complexes indicate that these possible artifacts are indeed small. The fluorescat cell can likewise be used to check how much light scattering exists. The use of the fluorescat cell is well documented in the article by Dorman *et al.* (52).

REFERENCES

1. Moscowitz, A. *In* "Optical Rotatory Dispersion" (C. Djerassi, ed.), p. 150. McGraw-Hill, New York. (1960).
2. Mommaerts, W. F. H. M. *In* "Methods in Enzymology" (L. Grossman and K. Moldave, eds.), Vol. 12, Part B, p. 302. Academic Press, New York. (1968).
3. Yang, J. T., and Samejima, T. *Prog. Nucleic Acid Res. Mol. Biol.* 9, 223. (1969).
4. Brahms, J., and Brahms, S. *Biol. Macromol.* 4, 191. (1970).
5. Gratzer, W. B. *Procedures Nucleic Acid Res.* 2, 3. (1971).
6. Bush, C. A., and Brahms, J. *Phys.-Chem. Prop. Nucleic Acids* 2, 147. (1973).
7. Bush, C. A. *Basic Prin. Nucleic Acid Chem.* 2, 99. (1974).
8. Tinoco, I., Jr. *J. Chim. Phys.* 65, 91. (1968).
9. Marvin, D. A., Spencer, M., Wilkins, M. H. F., and Hamilton, L. D. *J. Mol. Biol.* 3, 547. (1961).
10. Langridge, R., Wilson, H. R., Hooper, C. W., Wilkins, M. H. F., and Hamilton, L. D. *J. Mol. Biol.* 2, 19 and 38. (1960).
11. Fuller, W., Wilkins, M. H. F., Wilson, H. R., and Hamilton, L. D. *J. Mol. Biol.* 12, 60. (1965).
12. Davies, D. *Annu. Rev. Biochem.* 36, Part 1, 350. (1967).
13. Tunis-Schneider, M. J., and Maestre, M. F. *J. Mol. Biol.* 52, 521. (1970).
14. Ivanov, V. I., Minchenkova, L. E., Schyolkina, A. K., and Poletayev, A. I. *Biopolymers* 12, 89. (1973).
15. Lerman, L. S. *Proc. Natl. Acad. Sci. U.S.A.* 68, 1886. (1971).
16. Jordon, C. F., Lerman, L. S., and Venable, J. H., Jr. *Nature (London) New Biol.* 236, 67. (1972).
17. Mantiatis, T., Venable, J. H., Jr., and Lerman, L. S. *J. Mol. Biol.* 84, 37. (1974).
18. Shih, T., and Fasman, G. D. *J. Mol. Biol.* 52, 125. (1970).
19. Greenfield, N., and Fasman, G. D. *Biochemistry* 8, 4108. (1969).
20. Pardon, J., and Richards, B. *Biol. Macromol.* 6, 1. (1973).
21. Mandel, R., and Fasman, G. D. *In* "Organization and Expression of Eukaryotic Genome." Academic Press, New York (in press). (1977).
22. Fasman, G. D., Schaffhausen, B., Goldsmith, L., and Adler, A. *Biochemistry* 9, 2814. (1970).
23. Fasman, G. D. *In* "Conformation of Biological Molecules and Polymers, Jerusalem

Symposium on Quantum Chemistry and Biochemistry" (E. Bergmann and B. Pullman, eds.) p. 655. Academic Press, New York. (1973).

24. Henson, P., and Walker, O. J. *Eur. J. Biochem.* **16**, 524. (1970).
25. Mandel, R., and Fasman, G. D. *Biochem. Biophys. Res. Commun.* **59**, 672. (1974).
26. Mandel, R., and Fasman, G. D. *Nucleic Acids Res.* **3**, 1839. (1976).
27. Gratzer, W. B. *In* "Poly-α-Amino Acids" (G. Fasman, ed.), p. 177, Dekker, New York. (1967).
28. Beaven, G. H., Holiday, E. R., and Johnson, E. A. *In* "The Nucleic Acids" (E. Chargaff and J. N. Davidson, eds.), Vol. 1, p. 493. Academic Press, New York. (1955).
29. Fasman, G. D. (ed.) "CRC Handbook of Biochemistry and Molecular Biology. *Nucleic Acids,*" Vol. 1. CRC Press, Cleveland, Ohio. (1975).
30. Warner, R. C. *J. Biol. Chem.* **229**, 711. (1957).
31. Mii, D., and Warner, R. C. *Fed. Proc., Fed. Am. Soc. Exp. Biol.* **19**, 317. (1960).
32. Inman, R. B. *J. Mol. Biol.* **9**, 624. (1964).
33. Chamberlin, M. J., and Patterson, D. L. *J. Mol. Biol.* **12**, 410. (1965).
34. Michelson, A. M. "The Chemistry of Nucleosides and Nucleotides". Academic Press, New York. (1963).
35. Ames, B. N., and Dubin, D. T. *J. Biol. Chem.* **235**, 769. (1960).
36. Shih, T. Y., and Fasman, G. D. *Biochemistry* **10**, 1675. (1971).
37. Schneider, W. C. *In* "Methods in Enzymology" (S. P. Colowick and N. O. Kaplan, eds.), Vol. 3, p. 680. Academic Press, New York. (1957).
38. Adler, A. J., and Fasman, G. D. *In* "Methods in Enzymology" (L. Grossman and K. Moldave, eds.), Vol. 12, Part B, p. 268. Academic Press, New York.
39. Potts, W. J. "Chemical Infrared Spectroscopy," Vol. 1, p. 119. Wiley, New York. (1963).
40. Engel, J., Liehl, E., and Sorg, C. *Eur. J. Biochem.* **21**, 22. (1971).
41. Radding, W., Veyama, N., Gilon, C., and Goodman, M. *Biopolymers* **15**, 591. (1976).
42. Brand, E., Washburn, E., Erlanger, B. F., Ellenbogen, E., Daniel, J., Lippman, F., and Scheu, M. *J. Am. Chem. Soc.* **76**, 5037. (1954).
43. DeTar, D. F. *Anal. Chem.* **41**, 1406. (1969).
44. Krueger, W. C., and Pschigoda, L. M., *Anal. Chem.* **43**, 675. (1971).
45. Cassim, J. Y., and Yang, J. T. *Biochemistry* **8**, 1947. (1969).
46. Ottaway, C. A., and Wetlaufer, D. B. *Arch. Biochem. Biophys.* **139**, 257. (1970).
47. Urry, D. W., Hinners, T. A., and Masotti, L. *Arch. Biochem. Biophys.* **137**, 214. (1970).
48. Gordon, D. J., and Holzwarth, G. *Arch. Biochem. Biophys.* **142**, 481. (1971).
49. Schneider, A. S., Schneider, M. T., and Rosenheck, K. *Proc. Natl. Acad. Sci. U.S.A.* **66**, 793. (1970).
50. Duysens, L. N. M. *Biochim. Biophys. Acta* **19**, 1. (1956).
51. Schneider, A. *In* "Methods in Enzymology" (C. H. W. Hirs and S. N. Timasheff, eds.), Vol. 27, p. 751. Academic Press, New York. (1973).
52. Dorman, B. P., Hearst, J. E., and Maestre, M. F. *In* "Methods in Enzymology" (C. H. W. Hirs and S. N. Timasheff, eds.), Vol. 27, p. 767. Academic Press, New York. (1973).

Chapter 1 8

Intercalating Agents as Probes of Chromatin Structure

EDMOND J. GABBAY

Departments of Chemistry and Biochemistry,
University of Florida,
Gainesville, Florida

W. DAVID WILSON

Department of Chemistry,
Georgia State University,
Atlanta, Georgia

I. The Interaction Specificities of Intercalating Agents with Nucleic Acids (DNA)

Following the elucidation of the structure of double helical DNA in the early 1950s, numerous investigators began using this model in an attempt to determine the structure of the complex formed between DNA and planar aromatic cations. In the early 1960s, Lerman (*1–4*) proposed the intercalation model which involves the unwinding of the deoxyribose sugar phosphate chains and elongation of the DNA molecule to produce an approximately 3.4 Å-thick hydrophobic space to accommodate the insertion of planar aromatic molecules between base pairs. The intercalation process is depicted schematically in Fig. 1. Lerman provided extensive experimental support for the intercalation model, and other investigators have since contributed supporting experiments and refinement of the original model (*5–10*). Much of the earlier work is discussed in reviews (*11–13*). A number of planar aromatic molecules have been shown to form intercalated complexes with DNA; these include the acridine dyes (e.g., proflavine, acridine orange), antiparasitic drugs (e.g., chloroquine, quinacrine, ethidium bromide), antitumor agents (e.g., actinomycin D, adriamycin, daunorubicin), and a variety

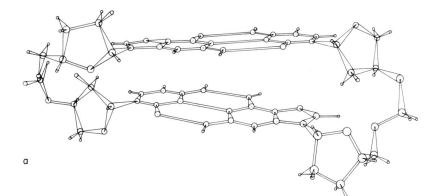

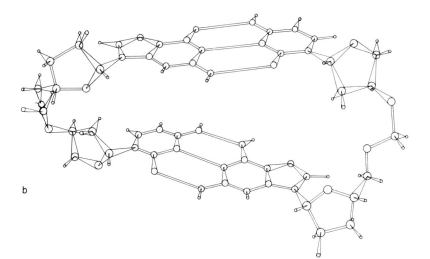

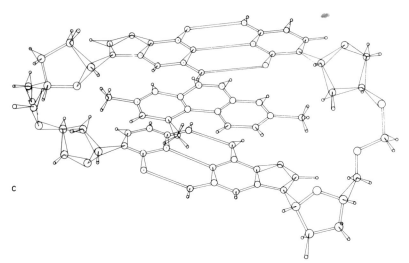

a

b

c

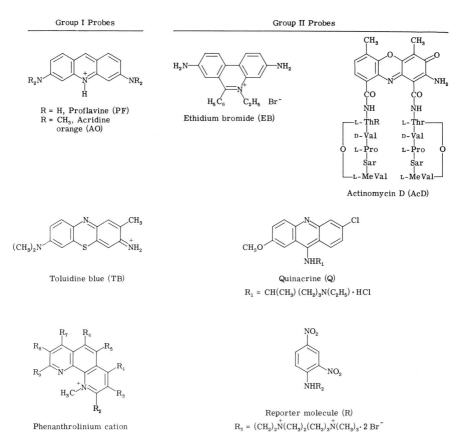

Group I Probes

Group II Probes

R = H, Proflavine (PF)
R = CH₃, Acridine
  orange (AO)

Ethidium bromide (EB)

Actinomycin D (AcD)

Toluidine blue (TB)

Quinacrine (Q)
R₁ = CH(CH₃) (CH₂)₃N(C₂H₅) · HCl

Phenanthrolinium cation

Reporter molecule (R)
R₂ = (CH₂)₂N(CH₃)₂(CH₂)₃N(CH₃)₃· 2 Br⁻

CHART 1. Structure of typical intercalating molecules.

of small molecules (e.g., reporter molecules, phenanthrolinium cations). The structure of some typical intercalating molecules is shown in Chart 1.

A brief description of the characteristic physical and chemical changes occurring in both DNA and the bound small molecule as a result of intercalation are presented here as an introduction to the more complex phenomena which occur with chromatin. This discussion is not intended to be a comprehensive review of the literature, but an overview taken from selected papers which together describe the current state of the intercalation model.

FIG. 1. Schematic illustration of a segment of DNA duplex showing (a) the right-handed stacking arrangements of adjacent base pairs, (b) the unwinding and separation of adjacent base pairs to accommodate (c) the intercalating agent, N-3, 8-trimethylphenanthrolinium cation.

A. Effects of Intercalation on the Physical Properties of DNA

1. CLASSICAL INTERCALATION MODEL

Owing to the elongation of the DNA molecule caused by intercalation, an increase in the intrinsic viscosity and a decrease in the sedimentation coefficient are observed for the DNA complex (1, 14, 15). Careful and thorough studies on the hydrodynamic properties of rodlike sonicated low-molecular-weight DNA (4 to 5 × 10⁵) in the presence and in the absence of proflavine (PF) have been carried out (14). The changes in the values of the intrinsic viscosity and sedimentation coefficient of DNA on binding PF are shown to be consistent with the intercalation model. For example, the ratio of the intrinsic viscosities, $[\eta]/[\eta]_0$ (where $[\eta]$ and $[\eta]_0$ are the intrinsic viscosity of solutions of DNA-PF complex and free DNA, respectively) are observed to be related to the helix length, L, as shown in Eq. (1), where $f(p)$ and

$$[\eta]/[\eta]_0 = (L/L_0)^3 \left[f(p)/f(p)_0 \right] \tag{1}$$

$f(p)_0$ are functions of the axial ratio of the complex, DNA–PF, and free DNA, respectively, Cohen and Eisenberg (14) showed that (i) the ratio of $f(p)/f(p)_0$ is close to unity (i.e., insensitive to binding) and (ii) the ratio of the intrinsic viscosity, $[\eta]/[\eta]_0$ is independent of the molecular weight of the sonicated DNA and is increased as predicted [Eq. (1)] with increasing $\bar{n}$ values (moles of bound ligand/moles DNA phosphate, see Section III,B,2) of $0 < \bar{n} \leq 0.1$. These findings provide strong support for the intercalation model and indicate that the intrinsic viscosity increases cannot be explained by other mechanisms, such as stiffening of the DNA molecule, which would cause increases varying with both molecular weight and $\bar{n}$. Because of the ready availability, and ease of use of viscometers, viscosity studies of potential intercalating molecules have been much more common than the equally valid but more laborious sedimentation experiments.

2. "PARTIAL" INTERCALATION MODEL

It should be noted that the absence of an increase in the value of the intrinsic viscosity of DNA upon binding a planar aromatic molecule does not necessarily preclude an intercalation mode of binding. The classical work of Muller and Crothers (5) has shown that the intrinsic viscosity of DNA solution in the presence of actinomycin D (AcD) is strongly dependent on the molecular weight (MW) of the polymer. For example, for the DNA–actinomycin complex, the ratio of the intrinsic viscosity, $[\eta]/[\eta]_0$ [see Eq. (1)] is observed to be less and more than unity for high and low MW DNA, respectively. The spectral characteristics of the complexes are identical, i.e., independent of the MW of the DNA polymer, and highly suggestive of an intercalation mode of binding of the phenoxazone ring of AcD. The anomalous viscosity reduction obtained with high MW DNA are thought to be due to extensive coiling of this long polymer, which is not possible for the more rodlike low MW DNA. Muller and Crothers (5) suggested that the cyclic peptides of AcD promote an intramolecular aggregation (or coiling) of high MW DNA, and they concluded that the phenoxazone ring of AcD is bound to DNA via an intercalation process.

Recent systematic studies on other types of aromatic cations indicate that an alternative explanation involving "bending" of the helix at the point of insertion of the aromatic ring between base pairs of DNA could also occur. For example Kapicak and Gabbay (8) have synthesized and studied the interaction specificities of the nitrophenyl-labeled diammonium salts (I), with DNA. Viscometric titration studies (at near infinite dilution of native salmon sperm DNA) showed that the effective length of DNA–(I) complexes is considerably diminished where $n = 1$, and enhanced with $n = 2, 3$, and 4. On the basis of the viscometric as well as the ^1H nuclear magnetic resonance (NMR) studies (Section I,B,5) of the DNA–(I) complexes, an intercalation model whereby the p-nitrophenyl ring of (I), where $n = 1$, is "partially" inserted between base pairs of DNA has been proposed (8). Figure 2 schematically illustrates the partial intercalation model which has also been found to adequately account for the interaction specificities of aromatic amino acids in oligopeptides (9, 10) and other sterically restricted aromatic cations (6, 7).

$$NO_2$$

$$H_A$$

$$H_B$$

$$(CH_2)_n \overset{+}{N}(CH_3)_2(CH_2)_3\overset{+}{N}(CH_3)_3 \cdot 2\,Br^-$$

$$n = 1\text{-}4$$

(I)

3. CLOSED CIRCULAR SUPERCOILED DNA

A sensitive method for detecting intercalation is the utilization of covalently linked circular supercoiled DNA (12, 16). Although there has been some initial speculation concerning the direction of supercoiling in naturally occurring circular DNA (17), present evidence now indicates that the natural supercoils are right-handed (18). Intercalation of an aromatic molecule between base pairs of DNA causes unwinding of the right-handed helix; if it occurs in closed circular supercoiled DNA, removal of the right-handed natural supercoils will result. For example, during the titration of closed circular DNA, a point is reached when the amount of intercalating

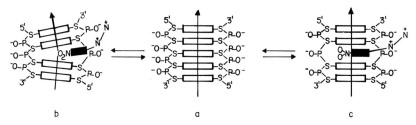

FIG. 2.   Schematic illustration of a segment of DNA duplex (a) which can either partially intercalate the nitrophenyl substituent of reporter molecules (I), where $n = 1$, to give structure (b) or fully intercalate the aromatic ring of (I), where $n = 2, 3$, and 4, to give structure (c). These two processes will either decrease or increase, respectively, the effective length of the DNA duplex.

ligand added to a constant amount of DNA causes the exact amount of unwinding necessary to remove all the supercoils and give an uncoiled closed circular DNA molecule. Addition of more ligand will cause continued unwinding and supercoiling in the reverse direction (i.e., left-handed vs right-handed) of the original supercoils (12). The amount of bound ligand required to exactly unwind the supercoils is found to be dependent on the nature of the ligand and is presumably also dependent on other molecular structural factors, e.g., (i) differences in the binding specificity of the ligand and/or (ii) the unwinding angles of the various intercalation sites (19).

The uncoiled form of the closed circular DNA is the most extended and therefore, exhibits the highest intrinsic viscosity and lowest sedimentation coefficient. Right-handed native supercoiled circular DNA and the left-handed supercoiled circular DNA (induced via an intercalation process) lead to a more compact DNA with a concomitant decrease in the intrinsic viscosity and an increase in the sedimentation coefficient. Figure 3 illustrates the changes in sedimentation coefficient that occur on titrating closed circular supercoiled PM2 DNA with ethidium bromide (12).

B.   Induced Changes in the Intercalated Molecules

In addition to predicting the changes described above for DNA, the intercalation model also predicts certain characteristic effects on the physical and chemical properties of the intercalated ligands. These changes are described briefly below.

1.   ABSORPTION AND FLUORESCENCE SPECTROSCOPY

The visible electronic spectrum of the DNA bound molecule is found to shift to longer wavelengths and to exhibit considerable hypochromism (lower extinction coefficients). Such observation is consistent with a stacking

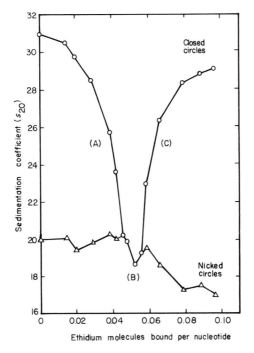

FIG. 3. The sedimentation behavior of closed (○) and nicked (△) circular duplex DNA (prepared from bacteriophage PM2) upon binding to ethidium bromide (EB). The curve for the closed circular DNA duplex reflects changes in supercoiling resulting from the accumulated unwinding of the double helix caused by binding of the drug. The three portions of the curve correspond to progressive removal (A), loss (B), and reversal (C) of the supercoils. From Waring (12).

interaction between the intercalating aromatic chromophore and the base pairs of DNA and is readily explainable on the basis of current theories (20). The fluorescence emission quantum yield of the DNA-bound molecules can be either reduced (quenched), as with proflavine (21) or enhanced, as with ethidium bromide (22). With some molecules, such as quinacrine (Q), base-pair specificity is observed; i.e., fluorescence quenching and enhancement is obtained upon binding of Q to G-C and A-T rich regions of DNA, respectively (23–25). On a percentage basis, the change in fluorescence quantum yield is typically greater than the change in extinction coefficient found by the absorption method. The fluorescence technique (if applicable) is a more sensitive method for detecting binding to DNA, but it must be remembered that neither the absorption nor the fluorescence spectral changes are unequivocal proof of an intercalation process. Nonintercalating planar aromatic cations also show spectral changes upon binding to DNA (26).

2. Reactivity of the Bound Ligand

It has been suggested (3,4) that a reduced reactivity of the functional groups of the aromatic ligand would be obtained upon intercalation of the latter between base pairs of DNA. For example, a careful analysis of the reaction rates of nitrous acid with the amino group of proflavine showed that the bimolecular process is considerably reduced in the presence of DNA. Unfortunately, such studies are difficult to evaluate since the observed lower reactivity of the DNA complex may be due to (i) different modes of binding of the ligand (i.e., intercalation and/or external electrostatic binding) and (ii) the relative rates of dissociation of the DNA complex as compared to the bimolecular reaction (e.g., nitrosation deamination reaction). Lerman (3,4) adequately discusses the pitfalls of this approach.

3. Induced Optical Activity

Induced circular dichroism (CD) in the electronic transition of proflavine (PF), acridine orange (AO), reporter molecules (R), ethidium bromide (EB), and many other planar aromatic cations is noted upon binding to DNA. According to present theories, the optical activity arises from the interaction of the transition moment(s) of the ligand with the asymmetrically distributed transition moments of the DNA base pairs (27). For example, a simple induced CD spectrum paralleling the absorption band of the ligands (e.g., EB, PF, R, etc.) is noted at low ligand to DNA phosphate ratios. However, at higher bound ligand-to-DNA ratio, the induced CD spectrum undergoes pronounced changes and becomes considerably more complex. Wetmur and co-workers (28) have carried out a careful analysis of the induced CD in the absorption bands of PF, EB, and methylene blue. On the basis of their results, it was concluded that at a low dye-to-DNA ratio only the ligand base-pair interactions contribute significantly to the induced CD spectra. The latter effect, i.e., asymmetric ligand–ligand interactions, may arise from external electrostatic type binding resulting in an asymmetric stacking of the ligands, which is expected to occur at high dye-to-DNA ratio (29).

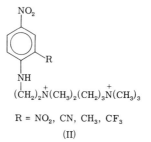

$$R = NO_2, \ CN, \ CH_3, \ CF_3$$

(II)

A systematic analysis of the induced CD in the absorption band of variously substituted reporter-type molecules (II) upon binding to DNA has been carried out in this laboratory (30). A large negative induced circular dichroism is observed in the nitroaniline transition of (II), where $R = NO_2$ and CN, upon binding to DNA which is independent of the DNA phosphate (P_t)-to-reporter (R) ratio $(P_t/R = 70-6.0)$. However, the induced CD soectra of the DNA–(II), where $R = CH_3$ and CF_3, is found to be strongly dependent on the DNA phosphate-to-reporter ratio; e.g., the molar ellipticity $[\theta]$ at 395 nm (for $R = CF_3$) at P_t/R of 70 and 6.0 is found to be -4000 and 2400, respectively. In order to account for the above observations, as well as the viscometric and ^1H NMR studies which show that the reporter molecules (II) form intercalated complexes at P_t/R ratio greater than 6.0, it has been suggested that the various possible intercalation sites in DNA may not necessarily induce the same sign and/or magnitude in the CD signal of the absorption band of the DNA-bound reporter molecule (30).

In summary, an induced CD spectrum in the absorption band of the DNA bound ligand is obtained. The origin of this effect is due to asymmetric stacking interactions, i.e., (i) ligand–base pairs (intercalation mode of binding) and (ii) ligand–ligand (external electrostatic interactions). The sign, magnitude, and complexity of the induced CD signal are found to be dependent on the ratio of the bound ligand to DNA phosphate and may be due to differences in the induced CD arising from (i) the 10 possible intercalation sites in DNA (30), and/or (ii) the two modes of binding, i.e., intercalation vs external electrostatic binding (28).

4. FLOW DICHROISM AND POLARIZED FLUORESCENCE

The technique of flow dichroism in DNA studies has been introduced and refined on a quantitative basis by Cavalieri et al. (31). Lerman (2–4) was the first to extend this method to determine the orientation of DNA-bound planar aromatic cations (e.g., proflavine and quinacrine) and to distinguish between external electrostatic and intercalation mode of binding. In order to accomplish this objective, at least two nonparallel in-plane transitions must be considered, e.g., two absorption transition or an absorption and an emission transition. To be useful these transitions must not be parallel, and to give optimum results they should be perpendicular to each other (2–4). For example, flow-oriented DNA–acridine (proflavine and acridine orange) complexes have been studied using the in-plane long and short axis transitions of the acridine. The results indicate that both transitions are in a plane parallel to the DNA base pairs. Similarly, Lerman (3,4) has examined floworiented DNA–quinacrine complex using an absorption (flow dichroism) and an emission transition (flow polarized fluorescence) of the bound quinacrine and showed that both are in plane parallel to the DNA base pairs.

The above results constitute highly supportive experimental data (and necessary requirements) in favor of the intercalation mode of binding. It is recognized, however, that similar results (although highly unlikely) may be obtained via external electrostatic parallel stacking of the aromatic cations. For this reason, an unequivocal conclusion concerning an intercalation mode of binding can be made only on the basis of combined hydrodynamic (e.g., viscosity and/or sedimentation) and flow orientation studies (2–4).

5. PROTON MAGNETIC RESONANCE TECHNIQUES

The line widths and chemical shifts of the ^{1}H NMR signal of a small molecule bound to a macromolecule often reveal considerable information concerning the nature of the binding process. For example, three types of binding may be distinguished. The first type of binding is characteristic of a rigid macromolecule–small molecule complex and leads to a total line broadening of the ^{1}H NMR signals. This effect has been observed for molecules which intercalate between base pairs of DNA (6–8,32). In such a case, the small molecule experiences strongly restricted tumbling in the DNA complex, leading to an unaveraged chemical shift of the individual protons and total line broadening (33,34). Examples of molecules exhibiting this behavior are the phenathrolinium cations (7) (Chart 1) and the reporter molecules (I), where $n = 2$–4, and (II), where R = NO$_2$, CH$_3$, CN (30) (see Sections I,A,2 and I,B,3).

The second type of binding is characteristic of molecules exhibiting a line-broadened ^{1}H NMR signal as well as a large upfield chemical shift. The latter effect (i.e., upfield chemical shift) experienced by the aromatic protons of the small molecule is indicative of close contact to the DNA base pairs and is presumably due to ring current anistropy (34). However, the observed ^{1}H NMR signal broadening of the aromatic protons of the DNA-bound small molecules may arise by distinctly different mechanisms: (i) weak restriction of molecular tumbling of the aromatic ring in the DNA complex; (ii) slow rate of exchange between the various DNA binding sites and the unbound state; (iii) large differences in the chemical shifts experienced by the aromatic protons in the different intercalation sites of DNA; and (iv) combination of (i)–(iii). Molecules which exhibit this behavior are found to have a profound effect on the tertiary structure of DNA and lead to shortening of the helix length. For example, the reporter molecule (I), where $n = 1$ (see Section I,A,2), and aromatic amino acid-containing peptides, which also exhibit a line-broadened ^{1}H NMR signal as well as large upfield chemical shifts of the aromatic protons, are found to decrease the intrinsic viscosity and the reduced dichroism of DNA solution (9,10). Such effects are consistent with the partial intercalation model (Fig. 2). It should be noted, that the reporter molecules (I), where $n = 2$, 3, and 4, are found to increase the effective length of the DNA helix and to exhibit total line broadening of the ^{1}H NMR signals of the aromatic protons upon binding to DNA. Presumably, the single methylene group (CH_2), between the 4-nitrophenyl ring of (I), $n = 1$, and the quaternary ammonium group of the side chain is not sufficient to allow for "full" insertion of the aromatic ring and hence lengthening of the helix (8).

The third type of binding is noted for molecules which have a high affinity to nucleic acids but exhibit no line broadening and upfield chemical shift of ^{1}H NMR signals. Among these are the polyammonium salts, $R_3N^+(CH_2)_n N^+R_3 \cdot 2Br^-$ (where $R = H, CH_3$; and $n = 2-6$), spermidine, spermine, and the oligopeptides which contain the simple aliphatic and polar amino acids, e.g., glycine, lysine, histidine, serine (35). Such molecules form external electrostatic complexes with DNA and undergo rapid tumbling leading to an efficient averaging of the chemical shift environment.

In general, ^{1}H NMR investigations have been attempted for only a few intercalating agents owing to the extensive signal line broadening that occurs when the small molecule experiences reduced tumbling in the DNA complex. In addition, with conventional NMR spectrometers, a high concentration of the ligand and DNA is required, which often leads to insolubility problems. However, with the present availability of Fourier-transform NMR instruments, it is routinely possible to detect ^{1}H NMR signals at relatively low concentration (0.5–1.0 mM).

II. Interaction Specificities of Intercalating Agents with DNA and Chromatin

The interaction of a variety of planar aromatic molecules with free DNA and chromatin have been reported. The objectives of these studies include (i) the estimation of the percentage of DNA in chromatin which is exposed and capable of binding the probe, (ii) the effect of the sequential removal of specific proteins (e.g., histones) on DNA exposure, (iii) the localization and determination of the binding site of the proteins in chromatin (i.e., minor or major groove), (iv) comparison of the structure of exposed DNA in chromatin with that of free DNA, and (v) the determination of the fidelity of chromatin reconstitution experiments.

A. The Probes

Chart 1 shows the names and chemical structures of the commonly used agents for probing the structure of chromatin. It is relevant to classify these probes into two groups based on structural considerations. Group I consists of the totally planar molecules (proflavine, PF; acridine orange, AO; and toluidine blue, TB). Group II consists of planar molecules containing non-planar and bulky side chain substituents (e.g., ethidium bromide, EB; actino-mycin D, AcD; quinacrine, Q; and dinitroaniline reporter molecule, R). With the exception of actinomycin D (which shows only type I binding), all these probes have been shown to exhibit two modes of binding to DNA: (i) intercalation (type I binding); and (ii) external weak electrostatic binding (type II). The latter type of intercalation is normally observed at low ionic strength (below 0.05) with an $\bar{n}_{max} \leq 1.0$ whereas the strong intercalative type binding is generally observed over a wide range of ionic strength with $\bar{n}_{max}$ values ≤ 0.25 (see Section III,B,2 for definitions of $\bar{n}_{max}$).

In order to avoid dissociation of the histones, the estimation of exposed DNA in "native" chromatin has been determined at low ionic strength. For example, Ohba and co-workers (36,37) and Itzhaki (38) using a precipitation titration technique with TB (at low ionic strength) showed that about "half" of the chromatin DNA reacted with the dye; i.e., the precipitated complex contained 0.30–0.48 drug per phosphate of calf thymus deoxyribonucleo-protein. Such studies, unfortunately, do not reveal the nature of the chroma-tin DNA binding sites, i.e., type I or II and/or both. In the same studies, for example, precipitation titration of free DNA with TB led to the formation of complexes containing 1.1–1.2 dyes per DNA phosphate, suggesting that type I and II complexes are formed.

In addition, it is recognized that the estimation of "exposed" DNA in chromatin via the utilization of the chemical probes shown in Chart 1 is dependent not only on the type of binding (discussed above), but also on other factors which are perhaps more critical. These include (i) the prepara-tion of the chromatin (e.g., the protein-to-DNA ratio), (ii) the stability of nucleoprotein preparation (e.g., susceptibility to degradation by nonhistone proteins), (iii) artificially induced protein exchange and/or protein re-organization of the deoxyribonucleoprotein, (iv) the heterogeneity of the chromatin (e.g., mitotic vs S-phase), and (v) induced dissociation of proteins from DNA by the agents which are used as probes. Clearly, therefore, the interpretation of the results obtained from chromatin interaction with the chemical probes are difficult to evaluate, and often seemingly conflicting data are obtained from different laboratories.

The interaction of the probes with chromatin DNA depends on the type and extent of protein coverage of the nucleic acid. It is possible to define three types of DNA coverage in chromatin: (1) total, (ii) partial, and (iii) no coverage (naked DNA). The presence of strongly bound proteins covering both the minor and major grooves of DNA would be characteristic of total coverage. Partial coverage of DNA is defined by the presence of proteins in one of the two grooves. It is recognized that important differences in the intercalative mode of binding of groups I and II type probes to deoxyribonucleoprotein may be observed. Owing to the absence of bulky side-chain substituents, group I probes may insert between base pairs of exposed DNA from the major and/or minor groove. The presence of DNA segments which are totally covered could lead to a lowering of the maximum number of intercalative binding sites, $\bar{n}_{max}$, for group I probes. However, it is theoretically possible that intercalative binding of type I probes may still occur to totally covered segments of DNA without a significant decrease in $\bar{n}_{max}$. This situation may occur via (i) local dissociation of the protein, (ii) insertion of the probe between base pairs of DNA, and (iii) subsequent reassociation of the protein to reform a totally covered DNA segment. On the other hand, the presence of either (i) partially covered or (ii) noncovered segments of DNA in chromatin is not expected to lead to a decrease in the maximal intercalative binding of type I probes. It may be argued, however, that the presence of proteins in only one of the grooves could prevent unwinding of the DNA helix and lead to lower values of $\bar{n}_{max}$. Such a situation is highly unlikely. For example, recent experimental data on the interaction of N-methylphenanthrolinium cation, 1 (a type I probe), with free DNA and DNA–polylysine complex have shown that the value of $\bar{n}_{max}$ remains the same for both systems (*39*).

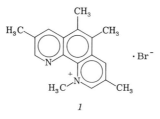

1

The intercalative mode of binding of group II probes to DNA would of necessity place the bulky side-chain substituents of the latter in one of the two grooves. A decrease in the value of the maximal DNA binding, $\bar{n}_{max}$, of group II probes will be obtained if (i) totally covered DNA segments are present and/or (ii) the presence of proteins is in the same groove as that of the binding site of the side chain substituents of group II probes.

Table I summarizes the above discussion with regard to the significance and evaluation of comparative binding studies of types I and II probes to free DNA and to chromatin. It should be noted that group II agents have been extensively utilized in connection with the determination of the site of binding of histones in chromatin. A brief survey of the literature concerning the interaction specificities of types I and II probes with DNA and chromatin is given below.

TABLE I

EFFECT OF VARIOUS TYPES OF DNA–PROTEIN
COVERAGE ON THE MAXIMAL INTERCALATIVE
BINDING, $\bar{n}_{max}$, OF GROUPS I AND II PROBES
TO DNA

DNA coverage	Group I	Group II
Total	a	Decrease
Partial	a, b	b, c
Naked	a, b	a, b

[a] May or may not decrease.
[b] Remains unchanged.
[c] Will decrease if side-chain substituents of group II are in the same groove as the proteins in partially covered DNA.

B.   Group I Probes

Dyes of the acridine class, such as proflavine (PF) and acridine orange (AO) exhibit significant changes in their absorption spectra and an induced optical activity upon binding to DNA (40–45). An increase in the viscosity of DNA solution (1) and in the apparent length of DNA molecules as determined by autoradiography (46) and low-angle X-ray scattering (47) accompany the binding of the acridines. Peacocke and Skerrett (41) proposed a two-site model for PF–DNA complexes based on the results of an extensive equilibrium dialysis investigation. The strong binding was shown to reach a maximal value ($\bar{n}_{max}$) of 0.22, while the weaker binding gave an $\bar{n}_{max}$ value of unity. On the basis of viscometric, flow dichroism, and polarized fluorescence evidence, Lerman (1,2) proposed that the strong binding site of Peacocke and Skerrett involves intercalation of the planar acridines between base pairs of DNA (i.e., type I binding). The weaker DNA binding of the acridines (type II binding) is attributed to external electrostatic self-stacking of the planar molecules adjacent to the phosphate groups of DNA. Additional evidence based on (i) fiber X-ray diffraction studies of DNA–acridine complexes (44) and (ii) unwinding of closed circular supercoiled DNA by PF (19,48) provide strong additional support of the intercalation model proposed by Lerman (1).

The interaction of PF with DNA and deoxyribonucleohistone from calf thymus has been studied (49–54). It has been concluded on the basis of the results of visible spectrophotometric and viscometric measurements that the presence of histones in DNA does not significantly decrease the number of strong binding sites in deoxyribonucleoprotein (50). In contrast, Dalgleish *et al. (54)* concluded from a similar binding study with PF that chromosomal proteins block about 40% of the total sites available in DNA. However, it is not clear from this work whether the decrease in maximal binding involves a decrease in the number of intercalation sites or external electrostatic-type binding or both. The results of Lawrence and Louis *(53)* based on fluorescence titration indicate that the availability of DNA binding sites is related to the protein content of the deoxyribonucleoprotein. For example, maximal binding, $\bar{n}_{max}$, of PF to chromatin increases from a value of 0.10 to 0.30 upon successive removal of histone proteins. The results, which appear to contradict the data of Gittelson and Walker *(50)*, are difficult to reconcile but could be explained. For example, the validity of the indirect spectrophotometric titration studies based on absorption *(50)* and fluorescence quenching *(53)* techniques for the determination of the binding isotherms of PF to DNA (and/or to chromatin) is highly questionable. The basic assumption of a linear dependence of the observed spectral change with the extent of binding is made without any justifications or corroboration via direct studies, such as equilibrium dialysis and/or solvent distribution techniques. Indeed, it has already been shown that the fluorescence characteristics of small molecules [e.g., quinacrine *(24, 25)*, daunorubicin *(55)*] is highly dependent on the nature of the DNA binding site. Insertion between specific base pairs and external electrostatic binding are found to induce different fluorescence properties (quenching and/or enhancement) in the DNA bound molecule.

Toluidine blue (TB), a cationic planar aromatic dye, is found to interact with DNA in a manner similar to that of the acridines *(36–38)*. Spectral titration studies indicate the presence of two types of binding sites of TB to DNA, i.e., the intercalative mode (type I) and a weaker external binding (type II), which is nearly abolished at high ionic strength. The two types of binding sites were estimated separately by Miura and Ohba *(36)*. Maximal binding values, $\bar{n}_{max}$, of 0.35 and 0.32 for the intercalative mode of binding of TB to DNA and calf thymus nucleohistone, respectively, were obtained. On the other hand, maximal binding for the external electrostatic type of interaction was found to be 1.0 and 0.48 for DNA and calf thymus nucleohistone, respectively. On this basis, it was concluded that approximately 50% of the DNA phosphate is exposed and available for TB binding (type II) in chromatin. It should be emphasized, however, that the studies of Ohba and his co-workers indicate that $\bar{n}_{max}$ for the intercalative mode of binding of TB to DNA and to chromatin is nearly identical. Such finding is consistent with earlier observation that group I type probes (lacking bulky side-chain substituents) may intercalate to the same extent to free DNA and to partially (or totally) covered DNA in chromatin.

The work of Itzhaki *(38)* based on rather inaccurate precipitation titration experiments (and corrected for various artifacts) indicates that 1.10–1.20 and 0.46–0.50 TB molecules per phosphate residue are obtained in the insoluble complexes of TB and calf thymus DNA and nucleohistone, respectively. On this basis, Itzhaki (38) concluded that 50% of the DNA phosphate is exposed in calf thymus chromatin. However, it is recognized that such conclusion is not warranted since two types of binding may be involved, i.e., intercalative and external electrostatic binding to DNA phosphate. For example, the precipitation titration technique does not distinguish between the two modes of bindings and therefore estimation of exposed DNA phosphate in nucleohistone (equivalent to $\bar{n}_{max}$ of type II binding) cannot and should not be made.

C.   Group II Probes

Group II probes, which are characterized by the presence of (i) a planar aromatic ring(s) capable of inserting between base pairs of DNA, and (ii) large bulky side-chain substituents, which may lie in either of the two grooves of DNA in the "intercalated" complex, have been studied extensively in a number of laboratories and are discussed separately below.

1.   ETHIDIUM BROMIDE (EB)

A considerable body of evidence exists for the intercalative mode of binding of the aromatic ring of EB to DNA (22,56,57) which is based on (i) the hydrodynamic behavior of DNA, i.e., increase in viscosity and decrease in sedimentation coefficient (5,58,59), and (ii) unwinding of closed circular supercoiled DNA (19,48,60). In addition to the intercalation mode of binding of EB to DNA, an external electrostatic weak binding process has also been detected at low ionic strength (56,60a). It should be noted that although a detailed molecular model of an intercalative DNA–EB complex has been described which places the phenyl ring substituent of EB in the major groove of the helix (61) there is no evidence (direct or indirect) for this assignment. The reported fiber X-ray data do not distinguish between minor and/or major groove binding of the out-of-plane phenyl substituent of EB (W. Fuller, private communication, 1975).

Comparative studies of the interaction specificity of EB with DNA, deoxyribonucleoproteins, and reconstituted histone–DNA complexes have been reported (62–68). For example, Angerer and Moudrianakis (65) examined the interactions of EB with whole, selectively deproteinized deoxyribonucleoproteins (DNP) and free DNA from calf thymus, using absorption and spectrofluorometric techniques. It is instructive to first compare the binding isotherm of EB to free DNA by the two spectral techniques. Scatchard plots of the absorption and fluorometric titration of

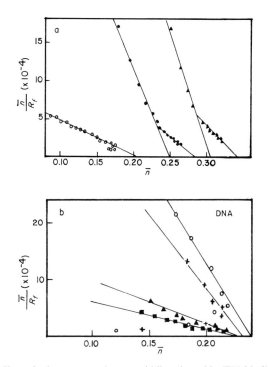

FIG. 4.   (a) Effect of salt concentration on ethidium bromide (EB) binding to calf thymus DNA as measured by absorption titrations. The Scatchard plots at the following three salt concentrations are shown: $\blacktriangle$, 1 mM sodium cacodylate; $\bullet$, 10 mM NaCl and 1 mM sodium cacodylate; $\bigcirc$, 100 mM NaCl and $10^{-3} M$ sodium cacodylate. From Angerer and Moudrianakis (65). (b)   Scatchard plots of fluorometric titrations with EB of calf thymus DNA in 1 mM sodium cacodylate at the following salt concentrations: $\bigcirc$, no NaCl; $+$, 10 mM NaCl; $\blacktriangle$, 50 mM NaCl; $\blacksquare$, 100 mM NaCl. Note the curvature of the Scatchard plots at the low salt concentrations. From Angerer and Moudrianakis (65).

DNA are shown in Figs. 4A and 4B, respectively. Several interesting points should be noted: (1) two types of binding sites are detected by the absorption method at the low ionic strength (Fig. 4A); (ii) the number of available strong binding sites in DNA is reduced by one-third, i.e., $\bar{n}_{max}$ values of 0.31 and 0.20 are obtained, as the salt concentration is raised from 1 mM to 100 mM, respectively (Fig. 4A); (iii) at low ionic strength, e.g., 1 mM sodium cacodylate, the fluorescence titration of DNA leads to an anomalous binding curve at low P/D which deviates from linearity in Scatchard plots (Fig. 4B); and (iv) at any given ionic strength, different values of maximal binding, $\bar{n}_{max}$, are obtained by the absorption and fluorometric methods. One possible explanation for the deviation from linearity of the

Scatchard plot by the fluorometric technique at low ionic strength which has been suggested (65) is that an EB molecule at one site influences the fluorescence yield of another EB molecule bound at a second site, perhaps by quenching of fluorescence or by dimer formation. However, this explanation is not likely since the fluorescence studies carried out by these workers indicate that varying the salt concentration influences neither the emission nor the quantum yield of fluorescence of EB bound to DNA at high P/D ratios. A second alternative is concluded to be more likely—i.e., that changes in the parameters of the helix induced by intercalation of EB may result in either facilitation or inhibition of further binding of EB depending on the ionic strength of the medium. It is recognized, however, that the extensive work of Angerer and Moudrianakis (65) still fails to offer any reasonable explanation for the discrepancy in the results of the binding isotherms obtained by the two spectral techniques. It is our opinion that such anomalous results, which are frequently obtained by the indirect spectral methods, can be clarified by direct-binding studies, i.e., equilibrium dialysis (see Section III,B).

Nonetheless, the careful and thorough work of Angerer and Moudrianakis (65) does establish the following: (i) there are at least two types of binding sites in both DNA and deoxyribonucleoprotein (DNP); (ii) the fluorescence properties (microenvironment) of these two kinds of binding sites appear to be very similar in DNP and DNA; (iii) the number of available intercalative binding sites is reduced by one-third in DNP as compared to DNA; (iv) the affinity of DNP for EB is reduced compared to that of DNA; and (v) the protein extracted from DNP by 0.6 M NaCl (shown to be histone f_1) has the greatest inhibiting effect both on the number of available sites and on the binding constant for EB.

Additional studies (69) on the interaction specificities of EB with DNA and DNP demonstrated that (i) there are at least two classes of highly fluorescent intercalation sites in DNP (i.e., causing EB fluorescence enhancement) compared to only one class in free DNA as shown by Scatchard plots of fluorescence binding isotherm data; (ii) the rotational relaxation time of EB bound to DNP is approximately 2-fold greater than in the DNA complex, indicating that the internal structure of DNA in DNP is more rigid than that of DNA (shown by fluorescence depolarization technique); (iii) on a weight basis, the lysine-rich histones are the most effective inhibitors of EB binding in DNP (shown by Scatchard plots); and (iv) no apparent differences in the nature of the intercalation sites of DNP and DNA are detected based on the similarity of the fluorescence lifetime and quantum yield of the bound ethidium bromide.

Lurquin and Seligy (66) have reported on a spectrophotometric study of the binding of EB to whole and partially deproteinized avian erythrocyte chromatin. Their chromatin preparations have a considerably lower EB-binding capacity than that observed for calf thymus DNP (69). For example, the maximal intercalative binding of EB to intact avian erythrocyte chromatin was found to be 0.033 as compared to 0.192 for free DNA. In contrast, the work of Angerer and Moudrianakis (65) on the binding of EB to DNA and DNP from calf thymus gave maximal intercalative binding of 0.31 and 0.21, respectively. The difference between the results with the whole chromatin (or DNP) may be due to the fact that the corresponding preparations were isolated from different tissues, by different methods, and presumably were assayed under slightly different ionic conditions. (Unfortunately, neither work reports the ionic conditions used in the spectrophotometric titration employed in the Scatchard analysis of the binding of EB to DNP.) In addition, it is very puzzling that the reported $\bar{n}_{max}$ values for the primary binding process (intercalation) of EB to calf thymus DNA and to avian erythrocyte DNA are found to be 0.31 and 0.19, respectively. Since the intercalative mode of binding of EB to DNA has been shown not to be dependent on base composition (22,56,70), the observed differences in the values of the maximal binding to calf thymus and avian erythrocyte DNA is probably due to the erroneous assumption upon which the spectral titration technique is based; i.e., the extinction coefficient of DNA bound EB is assumed to be constant and independent of (i) the base composition of the intercalation site and/or (ii) the extent of binding ($\bar{n}$). Nevertheless, the studies of Seligy and co-workers (66,67,71) do show that maximal intercalative binding of EB to avian DNP is approximately 16% of that found in free DNA. It is also found that successive deproteinization of the DNP leads to higher values of $\bar{n}_{max}$ and that histone V (a serine-rich histone specifically found in nucleated erythrocytes) has the greater inhibiting effect on the intercalative binding of EB to DNP.

Direct binding studies of EB with calf thymus DNP and DNA utilizing equilibrium dialysis technique have recently been reported (72). The maximal intercalative binding of EB to DNP ($\bar{n}_{max} = 0.20$) at low ionic strength (1 mM NaCl) is approximately 80% of that found with free DNA ($\bar{n}_{max} = 0.25$). These results are qualitatively consistent with the data obtained by the indirect spectral methods mentioned earlier. This work (72) also indicates that the classical spectrophotometric method of Peacocke and Skerrett (41) based on absorption titration is valid in the absence of externally bound species, i.e., at high ionic strength. However, an overestimation of the maximal intercalative mode of binding, $\bar{n}_{max}$, is obtained by the indirect spectral titration as compared to the direct equilibrium dialysis method at low ionic strength. Moreover, it is found that the binding curves derived from the fluorescence methods (based on the intensity) are either completely erratic or underestimate the results obtained by equilibrium dialysis. The latter finding together with the observation that the fluorescence quantum yield of DNA-bound EB is dependent on the binding ratio ($\bar{n}$) is in direct conflict with the data of Moudrianakis and co-workers (65).

It has been concluded by Seligy and co-workers (66, 67, 71) and Moudrianakis and co-workers (65, 69) on the basis of (i) the postulated model of Fuller and Waring (61) which places the phenyl ring of EB in the major groove and (ii) the lower maximal binding, $\bar{n}_{max}$, of EB to DNP as compared to DNA, that histone proteins may occupy the major groove of DNA. This conclusion may well be correct; however, as has been stated earlier, the proposed molecular model of Fuller and Waring (61) of DNA–EB complex should not be construed as evidence for the specific binding of the phenyl substituent of EB in the major groove of DNA. The reported fiber X-ray data do not distinguish between minor and/or major groove binding of the phenyl substituent (W. Fuller, private communication, 1975).

2. QUINACRINE (Q)

Quinacrine, an N,N-diethylaminopropyl-substituted 9-aminoacridine has been shown to bind to DNA in a manner characteristic of the other acridines, i.e., (i) strong intercalative binding and (ii) weak external electrostatic binding (2). The evidence for an intercalation mode of binding of Q to DNA is based on hydrodynamic studies (viscosity and flow dichroism) which indicate that (i) the aromatic ring of Q is in the same plane as the base pairs and (ii) lengthening of the DNA helix occurs upon formation of a DNA–Q complex.

The interaction of Q with DNA and deoxyribonucleoprotein (DNP) from calf thymus has been studied by direct equilibrium dialysis technique by Bontemps and Fredericq (72). At low ionic strength (1 mM NaCl) the value of maximal intercalative binding, $\bar{n}_{maxI}$, is observed to be 0.16 and 0.25 for DNP and DNA, respectively. The value of $\bar{n}_{maxII}$ for the external electrostatic binding process remains unchanged for DNA and DNP (i.e., $\bar{n}_{maxII} = 0.13$). At the higher ionic strength (100 mM NaCl), the external binding of Q to DNA is abolished, while $\bar{n}_{max}$ of the intercalative binding mode remains unchanged although the DNA affinity is decreased by 15-fold. These results indicate that the histone proteins in DNP lower the maximum number of DNA intercalative binding sites of Q. However, they appear to have no effect on the availability of DNA phosphate residues for the external binding of Q. The significance of these data in light of the chromatin structure is not clear at present.

Caspersson *et al.* (73) has shown that quinacrine-labeled mustards stain specific regions of chromosomes with a brilliant fluorescence intensity, leaving other areas of chromosomes relatively dark. It was originally thought that the linear differentiation of chromosomes into fluorescent bands was due to specific alkylation of guanine residues by the mustard function. The finding that Q itself produces identical banding patterns (74) suggested that other than covalent attachment to guanine is responsible for the observed differential fluorescence. Ellison and Barr (75) have suggested that enhancement of Q fluorescence might be a function of base ratio, with (A + T)-rich regions fluorescing brightly. Others (23,24) have investigated Q fluorescence in vitro with a series of natural and synthetic polynucleotides and found that A-T base pairs are responsible for fluorescence enhancement. Guanine residues were shown to give rise to quenching of Q fluorescence. These data, and several other lines of evidence (76,77), suggested that the fluorescent bands observed with Q-stained chromosomes arise from intercalative binding of Q to A-T clusters. On this basis, Bonner and co-workers (78) measured the fluorescence of Q in the presence of interphase chromatin and in the presence of chromatin which had been fractionated into extended and condensed regions (euchromatin and heterochromatin) in order to determine whether the Q fluorescence is due to intrachromosomal differences in DNA base composition or whether DNA–protein interactions in chromatin play a role in producing the banding patterns. It is found that Q fluorescence is quenched most effectively by the euchromatin fraction, intermediately by unfractionated chromatin, and least effectively by the heterochromatin fractions. Deproteinization abolishes this differential fluorescence, and, moreover, no differences in DNA base composition were found between the euchromatin and heterochromatin fractions. In addition, binding isotherms based on absorption titration studies in 10 mM Tris buffer (pH 8.0) indicate that the association constants for Q binding to the various chromatin fractions as well as free DNA differ by only a factor of 2, and that the values of the observed maximal intercalative binding, $\bar{n}_{\text{maxI}}$, are nearly identical, i.e., 0.047–0.051. On the above basis, it has been concluded (78) that differences in protein–DNA interactions (DNA conformation rather than base composition) along the chromatids of metaphase chromosomes are responsible for the observed fluorescence patterns obtained with Q. However, it should be noted that although the DNA base composition in the extended and condensed regions of chromatin are found to be similar, the differential presence and/or the differential binding accessibility of specific sequences (e.g., A–T clusters) in euchromatin and heterochromatin could also account for the observed Q fluorescence banding of chromosomes. In addition, it is apparent that the value of the maximal intercalative binding of Q to rat liver DNA obtained by the spectral titration by Gottesfeld et al. (78) is significantly lower ($\bar{n}_{\text{maxI}} = 0.05$) than that obtained by the direct equilibrium dialysis method of Bontemps and Fredericq (72) with calf thymus DNA ($\bar{n}_{\text{maxI}} = 0.25$). Indeed, it is strange that such a large discrepancy (i.e., maximal intercalative binding of 1 molecule of Q per 2 base pairs vs 1 molecule per 10 base pairs by the direct and indirect methods, respectively) is obtained, especially since both techniques employed similar ionic strength conditions and nucleic acids of similar base compositions. It is likely that artifacts associated with (i) the indirect spectral titration and/or (ii) the purity of the DNA preparation employed by Gottesfeld et al. (78) are involved (see also Section III,B).

A novel method of estimating the accessibility of DNA in chromatin from fluorescence measurements has recently been reported by Brodie, Giron, and Latt (79). The method involves the transfer of electronic excitation energy between pairs of fluorescent dyes bound to chromatin at low ratios of dye to phosphate. Quinacrine (as well as other dye molecules) was chosen as the energy donor since it exhibits a fluorescence emission spectrum that extensively overlaps the absorption energy of the acceptor used, ethidium bromide (EB). Following excitation of the donor, energy will be transferred with a high probability to EB molecules bound within a critical distance (approximately 25–35 Å) of the donor. Energy transfer efficiency decreases sharply with donor–acceptor separations above this value. The critical donor–acceptor distance for 50% efficient energy transfer, R_0, is determined by individual spectroscopic parameters and relative dye orientations (80,81). The relative efficiency of energy transfer between fixed amounts of donor and acceptor bound to either DNA or chromatin depends on the probability that an acceptor molecule will bind within R_0 of the donor, and thus, to a first approximation, this energy transfer varies inversely with the acceptor binding space. Hence it provides a comparative estimate of DNA accessibility in chromatin. For an accurate estimate of the acceptor binding space, the acceptor (i) should bind to all regions accessible to the donor and (ii) should not bind preferentially to particular regions of DNA. Q and EB constitute a suitable donor–acceptor pair, since they both bind to DNA by intercalation in a mutually competitive manner with little base-composition specificity (2,22,56,70). It is found that if EB is added to a solution containing DNA–Q complex, Q fluorescence is quenched by energy transfer to nearby EB molecules, which concomitantly exhibit sensitized fluorescence. At low donor and acceptor saturations and under conditions of total binding of both molecules, the donor fluorescence intensity (F) varies inversely with amount of acceptor (Fig. 5a). The amount of EB necessary for a 50% reduction in donor fluorescence EB_{50} increases in proportion to the total accessible DNA (Fig. 5b). Quenching of the fluorescence of Q bound to chromatin occurs at about one-third of the amount of added EB needed for a comparable effect on DNA–Q complexes. Direct spectrophotometric measurement of Q binding by Latt et al. (70) indicates that calf thymus chromatin ($\bar{n}_{maxI} = 0.07$) also possesses about one-third

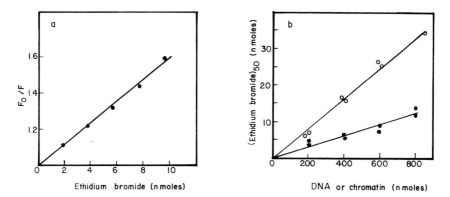

FIG. 5. (a) Energy-transfer quenching of quinacrine (Q) fluorescence by ethidium bromide (EB) in complexes with calf thymus DNA. The fluorescence of a 2-ml sample of a $0.7 \times 10^{-6} M$ solution of Q complexes with $2 \times 10^{-4} M$ calf thymus DNA (in 10 mM NaCl, 5 mM HEPES, pH 7) was measured after successive 20-μl additions of a 1×10^{-4} solution of EB. Excitation was at 420 nm. Q fluorescence (F), measured at 493 nm, was corrected for dilution by the EB solution and for a very small amount of quenching due to the trivial process of fluorescence and reabsorption. F_0 is the initial fluorescence in the absence of added ethidium. The ratio of F_0/F is plotted against the amount of added EB. From Brodie *et al.* (*79*). (b) Energy transfer titration of available phosphate in calf thymus DNA or chromatin. The amount of EB bromide at which Q fluorescence is reduced 50%, obtained from data, such as those of Fig. 5a, is plotted against the amount of calf thymus DNA (O) or calf thymus chromatin (●) present in solution. Data were obtained at ratios of ethidium to phosphate of 0.025 or less. From Brodie *et al.* (*79*).

the number of Q binding sites in an equal amount of DNA ($\bar{n}_{maxI} = 0.21$), consistent with the existence of an extensive overlap between Q and EB binding spaces in chromatin. It should be noted that energy transfer between dyes bound to chromatin in 5 M urea (conditions which are shown to cause negligible dissociation of proteins from DNA) was virtually identical to that with DNA. This finding is consistent with the idea that histone proteins in chromatin play an important role in determining the chromatin conformation which may govern interaction specificities and accessibility of the dyes (or probes) to DNA (*79*).

Finally, it should be mentioned that although extensive work has been carried out on the interaction of Q with DNA, there are no data available as to the DNA binding site (major or minor groove binding) of the *N,N*-diethylaminopropyl substituent upon intercalative binding of Q. It is highly probable that the positively charged diethylammonium group occupies the minor groove of DNA owing to the larger electrostatic repulsion between the polyanion chains in the minor as compared to the major groove.

3. ACTINOMYCIN D (AcD)

The interaction between AcD and DNA has been extensively investigated (5, 19, 82–89). It is now generally accepted that the strong DNA binding of AcD is due to intercalation of the phenoxazone ring between base pairs of DNA. This mode of binding is supported by the effect of AcD on (i) increasing the viscosity of solutions containing rodlike sonicated low MW DNA (5) and (ii) unwinding of supercoiled circular DNA (19). In addition, the X-ray data (89) on cocrystals of guanosine and AcD demonstrate a stacking interaction of the phenoxazone and guanine ring. The requirements for the presence of a G-C base pair (in natural DNAs) led Muller and Crothers (5) to postulate that the phenoxazone ring is intercalated into the helix adjacent to a G-C base pair and the peptide lactone rings project into the minor groove with concomitant H-bonding between the 2-amino group of guanine and the amino acids of the cyclic peptide. Sobel *et al.* (89) proposed a similar model involving insertion of the phenoxazone ring between G-C base pairs of DNA based on an extension of the X-ray studies of co-crystal of G and AcD. In an elegant study, Wells and Larson (87) demonstrated, however, that the presence of a 2-amino group (as in guanine) is not required for the intercalative binding of AcD to helical polydeoxyribonucleotides; moreover, they showed that the extent of AcD binding is strongly dependent on the base sequence. Additional evidence in favor of the Muller and Crothers' model whereby the peptide lactone rings of AcD are located in the minor groove of DNA comes from the study of Carroll and Botchan (90). They showed that the value of the maximal binding of AcD to calf thymus DNA and phage T4DNA (which has one or two glucose residues in the major groove attached to each hydroxymethylcytosine) are nearly identical.

Berlowitz *et al.* (91) studied the binding of [³H] AcD to extended (euchromatin) and condensed (heterochromatin) chromosomal preparations. The results indicate that maximal binding of AcD is 4 times greater in euchromatin as compared to heterochromatin. Olins (64) on the other hand, reported that the maximal binding of AcD to reconstituted nucleohistone (calf thymus DNA–histone I complex) and free DNA are similar in magnitude. Kleiman and Huang (88) studied the binding of AcD to calf thymus DNA, whole chromatin, and successively deproteinized chromatin. The results showed that maximum binding of AcD to chromatin is approximately one-third of that found in DNA. Selective removal of histone I from chromatin does not increase appreciably the number of AcD binding sites. However, removal of the remaining histone proteins (especially the arginine-rich histones) is found to lead to a significant increase in AcD binding sites. Similar results have also been obtained by Seligy and Lurquin (66, 71) who showed that avian erythrocyte chromatin had approximately 23% of the

AcD binding capacity of free DNA and that acid extraction of histones I and V had little effect on increasing the number of AcD binding sites. Significant increase in AcD binding sites is noted upon removal of the remaining histone proteins. On the other hand, it is observed that, in contrast to AcD, EB binding sites in chromatin are significantly increased upon removal of histones I and V. Based on these findings Seligy and Lurquin (66,71) and Olins (64) have suggested that the lysine-rich histones, i.e., histones I and V, do not bind in the minor groove of DNA and hence their removal would not lead to enhanced AcD binding capacity. On the other hand, assuming that the hypothetical model of Fuller and Waring (61) is correct (i.e., the phenyl substituent of EB occupies the major groove), then removal of histones I and V (presumably from the major groove of DNA) would lead to a significant enhancement of EB binding sites. It should be noted that an alternative explanation for the differential behavior of AcD and EB binding to partially deproteinized chromatin may involve a selective restriction of AcD binding to naked DNA regions by the remaining histone proteins via "telestability" mechanism similar to that described recently by Wells and co-workers (92–94).

4. DINITROANILINE REPORTER (R)

The dinitroaniline-labeled diamine (R) (and related derivatives) have been extensively studied in our laboratories and shown to form intercalated complexes with DNA at a base pair-to-reporter ratio greater than 3. This conclusion is based on several lines of evidence: (i) enhanced viscosity of DNA solution upon binding the reporter molecule; (ii) hypochromism and induced CD in the absorption transition of the bound molecules; (iii) total line broadening of the PMR signals of the aromatic protons of the bound molecules; and (iv) flow dichroism studies which indicate that the dinitro-aniline ring of the reporter molecule is in the same plane as the DNA base pairs.

Intercalation of the dinitroaniline ring of the reporter molecule (R) between base pairs of DNA could result in two types of complexes, i.e., with the diammonium side chain in the minor and/or the major groove. The

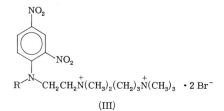

(III)

interaction specificity based on direct binding studies (equilibrium dialysis), hypochromism and induced circular dichroism in the 4-nitroaniline transition of N-substituted nitroanilines (III) with DNA of various base compositions showed that increasing the size of the R substituents led to selective binding to A–T rich DNA (95). In addition, it has been found that the binding of 2 to DNA is 10-fold greater than to synthetic double-stranded polyribonucleotides, e.g., poly(I)–poly(C) and poly(A)–poly(U) (96,97). Since the 2-hydroxyl group of double-stranded RNA and the additional H-bond found in G-C base pairs occupy the minor groove, it has been argued that (i) the greater binding to DNA as compared to RNA and (ii) the enhanced binding to A–T as compared to G–C sites, especially with increasing size of the R substituent in (III), provide indirect evidence for the binding of the diammonium side chain in the minor groove of the DNA helix. Such a model is also plausible from an electrostatic consideration, i.e., positively charged molecules would preferentially occupy the minor groove since the intrastrand repulsion between phosphate anions is greater across the minor groove.

The interaction of the reporter molecule, R (see Chart 1) with DNA and chromatin of rabbit liver has been studied by Simpson (98). It was found that (i) the reporter binds to chromatin without dissociation of the proteins of the complex, (ii) similar induced CD spectra are obtained in the absorption band of R upon binding to chromatin and to free DNA, and (iii) maximal binding of R to chromatin and DNA are found to be identical, i.e., 1 reporter per 3 base pairs. On this basis, Simpson (98) has suggested that the proteins of chromatin occupy only the major groove of DNA helix, leaving the minor groove available for the binding of R.

Unpublished work from our laboratories on the binding (via direct equilibrium dialysis) of R to HeLa S-3 cell mitotic chromatin (which has been subjected to minimal shear and contains 1.4:1.0 protein to DNA ratio) shows a maximal binding, $\bar{n}_{max}$, equal to 0.025 (i.e., 1 molecule per 20 base pairs) and a binding affinity, K_a, equal to 66,000. The value of $\bar{n}_{max}$, however, does not remain constant and approaches the value obtained with free HeLa S-3 cell DNA ($\bar{n}_{max} = 0.13$, $K_a = 80,000$) as the chromatin preparation becomes more aged, i.e., 4 days at $4\,°C$ (99). Identical results have also been obtained from DNA and chromatin isolated from rat liver. The instability of the chromatin, which may be due to the presence of proteolytic enzymes in the chromosomal proteins and/or self-association of the chromosomal proteins, may account for the results obtained by Simpson (98), i.e., that the maximum number of DNA binding sites for R remains constant for free DNA and chromatin.

Recent studies by Stein et al. (100) utilizing the reporter molecule, R, to examine the fidelity of chromatin reconstitution, showed that the maximal binding of R to reconstituted and native HeLa S-3 cell chromatin (freshly prepared) to be nearly identical ($\bar{n}_{max}$ equals 0.071 and 0.072, respectively). Approximately one reporter molecule is bound per seven base pairs, this indicates that in each case approximately 54% of the DNA binding sites ($\bar{n}_{max}$ is found to equal 0.13 for free DNA) are capable of interacting with the reporter molecule. The value of the maximal binding of R to both chromatin preparations ($\bar{n}_{max}$) approaches the value obtained with free DNA as the solution of chromatin is allowed to age at 4°C.

III. Methods in Chromatin Analysis

A. Buffer and Chromatin Preparation

Results from analyses of chromatin properties using intercalating agents have varied widely and are probably due to two principal factors: (i) use of different buffers, salts, ionic strengths, and pH values; and (ii) use of chromatin preparations in which the protein content, age, and stability are not determined. The cause of the first variation can be removed by appropriate control of buffer, ionic strength, and pH. Considering the published studies, a buffer containing 5 mM sodium phosphate (pH 7.0)–10 mM NaCl has been most frequently employed. These may not be the ideal conditions for studying the interaction of intercalating agents with DNA and/or chromatin, since the low ionic strength of the medium would lead to the formation of two types of complexes, i.e., the intercalative mode (type I) and external electrostatic mode (type II) of binding. Since the weaker binding mode (type II) can be eliminated at higher ionic strength, it appears appropriate to carry out the comparative binding studies at 10 and 100 mM NaCl concentrations. It is generally found that the binding isotherms obtained at the higher ionic strength (e.g., 100 mM NaCl) is solely due to an intercalation mode of binding. In addition, it should be emphasized that the values of the apparent binding constant, K_a and maximal binding, $\bar{n}_{max}$ (see Section III,B) for the intercalative binding process to free DNA are significantly reduced with increasing ionic strengths (see Fig. 4). This fact clearly demonstrates that, in all chromatin studies, purified DNA from the same species and preparation should be compared as a control under exactly the same conditions. Moreover, it is strongly recommended that the interaction specificity of the intercalating agent with highly purified calf thymus DNA (available commercially from Worthington Co.) be determined and compared to the results obtained with the DNA isolated from chromatin. Similar results in the binding isotherms (e.g., K_a and $\bar{n}_{max}$), spectral characteristics, etc., of the intercalating agent to the two DNA systems should be obtained. Failure to meet this criterion could indicate that the isolated DNA may contain protein contaminants and should not be used for comparative studies with chromatin.

The second cause of variation in chromatin experiments is difficult to remove owing to the labile nature of chromatin itself. The best solution is the use of freshly prepared chromatin in all investigations (e.g., less than 24 hours old). A standard reference intercalating agent, e.g., ethidium bromide, should be used to characterize different chromatin preparations for possible variations due to differences in methods of isolation. The protein content has a profound effect on binding studies and must be determined and carefully maintained or changed in a controlled desired manner to give reproducible results. Since chromatin preparations have a tendency to become turbid, they must be rigorously clarified before any spectrophotometric procedures are initiated. Some intercalating compounds which do not cause DNA to precipitate have caused turbidity in chromatin preparation and are presumably due to the neutralization of the charge on chromatin by the combined presence of the protein and the cationic molecule. Binding isotherms based on spectrophotometric studies with such compounds are subject to very large deviations, and meaningless data are obtained.

B. Binding Studies

1. DIRECT AND INDIRECT METHODS

The apparent binding constant, K_a, and the maximal binding, $\bar{n}_{max}$, of intercalating agents with DNA are obtained from binding isotherm data. The experimental technique and the mathematical analysis of the binding have been discussed in numerous papers (see Section III,B,2), and only the critical points are summarized below.

Binding isotherms may be determined from direct methods, such as equilibrium dialysis and solvent partition, or by indirect analysis involving spectrophotometric techniques, i.e., changes in absorption, fluorescence, and/or CD. The indirect method, which depends on spectral changes induced in the bound ligand, assumes a linear response which is independent of the nature of the DNA complex (i.e., intercalation or external electrostatic mode of binding). It is also assumed that (i) all of the 10 possible intercalation sites in DNA induce the same spectral change and (ii) ligand–ligand interactions do not contribute to the observed spectral change. Since these assumptions have been shown not to be valid (especially for the fluorescence method) for a number of intercalating agents (see Sections II,B and C), the indirect technique should be avoided. Equilibrium dialysis technique does not suffer from such assumptions, however, experimental problems, such as ligand binding to membranes, the Donnan effect, difficulty in determining concentrations of dilute solutions, and the required long equilibration times ($\geq$ 18 hours), complicate this method. The solvent-partition technique for direct binding studies, recently reviewed and experimentally analyzed by Waring and Henky (*101*), eliminates some of these problems, but to our knowledge it has not been used with chromatin. It is probably not applicable to nucleohistone studies owing to the possible denaturation of the proteins by the organic solvent.

The equilibrium dialysis technique then, must remain the method of choice for chromatin and DNA binding studies. Various types of apparatus and methods of membrane preparation have been described and proved satisfactory (*6,72,102*). The ligand concentration should be measured on both sides of the membrane in order to determine the bound and unbound concentration directly. Although the concentration of ligand on the macromolecule side of the membrane can, in principle, be determined by simply subtracting the amount on the free side from the total ligand added to the dialysis cell, this leads to considerable error because of ligand binding to the membrane, glass walls, etc. The concentration could easily be determined by radioactivity (if a labeled ligand is available) using standard techniques. If the concentration is determined using spectrophotometric or fluorescence methods, then the complex must first be dissociated (usually by the addition of an organic solvent) and the ligand concentration measured using appropriate standard concentration curves in the final solvent composition (*72,102*).

The indirect methods involving spectral titration studies whereby changes in the (i) absorption, (ii) relative flourescence, and (iii) induced CD of the intercalating agent are monitored may also be used for the determination of the binding characteristics of the latter to DNA and chromatin. However, as has been pointed out above and in the earlier discussion (see Sections II,B and C), the results of such studies are questionable and should be employed only if it can be shown that the direct methods, e.g., equilibrium dialysis, give similar data. For the absorption titration studies, the technique involves the determination of the extinction coefficient of the free and bound ligand at a wavelength where there is no interference from DNA (or chromatin). Assuming a linear relationship of the absorption change with the degree of binding, the determination of the absorbance at a known ligand concentration in the presence of DNA (or chromatin) will then allow the calculation of the concentration of the free and bound species. The data obtained are thus equivalent in form to those obtained by equilibrium dialysis, and the analysis of the binding process is carried out in the same manner (see below).

2. ANALYSIS OF THE BINDING DATA

Numerous methods of analyzing the binding data are available in order to evaluate the apparent binding constant, K_a, and the maximal binding, $\bar{n}_{max}$ (5,11,27,72,103). The most simple and frequently used method is the Scatchard (104) analysis based on Eq. (2)

$$\bar{n} = \bar{n}_{max} - (1/K_a)(\bar{n}/R_f) \tag{2}$$

where $\bar{n}$ = the number of moles of x bound per mole of DNA phosphate, $\bar{n}_{max}$ represents maximal binding, K_a is the apparent association constant for the DNA–x complex, and R_f is the concentration of unbound x. A plot of $\bar{n}/R_f$ (y axis) vs $\bar{n}$ (x axis) gives the values of $\bar{n}_{max}$ (x-axis intercept) and $\bar{n}_{max} K_a$ (y-axis intercept) (see Fig. 4).

The next simplest binding situation would involve two sets of homogeneous and independent binding sites. Most cationic intercalating agents show this behavior, i.e., strong intercalative mode and weak external electrostatic mode of binding. If the values of the apparent binding constants, K_{aI} and K_{aII} (for types I and II binding) are significantly different, then both values as well as the values of $\bar{n}_{maxI}$ and $\bar{n}_{maxII}$ could be determined (Fig. 4). Subtle differences in strong binding due to effects such as differences in ligand affinity to A-T and G-C base-pair sites are not resolved by the Scatchard analysis. Nonetheless, Eq. (2) remains the logical place to begin the analysis of the binding studies between DNA (or chromatin) and intercalating agents.

It should be noted that even at high ionic strength, where only the intercalation mode of binding is involved, nonlinear Scatchard plots are frequently obtained and the cause of this phenomenon remains a matter of some debate. For example, such curvature could be due to at least two factors including (i) noncooperative binding (103) and (ii) large differences in the binding affinity of the intercalating agent to the 10 possible intercalating sites of a right-handed DNA helix (30). Investigators desiring to use one of the more complicated methods of binding analysis should begin by consulting the references at the end of this article (27,72,103).

C. Spectral Studies

Electronic, fluorescence, and CD spectra for ligands bound to DNA can be determined using standard methods and equipments. The primary concern with these techniques is to ensure that the chromatin solutions are not turbid and do not become so on mixing chromatin with the intercalating molecules.

Changes in the absorption spectrum of a ligand on addition of chromatin can be monitored just as with addition of DNA, and the results should be qualitatively similar (e.g., red shift and hypochromism) if a common mode of binding is involved. Such changes may be quantitated to allow the determination of $\bar{n}_{max}$ and K_a (see Section III,B,2).

Changes in the fluorescence properties of a ligand can also be monitored with addition of chromatin just as with addition of DNA. Once again such changes should be qualitatively similar for chromatin and DNA in terms of shifting of peak wavelengths, and enhancement of reduction of quantum yield. However, equilibrium calculations based on quantitation of the fluorescence changes are frequently not reliable due to the nonlinear response of the fluorescence changes with respect to the extent of binding, $\bar{n}$.

Circular dichroism studies with chromatin are conducted as with DNA (once again using caution to prevent turbidity) and can in principle allow a comparison of the mode of binding of a ligand to DNA and to chromatin. The complicating effects observed with this technique prevent quantitative comparisons from being made between chromatin and DNA, but qualitative analysis of induced CD spectra from the standpoint of peak position, magnitude, and sign can give some indication of the similarity of the binding sites on the two DNAs.

Although no ^{1}H NMR spectra have been reported on the interaction of ligands with chromatin, considerable work has been done with purified, sonicated DNA (9, 10). The development of high-resolution pulsed NMR instruments should allow a sufficient reduction in the ligand concentration necessary for such experiments to bring these studies into a feasible range for binding analysis.

D. Methods Summary and Future Directions

Of the methods available for analysis of chromatin properties using intercalating agents, binding studies using the equilibrium dialysis technique have provided the most useful quantitative information. Binding studies using indirect methods have provided qualitative comparisons between purified DNA and chromatin. Spectral shifts, fluorescence changes, and induced CD in the bound ligand offer even more grossly qualitative information about chromatin structure and should only be used to determine whether ligand–chromatin interactions do exist. Fluorescence energy transfer studies (79) at low ratios of bound ligand, between carefully selected donor and acceptor molecules, have been successfully used to determine an approximate value for "free" DNA in chromatin and should be useful in future investigations requiring this type of information (e.g., in analyzing chromatin from various cellular reproductive states and comparing normal versus cancer cell chromatins). Circular dichroism studies, although in principle they offer information on the conformation of the chromatin ligand complex, have been difficult to interpret owing to (i) intercalated ligand–ligand interaction and (ii) asymmetric external stacking of ligands which obscure the interpretation of the data. Hydrodynamic and flow orientation techniques have been widely used in analyzing intercalation of ligands into purified DNA, but because of the complexity of chromatin and uncertainty in the structure of chromatin DNA, these techniques have not been nor are likely to be used in chromatin analysis. It should finally be emphasized that none of the more conclusive methods of demonstrating intercalation (e.g., viscosity, sedimentation, flow orientation) have been or can be applied to whole chromatin. Even our present conclusions regarding binding of ligands into chromatin DNA, therefore, rest on experiments giving only secondary support for an intercalative mode of binding.

REFERENCES

1. Lerman, L. S. *J. Mol. Biol.* **3**, 18 (1961).
2. Lerman, L. S. *Proc. Natl. Acad. Sci. U.S.A.* **49**, 94 (1963).
3. Lerman, L. S. *J. Cell. Comp. Physiol.* **64**, Suppl. 1, 1 (1964).
4. Lerman, L. S. *J. Mol. Biol.* **10**, 367 (1964).
5. Muller, W., and Crothers, D. M. *J. Mol. Biol.* **35**, 251 (1968).
6. Gabby, E. J., Sanford, K., Baxter, C. S., and Kapicak, L. *Biochemistry* **12**, 4021 (1973).
7. Gabbay, E. J., Scofield, R., and Baxter, C. S. *J. Am. Chem. Soc.* **95**, 7850 (1973).
8. Kapicak, L., and Gabbay, E. J. *J. Am. Chem. Soc.* **97**, 403 (1975).
10. Gabbay, E. J., Adawadkar, P. D., Kapicak, L., Pearce, S., and Wilson, W. D. *Biochemistry* **15**, 152 (1976).
11. Blake, A., and Peacocke, A. R. *Biopolymers* **6**, 1225 (1968).
12. Waring, M. J. *In* "The Molecular Basis of Antibiotic Action" (E. F. Gale *et al.*, eds.), p. 173. Wiley, New York.
13. Lang, M., Dourlent, M., and Hélène, C. *Phys.-Chem. Prop. Nucleic Acids* **3**, 19 (1973).
14. Cohen, G., and Eisenberg, H. *Biopolymers* **8**, 45 (1969).
15. Eisenberg, H. *Basic Prin. Nucleic Acid Chem.* **2**, 43 (1974).
16. Bauer, W., and Vinograd, J. *Basic Prin. Nucleic Acid Chem.* **2**, 265 (1974).
17. Saucier, J. M., Festy, B., and LePecq, J. B. *Biochimie* **53**, 973 (1971).
18. Pulleyblank, D. E., and Morgan, A. R. *J. Mol. Biol.* **91**, 1 (1975).
19. Waring, M. J. *J. Mol. Biol.* **54**, 247 (1970).
20. Bush, C. A. *Basic Prin. Nucleic Acid Chem.* **2**, 91 (1974).
21. Ellerton, N. F., and Isenberg, I. *Biopolymers* **8**, 767 (1969).
22. LePecq, J. B., and Paoletti, C. *J. Mol. Biol.* **27**, 87 (1967).
23. Weisblum, B., and deHaseth, P. L. *Proc. Natl. Acad. Sci. U.S.A.* **69**, 629 (1972).
24. Pachmann, U., and Rigler, R. *Exp. Cell. Res.* **72**, 602 (1972).
25. Krey, A. K., and Hahn, F. E. *Mol. Pharmacol.* **10**, 686 (1974).
26. Festy, B., and Daune, M. *Biochemistry* **12**, 4827 (1973).

27. Bloomfield, V. A., Crothers, D. M., and Tinoco, I., Jr. *In* "Physical Chemistry of Nucleic Acids," p. 373. Harper, New York. (1974).
28. Lee, C. H., Chang, G. J., and Wetmur, J. *Biopolymers* **12**, 1099 (1973).
29. Li, H. J., and Crothers, D. M. *Biopolymers* **8**, 217 (1969).
30. Gabbay, E. J., DeStefano, R., and Sanford, L. *Biochem. Biophys. Res. Commun.* **44**, 155 (1972).
31. Cavalieri, L. F., Rosenberg, B. H., and Rosoff, M. *J. Am. Chem. Soc.* **78**, 5235 (1956).
32. Gabbay, E. J., and DePaolis, A. *J. Am. Chem. Soc.* **93**, 562 (1971).
33. Pople, J. A., Schneider, W. G., and Bernstein, H. J. "High Resolution Nuclear Magnetic Resonance." McGraw-Hill, New York. (1959).
34. Jardetsky, O., and Jardetsky, C. D. *Methods Biochem. Anal.* **9**, 235 (1962).
35. Gabbay, E. J., Sanford, K., and Baxter, C. S. *Biochemistry* **11**, 3429 (1972).
36. Miura, A., and Ohba, Y. *Biochim. Biophys. Acta* **145**, 436 (1967).
37. Kurashina, Y., Ohba, Y., and Mizuno, D. *J. Biochem. (Tokyo)* **67**, 661 (1970).
38. Itzhaki, R. F. *Biochem. J.* **122**, 583 (1971).
39. Adawadkar, P. D., Sanford, K., Scofield, R. E., and Gabbay, E. J. Unpublished results (1974).
40. Michaelis, L. *Cold Spring Harbor Symp. Quant. Biol.* **12**, 131 (1947).
41. Peacocke, A. R., and Skerrett, J. N. H. *Trans. Faraday Soc.* **52**, 261 (1956).
42. Bradley, D. F., and Wolf, M. K. *Proc. Natl. Acad. Sci. U.S.A.* **45**, 944 (1959).
43. Neville, D. M., Jr., and Bradley, D. F. *Biochim. Biophys. Acta* **50**, 397 (1961).
44. Neville, D. M., Jr., and Davies, D. R. *J. Mol. Biol.* **17**, 57 (1966).
45. Armstrong, R. W., Kurucsev, T., and Strauss, U. P. *J. Am. Chem. Soc.* **92**, 3174 (1970).
46. Cairns, J. *Cold Spring Harbor Symp. Quant. Biol.* **27**, 311 (1962).
47. Luzzati, V., Masson, F., and Lerman, L. S. *J. Mol. Biol.* **3**, 634 (1961).
48. Bauer, W., and Vinograd, J. *J. Mol. Biol.* **33**, 141 (1968).
49. Walker, I. O. *Biochim. Biophys. Acta* **109**, 585 (1965).
50. Gittelson, B. L., and Walker, I. O. *Biochim. Biophys. Acta* **138**, 619 (1967).
51. Houssier, C., and Fredericq, E. *Biochim. Biophys. Acta* **88**, 450 (1964).
52. Houssier, C., and Fredericq, E. *Biochim. Biophys. Acta* **120**, 434 (1966).
53. Lawrence, J. and Louis, M. *Biochim. Biophys. Acta* **272**, 231 (1972).
54. Dalgleish, D. G., Dingsoyr, G., and Peacocke, A. R. *Biopolymers* **12**, 445 (1973).
55. Quadrifoglio, F., and Crescenzi, V. *Biophys. Chem.* **2**, 64 (1974).
56. Waring, M. J. *J. Mol. Biol.* **13**, 269 (1965).
57. Waring, M. J. *Biochim. Biophys. Acta* **114**, 234 (1966).
58. Kersten, W., Kersten, H., and Szybakski, W. *Biochemistry* **5**, 236 (1966).
59. O'Brien, R. L., Allison, J. L., and Hahn, F. E. *Biochim. Biophys. Acta* **129**, 622 (1966).
60. Crawford, L. V., and Waring, M. J. *J. Mol. Biol.* **28**, 455 (1967).
60a. Aktipis, S., and Kindelis, A. *Biochemistry* **12**, 1213 (1973).
61. Fuller, W., and Waring, M. J. *Ber. Bunsenges. Phys. Chem.* **68**, 805 (1964).
62. Bollund, L., Ringertz, N. R., and Harris, H. J. *J. Cell Sci.* **4**, 71 (1969).
63. Ringertz, N. R., and Bollund, L. *Exp. Cell Res.* **55**, 55 (1969).
64. Olins, D. E. *J. Mol. Biol.* **63**, 505 (1969).
65. Angerer, L. M., and Moudrianakis, E. N. *J. Mol. Biol.* **63**, 505 (1972).
66. Lurquin, P. F., and Seligy, V. L. *Arch. Int. Physiol. Biochim.* **80**, 606 (1972).
67. Williams, R. E., Lurquin, P. F., and Seligy, V. L. *Eur. J. Biochem.* 426 (1972).
68. Seligy, V. L., and Neelin, J. M. *Biochim. Biophys. Acta* **213**, 380 (1970).
69. Angerer, L. M., Georghiou, S., and Moudrianakis, E. N. *Biochemistry* **13**, 1075 (1974).
70. Latt, S. A., Brodie, S., and Munroe, S. H. *Chromosoma* **49**, 17 (1974).

71. Seligy, V. L., and Lurquin, P. F. *Nature (London)*, *New Biol.* **243**, 20 (1973).
72. Bontemps, J., and Fredericq, E. *Biophys. Chem.* **2**, 1 (1974).
73. Caspersson, T., Farber, S., Foley, G. E., Kudynowski, J., Modest, E. J. Simonsson, E., Wagh, V., and Zech, L. *Exp. Cell Res.* **49**, 219 (1968).
74. Caspersson, T., Zech, L., Modest, E. J., Foley, G. E., Wagh, V., and Simonsson, E. *Exp. Cell Res.* **58**, 128 (1969).
75. Ellison, J. R., and Barr, H. J. *Chromosoma* **36**, 375 (1972).
76. Schreck, R. R., Warburton, D., Miller, O. J., Beiser, S. M., and Erlanger, B. F. *Proc. Natl. Acad. Sci. U.S.A.* **70**, 804 (1973).
77. Lomholt, B., and Mohr, J. *Nature (London)*, *New Biol.* **234**, 109 (1971).
78. Gottesfeld, J. M., Bonner, J., Radda, G. K., and Walker, I. O. *Biochemistry* **14**, 2937 (1974).
79. Brodie, S., Giron, J., and Latt, S. A. *Nature (London)* **253**, 470 (1975).
80. Forster, T. *In* "Modern Quantum Chemistry" (O. Sinanoğlu, ed.), p. 93. Academic Press, New York (1965).
81. Stryer, L. *Science* **162**, 562 (1968).
82. Kahan, E., Kahan, F. M., and Hurwitz, J. *J. Biol. Chem.* **238**, 2491 (1963).
83. Reich, E., and Goldberg, I. H. *Prog. Nucleic Acid Res. Mol. Biol.* **3**, 183 (1964).
84. Cavalieri, L. F., and Nemchin, R. G. *Biochim. Biophys. Acta* **87**, 641 (1964).
85. Gellert, M., Smith, C. E., Neville, D., and Felsenfeld, G. *J. Mol. Biol.* **11**, 445 (1965).
86. Cerami, A., Reich, E., Ward, D. C., and Goldberg, I. H. *Proc. Natl. Acad. Sci. U.S.A.* **57**, 1036 (1967).
87. Wells, R. D., and Larson, J. E. *J. Mol. Biol.* **49**, 319 (1970).
88. Kleiman, L., and Huang, R. C. C. *J. Mol. Biol.* **55**, 503 (1971).
89. Sobell, H. M., Jain, S. C., Sakore, T. D., and Nordman, C. E. *Nature (London)*, *New Biol.* **231**, 200 (1971).
90. Carrol, D., and Botchan, M. H. *Biochem. Biophys. Res. Commun.* **46**, 1681 (1972).
91. Berlowitz, L., Paollotta, D., and Sibley, C. H. *Science* **164**, 1527 (1969).
92. Wartell, R. M., Larson, J. E., and Wells, R. D. *J. Biol. Chem.* **249**, 6719 (1974).
93. Burd, J. F., Larson, J. E., and Wells, R. D. *J. Biol. Chem.* **250**, 6002 (1975).
94. Burd, J. F., Wartell, R. M., Dodgson, J. B., and Wells, R. D. *J. Biol. Chem.* **250**, 5109 (1975).
95. Gabbay, E. J., and Sanford, K. *Bioorg. Chem.* **3**, 91 (1974).
96. Gabbay, E. J. *J. Am. Chem. Soc.* **91**, 5136 (1969).
97. Passero, F., Gabbay, E. J., Gaffney, B. L., and Kurucsev, T. *Macromolecoules* **3**, 158 (1970).
98. Simpson, R. T. *Biochemistry* **9**, 4814 (1970).
99. Scofield, R., Sanford, K., Adawadkar, P. D., and Gabbay, E. J. Unpublished results. (1974).
100. Stein, G. S., Mans, R. J., Gabbay, E. J., Stein, J. L., Davis, J., and Adawadkar, P. D. *Biochemistry* **14**, 1859 (1975).
101. Waring, M. J., and Henley, S. M. *Nucleic Acids Res.* **2**, 567 (1975).
102. Muller, W., Crothers, D. M., and Waring, M. J. *Eur. J. Biochem.* **39**, 223 (1973).
103. Crothers, D. M. *Biopolymers* **6**, 1391 (1968).
104. Scatchard, G. *Ann. N. Y. Acad. Sci.* **51**, 660 (1949).

Chapter 19

Thermal Denaturation Analysis of Chromatin and DNA–Nuclear Protein Complexes

HSUEH JEI LI

Division of Cell and Molecular Biology,
State University of New York at Buffalo,
Buffalo, New York

I. Introduction

Thermal denaturation of DNA complexed with proteins provides a means for measuring the transition of base pairs from a native, double-helical state to a denatured, unordered-coil state, as a function of temperature. A thermal denaturation profile (or melting profile) measures this transition through a plot of hyperchromicity against temperature. The melting profile of a DNA–protein complex is sensitive to a variety of parameters, such as type of DNA and protein, input ratio of protein to DNA, method of complex formation, type of ions, ionic strength of solutions, pH. Because of this sensitivity, thermal denaturation has been used extensively in obtaining detailed information about chromatin, DNA–histone, and other DNA–protein complexes. Both the principles and the applications of thermal denaturation to the study of chromatin structure have been reviewed recently (*1*). The present discussion emphasizes the methodology of thermal denaturation techniques in studies of chromatin and DNA–nuclear protein complexes and includes a brief description of applications.

II. Experimental Methods

A. General Considerations

In principle, any physical or biochemical assay of helix–coil transition in a DNA–protein complex as a function of temperature can be used for

385

measuring thermal denaturation. For instance, absorbance, circular dichroism (CD), optical rotary dispersion (ORD), nuclear magnetic resonance (NMR), viscosity, sedimentation, and other properties of a DNA–protein complex are changed when the native DNA is denatured. One can then follow the changes of these properties as a function of temperature and obtain a melting profile of a DNA–protein complex in terms of these parameters. Two practical questions should be considered before a particular instrument for measurement is selected: (a) Which method can provide the most accurate and detailed information about helix–coil transition of DNA segments, whether free or complexed with proteins? and (b) Which method costs less, is most readily available for use, and can be handled most easily.

Consideration of both factors suggests absorbance as the best technique for obtaining melting profiles, through measurement of the characteristic increase in hyperchromicity during denaturation. CR, ORD, NMR, viscosity, and sedimentation properties of a DNA–protein complex depend not only upon local helix–coil transition, but also upon changes in the tertiary structure of the whole complex. Although they can provide some information not obtainable by absorbance, interpretation of the melting profile is often more complicated when measured by these other techniques.

B. Instruments

Absorbance profiles which describe thermal denaturation of DNA–protein complexes can be obtained from any spectrophotometer to which a constant-temperature circulator can be attached. Both Gilford spectrophotometer Model 2400-S and Beckman Acta Century spectrophctometer with accessories are suitable. Since our laboratory has used the Gilford spectrophotometer, this model will be described as an example. Basic components of the Gilford Model 2400-S when used for thermal denaturation measurement are as follows:

1. An automatic recording spectrophotometer, which includes a monochromator, a light source, a photometer, a recorder, an automatic multiple-sample handling system with dual thermoplates

2. An automatic reference compensator, which maintains a constant zero reference line on a recording chart paper from the buffer solution in the reference cuvette at various temperatures

3. A thermosensor, which measures the temperature of the sample chamber in the automatic cuvette positioner

4. A constant temperature circulator, which controls the temperature in the sample chamber

5. A temperature programmer (Nelslab Instruments, Inc.), which controls the speed of temperature change in a melting experiment

The cuvette holder can accommodate four cuvettes, one for the buffer solution and the other three for samples. In each experiment, the recording chart reads the temperature, the reference line (from the reference cuvette), and the absorbance change in each of the three sample cuvettes.

C. Measurements of Thermal Denaturation

1. SELECTION OF BUFFERS

Various buffers have been used for thermal denaturation of chromatin DNA–protein complexes, such as $2.5 \times 10^{-4}M$ EDTA (pH 8.0) (2), 1 mM cacodylate buffer (pH 7.0) (3), and 3.6 M urea–5 mM cacodylate buffer (4). Since the melting temperature of a DNA is extremely sensitive to the presence of even a trace of divalent cation in the solution medium (Mg^{2+}, for example) (5), the use of EDTA in the buffer is recommended. In order to get a wider separation between the melting temperature of protein-free and protein-bound segments of a DNA–protein complex, it is desirable to use a buffer of lower ionic strength, since this allows the free DNA to melt at a lower temperature and permits a wider separation between the melting temperatures of these segments and those bound by proteins. As illustrated later, $2.5 \times 10^{-4} M$ EDTA (pH 8.0) satisfies this requirement and is satisfactory for melting experiments of both chromatin and DNA–protein complexes. Since the ionic strength of this EDTA buffer is extremely low, extensive dialysis with several changes of buffer solution is needed before melting. Since slight variation in buffer conditions is possible, despite careful preparation, it is recommended that those samples whose melting profiles are to be compared with one another be dialyzed against the same buffer solution and measured simultaneously on the spectrophotometer (three samples per run).

2. MEASUREMENT IN A SPECTROPHOTOMETER

Because of differences in the solubility of air at various temperatures, the solution may be degassed under vacuum before initiating a denaturation experiment. Such a procedure is sometimes undertaken in order to avoid the formation of air bubbles at higher temperatures. In our experiments, however, we have experienced no problem with air bubbles if the solutions are simply left in test tubes at room temperature for about half an hour before being transferred to the cuvettes for denaturation experiment.

Because of the extremely low ionic strength of the buffer, it is suggested that disposable pipettes which are used for transferring solutions be washed before use.

After the samples are transferred to the cuvettes and stoppered, the absorbance is measured, both at 260 (A_{260}) and 320 nm (A_{320}). The 320 nm measurement is made in order to be able to estimate the contribution to the measured A_{260} of light scattering and turbidity in the sample solution. Any significant change in light scattering and turbidity during the run can increase the total hyperchromicity measure at 260 nm. Therefore, in an ordinary experiment, it is recommended that the absorbance ratio, A_{320}/A_{260}, be calculated both before and after the sample is thermally denatured. If this ratio remains below 0.04, the scattering contribution is considered to be small; if the ratio is higher than 0.04 but remains constant before and after melting, the melting profile can still be considered to derive primarily from denaturation of DNA. On the other hand, if the A_{320}/A_{260} ratio is significantly changed after denaturation at high temperature, the apparent melting profile would be a superposition of temperature-dependent light scattering on the genuine denaturation of DNA, in which case the following analysis would not be applicable.

The hyperchromicity of DNA is about 35% at 260 nm, the wavelength at which runs are customarily made. The error of absorbance measurement by a spectrophotometer is about 0.001. To obtain a reliable melting profile, it is necessary to maintain an average absorbance change of this order of magnitude per degree, which gives a total absorbance change of about 0.1 or higher after full denaturation. For best results, the lower limit of the initial absorbance at 260 nm for a chromatin or a DNA–protein complex should be about 0.3. If there is an adequate supply of material, a higher concentration, with a starting absorbance around 1.0 optical density for each sample is recommended. Since three samples can be thermally denatured simultaneously in each run, and the melting profiles recorded on the same chart, it is desirable to prepare these three samples with A_{260} absorbance reading as close as possible to each other, so that all three samples can benefit from maximum use of the full scale.

Theoretically, the heating rate during a thermal denaturation experiment should be slow so that, at each temperature, the helical and coiled states of the complex are in equilibrium before the absorbance is recorded. No extensive research on the kinetics of denaturation of nucleoproteins, has yet been made, so one cannot tell exactly what would be an appropriate heating rate for each type of complex in order to permit equilibrium readings. Nevertheless, in chromatin and in DNA–protein complexes, it has been possible to obtain reproducible melting profiles if a constant heating rate with increments of about $1/2°–2/3°$ per minute is maintained. Maintenance of a linear increment of temperature as a function of time has the additional advantage of making it possible to extrapolate the temperature beyond 100°C from the straight portion of the line below this temperature; this can be calibrated between 0° and 100°C by use of a thermosensor. Since the cuvettes are sealed with a stopper, the solution can be heated above 100°C without boiling. When no further increase in absorbance is detected over a span of about 5°–10°C, and if the total hyperchromicity is reasonably close to what is expected for the uncomplexed DNA (about 30%, for example), the melting experiment is considered to be completed.

With complexed DNA, the kinetics of denaturation (6–8) and renaturation (9–14) is complicated. It is expected that the kinetics of denaturation and renaturation of chromatin and DNA–protein complexes will prove to be much more complicated. Because of its extreme complications, a quantitative analysis of renaturation has not been seriously attempted, but a qualitative experiment on the renaturation of chromatin and DNA–protein complexes can be performed by measuring absorbance changes as the temperature is lowered slowly, from the high temperature at which full denaturation was accomplished, back to the temperature at which the experiment was begun. Although there is no exact rule to suggest the most effective cooling rate, 1 °C per few minutes seems to provide adequate results. If the renaturation profiles for a variety of complexes are to be compared, it is advisable to maintain the same cooling rate for all of them. After renaturation has occurred, through returning from high to the initial temperature, the samples may be redenatured and their redenaturation profiles recorded.

3. Analysis of Thermal Denaturation Data

If A_0 is the absorbance at 260 nm for a sample in its native state, at room temperature, for example, and $A(T)$ is the absorbance at temperature T, the hyperchromicity at temperature T, $h(T)$, is

$$h(T) = \frac{[A(T) - A_0]}{A_0} \times 100\% \tag{1}$$

A plot of $h(T)$ against T gives the ordinary melting profile of the sample. The derivative of a melting profile, which makes slight changes more apparent, can be obtained by the following equation (15).

$$\frac{dh(T)}{dT} = \frac{h(T + 1) - h(T - 1)}{2} \tag{2}$$

or by other appropriate equations. The plot of $dh(T)/dt$ against T is termed the derivative profile. Melting data can be analyzed easily by use of a simple calculator; use of a computer does not seem warranted, since the equations involved are simple and the further resolution of derivative melting curves into many bands and subbands is limited both by the error of absorbance reading from a spectrophotometer (about 0.001) and by the reproducibility among samples (see Section III,A).

III.   Applications of Thermal Denaturation Methods to Studies of Chromatin and DNA–Protein Complexes

A.   Characteristic Melting Bands in Chromatin and Other DNA–Protein Complexes

Figure 1 shows three melting profiles of pea bud chromatin and partially dehistonized chromatin by 0.6 M and 2.0 M NaCl. The data points in the figure were obtained from Eq. (1). These melting profiles are not monophasic, but show at least three phases of transitions. The advantage of a derivative melting profile over an ordinary profile is seen in Fig. 2, wherein the curve is calculated from data points of Fig. 1 using Eq. (2) (*15*). When a smooth line is drawn through the data points, several characteristic bands appear. $T_{m,I}$ at 42°C, has been assigned as the melting temperature of free DNA segments; $T_{m,II}$, at about 52°C, as that of DNA segments bound by nonhistone proteins or short free DNA gaps between two histone-bound seg-

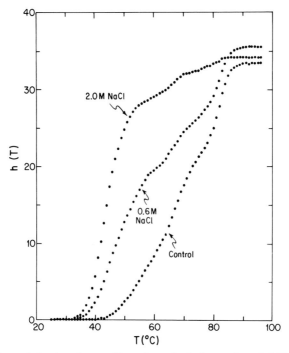

FIG. 1.   Thermal denaturation profiles of pea bud chromatin and 0.06 M and 2.0 M NaCl-treated chromatin in 2.5 × 10⁻⁴ M EDTA (pH 8.0). The data points are included in the figure (*15*).

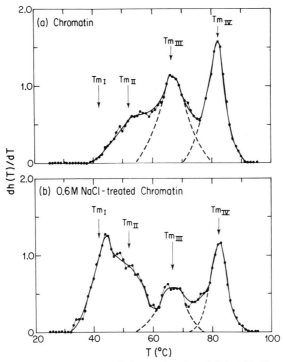

FIG. 2. Derivative melting profiles of chromatin (a) and 0.6 M NaCl-treated chromatin (b) from Fig. 1. The data points are calculated from those of Fig. 1 using Eq. (2). Derivative melting profiles emphasizing the major melting bands (————), those emphasizing sub-bands (————), and resolved melting bands for histone-bound base pairs (- - -). For the assignment of melting temperature of each band, see the text (15).

ments (15). The histone-bound segments also show two bands, centered at 66°–67°C ($T_{m,\,III}$) and 82°–83°C ($T_{m,\,IV}$): these have been assigned, res-pectively, as the melting temperatures of DNA segments bound by the less basic and the more basic regions of histones.

The following comments explain the method used in assigning melting bands of chromatin or DNA–protein complexes.

If the data points in Fig. 2 are followed carefully, subbands can be seen in each melting band. The obvious question is: Should we ignore these subbands and draw a smooth curve or should we assign more melting bands? The basic criteria for such assignment are (i) instrumental reli-ability and (ii) reproducibility among samples. If A_0 of the sample is 1.0, an absorbance change of 0.001, which is equal to the magnitude of instru-mental error, would be equivalent to a change of 0.1 in a dh/dt plot, since h is recorded on a percentage basis. Therefore, any variation of about 0.1 in the dh/dT plot should not be considered seriously. Substantial variation in these submelting bands may occur when different samples of chromatin or partially dehistonized chromatin are used. The recommended strategy is to put the derivative melting profiles of these samples together and draw a curve through the points which show the same characteristic bands in each of several different preparations. For a new type of chromatin that has never been studied before, an examination of results from about five different preparations should be adequate; for a simpler system that exhibits only two or three phases of transition, such as a DNA–histone or a DNA–polypeptide complex, the number of preparation can be reduced to about three with reasonable confidence that a reliable melting pattern has been obtained.

Derivative melting profiles have been used in the analysis of various chromatins (4, 15–20), histone–DNA complexes (15, 21–25) and other protein–DNA complexes (1, and references therein).

In thermal denaturation studies of nonhistone proteins, the reconstituted complexes of calf thymus DNA complexed with calf thymus HMG1 and HMG2 nonhistone proteins, in $2.5 \times 10^{-4} M$ EDTA (pH 8.0), showed two melting bands, one band at 47°C corresponding to free DNA, and another at 60°–65°C corresponding to nonhistone protein-bound DNA (25a). In calf thymus chromatin, the two melting bands of histone-bound base pairs are at 73° and 82°C (17). In other words, nonhistone proteins HMG1 and HMG2 when bound stabilize calf thymus DNA to a temperature about 10°C lower than that do the less basic regions of calf thymus histones.

B.   Quantitative Analysis of Derivative Melting Profiles of Chromatin and DNA–Protein Complexes

If a DNA–protein complex shows only two melting bands, T_m for free DNA and T_m' for protein-bound DNA, the fraction (F) of DNA base pairs in the complex bound by the protein can be measured by Eq. (3) (26)

$$F = A_{T_{m'}}/A_T \qquad (3)$$

where A_{T_m}' and A_T are, respectively, the area under the melting band at T_m' and the total area under the whole melting curve. If r is the ratio of protein to DNA (in amino acid residues/nucleotide), there is a linear relationship between r and F.

$$r = \beta F = \beta(A_{T_{m'}}/A_T) \qquad (4)$$

where slope β is the number of amino acid residues per nucleotide in protein-bound regions. If N is the number of amino acid residues in a protein molecule, the length of DNA bound by one protein molecule will be

$$m\text{(base pairs)} = N/2\beta \tag{5}$$

The introduction of 2 in the denominator is needed since the length of DNA is calculated in terms of base pair rather than nucleotide.

In chromatin, the DNA base pairs bound by histones melt at $T_{m,\mathrm{III}}$ and $T_{m,\mathrm{IV}}$; therefore, the $A_{T_{m'}}$ in Eqs. (3) and (4) can be replaced by the term $A_{T_{m,\mathrm{III}}} + A_{T_{m,\mathrm{IV}}}$. The areas of these melting bands can be obtained by assuming that each band is symmetric with respect to its peak position. On the basis of this assumption, melting band IV can be resolved first, then melting band III (Fig. 2). The rest of the total melting area belongs to the melting bands I and II.

By means of Eq. (3), it was determined that in pea bud and calf thymus chromatin 75–80% of DNA base pairs are directly bound by histones. This value dropped to 50–60% when histone H1 was removed by 0.6 M NaCl. In chromatin, histone H1 presumably binds directly about 20% of the DNA base pairs not bound by other histones, (17,26).

Using Eq. (4), it was determined that histone-bound regions in both pea bud and calf thymus chromatin contain 3.5 amino acid residues per nucleotide. From Eq. (5) it was calculated further that one histone H1 molecule alone would bind 30–40 pairs, and one molecule each of the other histones (H2A, H2B, H3, and H4) would bind 15–18 base pairs. If histones bind DNA as dimers, the above numbers should be doubled for each histone dimer. Combining the octamer hypothesis of Kornberg (27) with the results of the above analysis, Li (28) concluded that a chromatin subunit would contain 130–150 base pairs bound by an octamer of (H2A)₂ (H2B)₂ (H3)₂ (H4)₂ adjacent to a segment of 30–40 base pairs bound by one histone H1 molecule. This conclusion was recently confirmed by experiments on nuclease digestion of chromatin (29–32).

The above equations have also been applied to other DNA–protein complexes as well as to chromatin. For a recent review of these results, see Li (1).

C. Use of Renaturation to Study DNA–Protein Complexes

Theoretically, if a free DNA segment between two protein-bound segments is thermally denatured and renatured, the amount of renaturation is expected to be greater, the shorter the segment. Thus, when the same conditions are maintained for both denaturation and renaturation experiments, the amount of renaturation of a free DNA segment can be used as a measure for the length of these segments in DNA–protein complexes.

If T_1 and T_2 are two temperatures in the first denaturation, selected arbitrarily from within the free DNA melting band of a DNA–protein complex $(T_2 > T_1)$, a difference in hyperchromicity can be calculated using the equation

$$\Delta h_{T_2 - T_1} = h(T_2) - h(T_1) \tag{6}$$

The complexes are denatured up to a temperature at which the free DNA segments are fully denatured but the protein-bound segments are not. They are renatured by slowly lowering the temperature to well below the initial melt (e.g., to about 25°C for calf thymus DNA), then re-denatured as before.

Calculating $\Delta h'_{T_2 - T_1} = h'(T_2) - h'(T_1)$ this time in the re-denaturation, the amount of renaturation (R) of free DNA regions in the complex can be measured by use of the following equation.

$$R = (\Delta h'_{T_2 - T_1} / \Delta h_{T_2 - T_1}) \times 100\% \tag{7}$$

By means of Eq. (7) it was shown that the renaturation of free DNA regions was observed with 0.6 M NaCl and 1.6 M NaCl-treated chromatin, polylysine bound to DNA and to the decrease of DNA chain length in polylysine-free regions. A high degree of renaturation in histone-free regions was observed with 0.6 M NaCl- and 1.6 M NaCl-treated chromatin, implying that NaCl removes histones from chromatin DNA not in large clusters, but by individual histone molecules or small groups of histones (33). This conclusion is consistent with the recent view of histone distribution in chromatin, especially with respect to histone H1, as mentioned earlier in Section III, B.

Equations (6) and (7) can also be used for renaturation of protein-bound DNA, in chromatin or in a DNA–protein complex. In this case T_2 and T_1 are selected at two points in the melting band of protein-bound DNA. A high degree of renaturation is expected if both strands of a DNA segment are still tightly bound by a protein, such as polylysine. In the case of chromatin, a very low degree of renaturation was observed for histone-bound DNA, implying release from DNA of the main portions of histones after denaturation (33).

D. Use of Thermal Denaturation to Study Competition between Two DNAs for Protein Binding

Thermal denaturation has also been used to study the selectivity of a protein for binding a particular sequence of DNA. A mixture of two DNAs of different G+C (guanine + cytosine) content is expected to show two separate melting bands if the difference in G+C content is large, or to show broad, overlapping melting bands if the DNAs are close to each other in their G + C content. If a particular type of DNA is selectively chosen for binding by proteins, such as histones, the free melting band of that particular DNA would be reduced at a faster rate than that of the competing DNA. By means of this technique, it was shown that (A + T)−rich DNA is preferentially bound by polylysine (*34*), in agreement with results of Leng and Felsenfeld (*35*) using another method, (A + T)−rich DNA is also preferentially bound by histone H5 (V or f2c) (*24*), by H1 (I_2 or f) (352), and by slightly lysine-rich histones H2A (IIbl or f202) (IIb_2 or f_2b). This technique is simple and can be used for many proteins. (*35b*).

E. Other Possible Applications

As mentioned in Section II,A, temperature dependence of CD, ORD, NMR, viscosity, sedimentation, and other other physical properties of chromatin and DNA–protein complexes depend not only upon the helix–coil transition of protein-free and protein-bound regions, but also upon changes in the tertiarty structure. If both the temperature-dependent profiles and the absorbance melting curves are measured, it may be possible to separate the contributions to these profiles of CD, NMR, sedimentation, etc., made by local helix–coil transition from those made by changes in tertiary structure. Numerous reports have dealt with the temperature dependence of CD in DNA (*36–38*), in chromatin (*39,39a*), and in DNA–model protein complexes (*40–45*). The interpretation of these CD melting curves, however, is inconclusive without the assistance of the absorbance melting data.

ACKNOWLEDGMENT

This research was supported, in part, by National Institutes of Health Grants GM 23079 and 23080 and National Science Foundation Grant PCM 76–03268.

REFERENCES

1. Li, H. J. *In* "Chromatin and Chromosome Structure" (H. J. Li and R. Eckhardt, eds.). Academic Press, New York (in press). (1976).
2. Ohlenbusch, H. H., Olivera, B. M., Tuan, D., and Davidson, N. *J. Mol. Biol.* **25**, 229 (1967).
3. Olins, D. E., Olins, A. L., and von Hippel, P. H. *J. Mol. Biol.* **24**, 157 (1967).
4. Ansevin, A. T., Hnilica, L. S., Spelsberg, T. C., and Kehn, S. L. *Biochemistry* **10**, 4793 (1971).
5. Dove, W. F., and Davidson, N. *J. Mol. Biol.* **5**, 467 (1962).
6. Crothers, D. M. *J. Mol. Biol.* **9**, 712 (1964).
7. Spatz, H. C., and Crothers, D. M. *J. Mol. Biol.* **42**, 191 (1969).
8. Massie, H. R., and Zimm, B. H. *Biopolymers* **7**, 475 (1969).
9. Britten, B. J., and Kohne, D. E. *Carnegie Inst. Washington, Year.* (1966).
10. Wetmur, J. G., and Davidson, N. *J. Mol. Biol.* **31**, 349 (1968).
11. Studier, F. W. *J. Mol. Biol.* **41**, 199 (1969).

12. Cohen, R. J., and Crothers, D. M. *Biochemistry* **9**, 2533 (1970).
13. Wetmur, J. G. *Biopolymers* **10**, 601 (1971).
14. Hoff, A. J., and Roos, A. L. M. *Biopolymers* **11**, 1289 (1972).
15. Li, H. J., and Bonner, J. *Biochemistry* **10**, 1461 (1971).
16. Hanlon, S., Johnson, R. S., Wolf, B., and Chan, A. *Proc. Natl. Proc. Acad. Sci. U.S.A.* **69**, 3263 (1972).
17. Li, H. J., Chang, C., and Weiskopf, M. *Biochemistry* **12**, 1763 (1973).
18. Subirana, J. A. *J. Mol. Biol.* **74**, 363 (1973).
19. Hjelm, R. P., and Huang, R. C. C. *Biochemistry* **13**, 5275 (1974).
20. Tsai, Y. H., Ansevin, A. T., and Hnilica, L. S. *Biochemistry* **14**, 1257 (1975).
21. Shih, T. Y., and Bonner, J. *J. Mol. Biol.* **48**, 469 (1970).
22. Ansevin, A. T., and Brown, B. W. *Biochemistry* **10**, 1133 (1971).
23. Leffak, I. M., Hwan, J. C., Li, H. J., and Shih, T. Y. *Biochemistry* **13**, 1116 (1974).
24. Hwan, J. C., Leffak, I. M., Li, H. J., Huang, P. C., and Mura, C. *Biochemistry* **14**, 1390 (1975).
25. Yu, S. S., Li, H. J., and Shih, T. Y. *Biochemistry* **15**, 2027 (1976).
25a. Yu, S. S. *et al.* In preparation. (1977).
26. Li, H. J. *Biopolymers* **12**, 287 (1973).
27. Kornberg, R. D. *Science* **184**, 868 (1974).
28. Li. H. J. *Nucleic Acids Res.* **2**, 1275 (1975).
29. Sollner-Webb, B., and Felsenfeld, G. *Biochemistry* **14**, 2915 (1975).
30. Simpson, R. T., and Whitlock, J. P., Jr. *Nucleic Acids Res.* **3**, 117 (1976).
31. Varshavsky, A. J., Bakayev, V. V., and Georgeiv, G. P. *Nucleic Acids Res.* **3**, 477 (1976).
32. Shaw, B. R., Harman, T. M., Kovacic, R. T., Beaudreau, G. S., and Van Holde, K. E. *Proc. Natl. Acad. Sci. U.S.A.* **73**, 505 (1976).
33. Li, H. J., Chang, C., Weiskopf, M., Brand, B., and Rotter, A. *Biopolymers* **13**, 649 (1974).
34. Li, H. J., Brand, B., and Rotter, A. *Nucleic Acids Res.* **1**, 257 (1974).
35. Leng, M., and Felsenfeld, G. *Proc. Natl. Acad. Sci. U.S.A.* **56**, 1325 (1966).
35a. Polacow *et al.* Submitted for publication. (1977).
35b. Leffak, I. M., and Li, H. In preparation. (1977).
36. Gennis, R. B., and Cantor, C. R. *J. Mol. Biol.* **65**, 381 (1972).
37. Studdert, D. S., Patronic, M., and Davis, R. C. *Biopolymers* **11**, 761 (1972).
38. Usatyi, A. R., and Shylakhtenko, L. S. *Biopolymers* **12**, 45 (1973).
39. Wilhelm, F. X., DeMurcia, G. M., Champagne, M. H., and Duane, M. P. *Eur. J. Biochem.* **45**, 431 (1974).
39a. Li, H. J. *et al.* (1977). Submitted for publication.
40. Santella, R. M., and Li, H. J. *Biopolymers* **13**, 1909 (1974).
41. Santella, R. M., and Li, H. J. *Biochemistry* **14**, 3604.
42. Mandel, R., and Fasman, G. D. *Biochem. Biophys. Res. Commun.* **59**, 672 (1974).
43. Ong, E. C., and Fasman, G. D. *Biochemistry* **15**, 477 (1976).
44. Chang, C. *et al.* Submitted for publication. (1977).
45. Pinkston *et al.* In preparation. (1977).

Chapter 20

Thermal Denaturation Analysis of Chromatin and DNA–Nuclear Protein Complexes

ALLEN T. ANSEVIN

Department of Physics,
University of Texas System Cancer Center,
M.D. Anderson Hospital and Tumor Institute,
Texas Medical Center, Houston, Texas

I.   Introduction

When a double-stranded nucleic acid is gradually heated, it denatures at a temperature that is characteristic of the base composition of the nucleic acid and the cation concentration of the medium. This denaturation is easily detected by means of an increase of UV absorption that accompanies the local unwinding of the two strands of the double helix. Because the transition from an ordered helix to denatured, randomly coiled strands might be regarded as a phase change analogous to the disordering of a crystal, the process frequently is referred to as melting, and the midpoint of the transition (T_m) (i.e., the temperature for 50% completion) often is described as the melting temperature of the nucleic acid. If a native nucleic acid is allowed to interact with a basic polypeptide or protein to form a complex, the T_m of the nucleoprotein usually is found to be significantly higher than that of the free nucleic acid. Thus, thermal denaturation can serve as a convenient method to detect the formation of complexes between nucleic acids and basic molecules. When observations of this type are taken over an extended range of temperatures, the data can be used to characterize even the complicated nucleoproteins of chromatin and to detect modifications from their normal composition. Two types of plot are commonly employed to analyze these thermal transitions: the first will be described as a hyperchromicity curve, or "melting" curve; the second, as a derivative denaturation profile. Both types of graph may be used to examine the broad range of thermally stabilized complexes in chromatin preparations, but the derivative denaturation profile appears to be the more easily interpreted of the two.

II. Background and Theory

As double-stranded nucleic acids, whether DNA or RNA, denature from
a highly ordered helix to a pair of "random" coils, the bases along each
chain change from a totally stacked arrangement to a semiflexible, predominantly unstacked conformation. As a result, the average base–base
electronic interaction is greatly diminished. This is accompanied by an
increase in absorbance to give an optical extinction that is more nearly
like that of free nucleotides; for a given DNA, the magnitude of the increase
is proportional to the extent of denaturation.

In establishing a reference state for physical studies on DNA, one might
consider it desirable from a chemical standpoint to choose free nucleotides
or denatured single strands as a standard state. However, in the following
discussion, a biological orientation will be adopted by taking the reference
state of DNA as the native double helix at room temperature. This choice
has the advantage of reflecting several experimental realities: (a) double-
stranded DNA is the form most easily obtained in a pure and relatively
reproducible state, (b) single-stranded nucleic acids are undesirable as
standards because they change their optical extinction over a wide range of
temperatures, including all accessible temperatures above the main helix-
to-coil transition,[1] and (c) nucleic acids are chemically unstable to a
measurable extent at the high temperatures or pH extremes which generate
single strands. Therefore, in the remainder of this chapter, single strands
will be considered to be *hyper*chromic with respect to native DNA, and
these comments will be concluded by pointing out that the magnitude of the
*hyper*chromicity that results from the thermal denaturation of double-
stranded nucleic acids is a function of the base composition of the nucleic
acids, the wavelength of light employed, and to some extent, the final
observed temperature.

That the wavelength, as well as the base composition, is an important
variable in thermal denaturation was shown a number of years ago by Inman
and Baldwin (*1*). For greatest sensitivity in thermal denaturation measurements, it would therefore seem desirable to examine the optical change at
the wavelength where the hyperchromicity has its maximum value. However, since there is no single wavelength for which DNAs of all compositions have a common hyperchromic maximum for the helix-to-coil transition, the best experimental approach is to choose a wavelength at which
the hyperchromic increase during denaturation is the same for $A + T$ as for

---

[1] Although free native DNA often displays an initial small hyperchromicity gradient prior
to the main transition, this ordinarily is smaller than $+6 \times 10^{-4}$ deg, and usually is not observed for nucleoproteins.

G + C. Recent observations by D. L. Vizard (unpublished results) show that this wavelength for hyperchromic isoextinction is close to 270 nm; here the hyperchromic increase for the main portion of the transition is around 40%.

A number of observations support the concept that the thermal stabilization of DNA within nucleoproteins strongly reflects the partial neutralization of negative charges on the DNA backbone by positive charges introduced by the ligand molecules. Presumably, the close proximity of positive charges to the negatively charged phosphates reduces the usual electrostatic repulsion between the two strands of the double helix so that higher temperatures are required for the chains to be driven apart by thermal energy. That such an interpreation is reasonable, is suggested by the fact that double-helical nucleic acids are stabilized by even subequivalent amounts of small, divalent cations (2) and that the thermal stability of native DNA in dilute sodium salts of various acids is a function of the log of the sodium ion concentration, rather than the ionic strength of the medium (3). Furthermore, it is generally observed that the acidic proteins which remain bound to DNA after the histones have been extracted from chromatin contribute little, if any, thermal stability to the DNA. These statements should not be taken to indicate that only ionic interactions are involved in protein–DNA complex formation, but merely that electrostatic interactions usually are dominant in the thermal stabilization that can result from ligand binding.

In recent years, the analysis of thermal denaturation data for nucleoproteins has been greatly facilitated by the use of derivative denaturation profiles, which display the first derivative of hyperchromicity (dH/dT) as a function of temperature (T). The relationship between a hyperchromicity curve and a derivative denaturation profile is illustrated in Fig. 1, which shows the denaturation transitions for purified rat DNA and rat thymus chromatin. The derivative denaturation profiles of the simpler complexes formed by binding pure histones to DNA are predictably less complicated that chromatin profiles. This is demonstrated in Fig. 2 and has been shown also by Shih and Bonner (4), and Li and Bonner (5) for simple nucleohistones examined under slightly different denaturation conditions.

The derivative profile for DNA in Fig. 1b appears smooth and almost symmetrical and it is now known that rat DNA has an unusually homogeneous melting profile; in this respect, rat DNA is not typical for eukaryotic DNAs. In fact, eukaryotic DNAs contain substantial proportions of repetitious sequences in addition to a large variety of unique sequences; thus, there is no reason to expect Gaussian denaturation profiles or even similar profiles for different DNAs. The complexity of the denaturation for protein-free bovine DNA is illustrated in Fig. 3 (6), where most of the data have been taken at intervals of about 0.3°C, so that the finer features of the

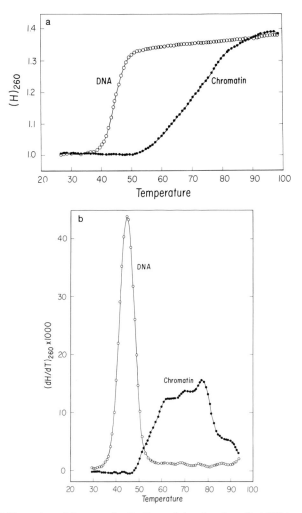

FIG. 1. (a) Hyperchromicity curve for the thermal denaturation of rat DNA and rat thymus chromatin; hyperchromicity at 260 nm vs temperature. (b) Derivative denaturation profile for the same experiment: first derivative of hyperchromicity with respect to temperature vs temperature. The heating rate was approximately 0.4 °/minute; the medium contained 3.5 M urea–5 mM sodium cacodylate buffer (pH 7)–0.15 mM sodium EDTA; initial optical absorbance was about 0.7 for both samples; the derivative, was calculated by 11-point quadratic fit of hyperchromicity data.

profile could be resolved. It is pertinent to ask whether such irregularity or broad spread of a DNA profile can be reflected in the thermal denaturation of corresponding nucleoproteins. Part of the answer can be deduced from an examination of nucleoprotein profiles in Figs. 1, 2, and 4, where it

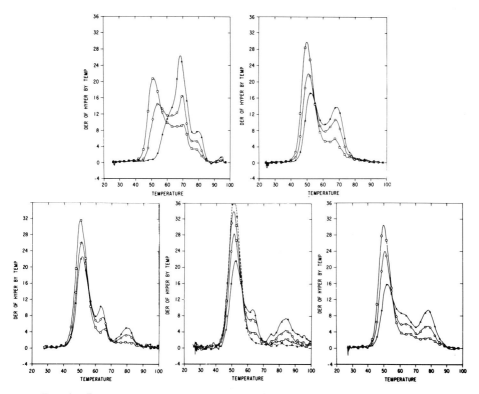

FIG. 2. Derivative thermal denaturation profiles of nucleohistone complexes formed by the "reconstitution" of single histone fractions with DNA. Upper row: histone H1 (left), histone H2B (right); lower row: histone H2A (left), histone H4 (center), histone H3 (right). Each graph shows complexes containing 0.25, 0.5, and 0.75 weight parts of calf thymus histone per unit weight of rat DNA. Same conditions of heating rate, medium, and calculation as for Fig. 1. The peak near 50° is due to free or weakly complexed DNA. Data points in all graphs are connected by straight-line segments; symbols appear only at every fifth point of the data. Reprinted with permission from Ansevin, A. T., and Brown, B. W. *Biochemistry* **10**, 1133 (1971). Copyright by the American Chemical Society.

is apparent that the major components ordinarily resolved in nucleoprotein profiles are considerably coarser that the irregularities in the DNA curve of Fig. 3. More important, the results of Fig. 2 suggest that the major features of the nucleoprotein profiles must be determined by the proteins, rather than the DNA, since the DNA in this case (rat) has a homogeneous-appearing profile (Fig. 1) and different histones yield different derivative profiles. Although it is probably true that the finer features of DNA melting are unimportant here, it remains to be established whether or not a broadly spread DNA denaturation profile, such as that of bovine DNA, introduces additional overlap among major peaks in corresponding nucleoprotein profiles.

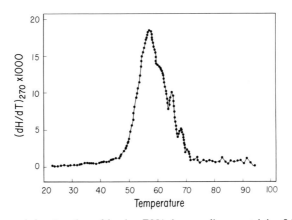

Fig. 3. Thermal denaturation of bovine DNA in a medium containing 0.003 M sodium ions (chloride–cacodylate buffer, pH 7). Readings were taken about 0.3° apart within the main transition; heating rate was 0.4°/minute; DNA was type V, Sigma Chemical Co. (St. Louis, Missouri). The derivative was calculated from the parameters of a 9-point cubic fit of hyperchromicity data.

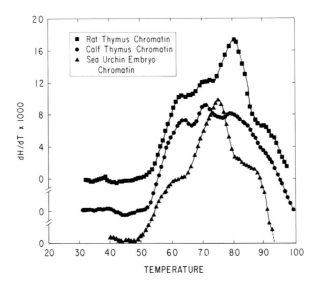

Fig. 4. Derivative denaturation profiles of chromatin preparations from different sources, showing distinctive shapes. Conditions as indicated for Fig. 1. Redrawn, from "The Structure and Biological Function of Histones," by L. S. Hnilica, The Chemical Rubber Co., 1972. Used by permission of The Chemical Rubber Co.

III. Sample Treatment and Denaturation Media

It is appropriate that most of the details of sample preparation should be determined by the investigator according to the requirements of his study. However, a few general comments can be made. Because most divalent and multivalent cations very effectively stabilize double-stranded nucleic acids, they must be absent from the preparations. These cations, if present at an earlier stage of the preparation, should be removed by thorough dialysis in the presence of excess sodium ethylenediaminetetra-acetate (EDTA). In addition, the final solution should contain a low level of EDTA (0.1 mM) to chelate trace ions that may be introduced adventitiously.

Preparations to be denatured must be soluble at the start of the determination and remain free of particulates throughout the melting experiment. It is usually necessary to remove small aggregates or dust particles at the beginning by centrifuging preparations at about 3000 g, so that absorbance at 320 nm (due to turbidity) is reduced to 5% or less of the reading at 260 or 270 nm. One should note carefully at this point whether the removal of particulates has resulted in a significant loss of sample, which could produce an inadvertent fractionation. If the sample initially has much turbidity, it is advisable to assay for 260 nm absorbance in a solution containing 0.2% sodium dodecyl sulfate (SDS); this usually is adequate to dissolve aggregates and to dissociate histone–DNA complexes, thereby obviating most errors due to light scattering.

Shortly before the clarified samples and blank solution are put into spectrophotometer cuvettes, they should be degassed to prevent bubble formation during subsequent heating. This may be done by holding the samples at room temperature in a vacuum chamber for about 10 minutes, or by exchanging dissolved oxygen and nitrogen for helium (which is more soluble at high temperatures than at low). As long as foaming is not a problem, the latter is the more effective of the two treatments and can be accomplished by a gentle but steady bubbling for about 5 minutes.

The cuvettes preferably should be closed by stoppers made of Teflon, which expands with heat to provide a tight seal. This type of positive closure permits the development of a limited pressure that opposes the generation of bubbles and makes it possible to raise the temperature of aqueous samples as high as 115°C. Semimicro cells, type 290, manufactured by Precision Cells, Inc. (Hicksville, New York) have a sufficiently rugged construction to give good service under these conditions. They permit the routine use of 1-ml samples. When care is taken about proper positioning and masking, the sample size may be reduced to 0.6 ml, or even to 0.3 ml if a cuvette with

2 mm width is used, such as type 115 of Helma Cells (Jamaica, New York). In all cases, optical path lengths conventionally are 10 mm.

Although Teflon-stoppered cuvettes do not leak, the samples may become slightly more concentrated during the determination because of evaporation and subsequent condensation of vapor in the upper portions of the cuvettes. This difficulty is especially likely when cuvettes extend above the limits of a standard holder so that the upper portion is heated less efficiently. The magnitude of the problem may be reduced by constructing a special cuvette holder or a metal adapter which increases the height of the holder, or by using shorter cuvettes (available on special order from Helma Cells). The condensation can be circumvented by placing a 2-mm layer of silicone oil (Dow-Corning, type 200) on top of each sample.

For a number of reasons, denaturation solvents for nucleoproteins are limited to solutions with relatively low salt concentrations. For instance, sodium chloride concentrations approaching 0.1 M tend to precipitate chromatin and many other nucleoproteins, while slightly higher salt concentrations begin to dissociate nucleohistones. Furthermore, since the T_m of DNA increases with the log of cation concentration, even moderately high salt molarities (e.g., 50 mM) decrease the potential for observing a difference between uncomplexed DNA and protein-bound DNA and also disfavor the detection of differences among various complexes. Cacodylic acid buffers often have been used because of their low coefficient of pH change with respect to temperature.

Probably the most popular medium for the thermal denaturation of nucleoproteins is 0.25 mM sodium EDTA (pH 8) (5,7). This solution provides a reasonable chelating capacity, and by its relatively low cation concentration (slightly less than 1 mM Na$^+$) it permits a practical maximum of separation between the T_m of free DNA and the denaturation temperatures of stabilized complexes. Samples must be well dialyzed to assure that all have the same ion concentration. The EDTA solution has practically no buffering capacity, but there may be little need for it since DNA denaturation is rather insensitive to pH over the range from pH 5 to pH 10 (8).

Another medium that has been used for nucleoprotein denaturation studies contains 3.6 M urea, in addition to a dilute buffer and a chelant (pH 7, 5 mM sodium cacodylate buffer–0.15 mM sodium EDTA) (6). This medium was adopted because both reconstituted nucleohistones and frozen chromatin preparations could be dissolved more completely than in dilute EDTA alone. Also, it was reasoned that denaturation in the presence of a hydrogen bond breaker should be a simpler process if the protein–protein interactions within nucleoproteins could thereby be reduced. The presence of urea in this medium lowers the denaturation temperature of free DNA below that predicted from the cation concentration of the solution. Although the derivative denaturation profiles of some nucleohistone complexes were found to be better resolved in the urea-containing medium than in dilute EDTA (9), this solution cannot be considered ideal because the urea undergoes a gradual decomposition during the experiment (6). Nonetheless, the urea-containing solution has considerable utility for nucleoprotein characterization because, in practice, the profiles are both detailed and reproducible. Derivative profiles recorded in this medium may be highly characteristic of the nucleoprotein source, as illustrated in Fig. 4 (10). Comparisons of results between urea-medium and dilute EDTA were considered by Ansevin et al. (9), and Van and Ansevin (11).

The importance of the rate of heating during the denaturation of nucleoproteins has not been well established. Many experiments have been conducted with a temperature rise in the sample of about 0.5° per minute; studies in this laboratory often have been made at a heating rate as high as 0.8° per minute. It is entirely possible that this is faster than what could be described as an equilibrium heating rate for DNA strand unwinding. For this and other reasons, it would seem only prudent to maintain a standard heating rate when close comparisons are to be made among several samples.

IV.   Equipment

The major piece of equipment required for thermal denaturation studies is a research-quality UV spectrophotometer. It must accommodate a system for regulating the temperature of the samples and the corresponding reference solution. Such a system may involve one of three possible arrangements: a heated sample compartment and adjustable bath from which fluid is circulated to the walls of the compartment, jacketed cuvettes supplied by a similar circulating bath system, or a regulated, electrically heated sample holder. The most common of these is a heated sample compartment supplied by a regulated external bath. Experience indicates that DNA denaturation can be accomplished fully as satisfactorily in an experiment having a slow, steady temperature rise as in one in which the temperature is adjusted by small increments that are followed by equilibrium periods (9, 12). For slow, steady heating, the bath temperature can be controlled with good precision by means of an inexpensive, adjustable-contact mercury thermoregulator driven by a small, fractional rpm synchronous motor. Possibly the most elegant solution to temperature control is an electrically heated sample holder (Thermo-Programmer, manufactured by Gilford Instrument, Oberlin, Ohio).

Sample temperature should be measured directly within one or more of the cuvettes. This is conveniently accomplished with a small thermistor encased in a waterproof jacket and dipped into the liquid above the optical path of the "blank" cell. Voltages read from a bridge circuit containing a single-element thermistor should be calibrated over the entire range of interest (ca. 20°C to 100°C) using standard thermometers to arrive at the linearizing corrections for each temperature. If a dual-element Thermilinear thermistor (Yellow Springs Instrument Co., Yellow Springs, Ohio) is used with the proper bridge circuit, voltages are directly proportional to temperature within an accuracy of ± 0.2°C.

If thermal denaturation is to be used as a routine method, it is readily apparent that a rather advanced degree of automation is appropriate. However, if it is to be applied only occasionally, there is little reason most measurements on nucleoproteins could not be made with a standard temperature-controlled spectrophotometer. In either case, the apparent resolution of optical absorbance readings for nucleoprotein denaturation experiments should be about 1 part in 500, or better. Temperature increments of 0.5° to 0.75°C between successive observations appear satisfactory for most nucleoproteins. An apparent resolution in temperature readings that is close to 0.1°C, or better, favors good results in calculations of derivatives, even if the absolute accuracy of the reading is less precise. [However, high-resolution studies of DNA denaturation are considerably more demanding (12).] Adequate precision of absorbance readings is attainable with most commercial research spectrophotometers (including that manufactured by Gilford Instrument, Oberlin, Ohio, with which the author has had extensive, favorable experience). When the preparations are free of particulates, reasonable data can be obtained on samples with absorbance at 260 nm ranging from 0.2 to above 1; at the lower concentration, maintenance of the precision recommended earlier would require at least 4-digit accuracy; this can be achieved with a digital voltmeter having a relatively long integration time (ca. 0.1 second).

To secure reliable results on nucleoproteins, one should take regular data not only at 270 nm (or 260 nm, if it is acceptable to weight A · T base pairs more than G · C pairs) but also periodically at 320 nm, or at this and several higher wavelengths, to monitor any change in the turbidity of nucleoprotein samples. Such turbidity readings are essential for reliable conclusions because nucleoproteins frequently undergo turbidity changes as a function of temperature. Automatic wavelength scanning is available in several commercial spectrophotometers, or can be achieved at very modest expense by adaptations to a manual monchromator. Highly automated spectrophotometer systems that should require little or no modification for use in thermal denaturation studies of nucleoproteins are now offered commercially.[2]

V. Numerical Analysis

Thermal denaturation experiments on DNA or nucleoproteins are customarily analyzed in terms of the parameter, hyperchromicity (H), which will be defined here as the ratio of optical absorbance at any elevated temperature to that at ambient temperature (frequently taken to be 25°C). Plotting the hyperchromicity as a function of temperature is more useful than displaying the absorbance, because hyperchromicity is a type of normalized parameter which permits direct comparisons to be made among samples with different concentrations. In theory, all samples containing the same DNA ought to attain the same *final* hyperchromicity when all the DNA has been denatured, whether the sample is free DNA or nucleoprotein, assuming the wavelengths are identical. However, in practice, final hyperchromicity values may vary for several reasons. These include the presence of RNA, denatured DNA, protein with a high content of aromatic amino acids or other absorbing contaminant, blank mismatch, turbidity at the start or at the end of the determination, the possible occurrence of helix distortions in chromatin, and the temperature at which measurements were discontinued.[3] For this reason, many investigators prefer to standardize

[2] Gilford Instrument, Oberlin, Ohio 44074; GCA/McPherson Instrument, Acton, Massachusetts 01720; Cary Instruments, Monrovia, California 91016; Varian Techtron, Walnut Creek, California; Instrument Division, Perkin-Elmer Corporation, Norwalk, Connecticut 06856; GCA/Precision Scientific, Chicago, Illinois 60647.

[3] Observations in this laboratory suggest that the optical absorbance of DNA within a typical chromatin, to a first approximation, is uninfluenced by the protein components of the complex, provided the turbidity is low; this means that any helix distortion introduced by the proteins of calf thymus chromatin must not cause more than a 5% increase in the absorbance of the DNA, although earlier literature suggested considerably larger values (13). While the absorbance of aromatic amino acids in chromosomal proteins is a small fraction of that of DNA, the absorbance due to RNA components of chromatin can vary from zero, to values approaching that of the DNA, depending on the source and the preparation method (10). It is worth noting that when the derivative profile is normalized to a standard area above the base line, the analysis of the profile is rendered independent of an absorbing contaminant that has either a constant absorbance or a linearly increasing absorbance. When the concentration of the RNA is known and the RNA can be isolated and denatured by itself, its expected contribution to the derivative profile of chromatin can be subtracted. In this way, it could be shown that various RNAs could cause a significant thermal destabilization of chromatin (14).

hyperchromicity curves to 100% at the highest temperature recorded, setting the ambient temperature level of the plot to 0%. If the experimenter is aware of the factors that might cause nonstandard final hyperchromicity and has evidence that turbidity is not changing during the experiment, this form of normalization may, indeed, aid comparisons. However, because information is lost in the process, the normalization of hyperchromicities should not be misapplied so as to mask totally unknown parameters that might have some pertinence to the proper evaluation of the experiment. Derivative profiles are a far more effective means to facilitate comparisons and are discussed below.

Some aspects of numerical analysis depend strongly on the type of apparatus in use. For instance, if only manually operated equipment is available, the investigator may wish to carry out the simplest permissible analysis. If absorbance data have been collected with a double-beam spectrophotometer or, if the blank has been adjusted to zero absorbance before reading samples in a single-beam instrument, any changes in the solvent absorbance are corrected automatically; however, the decrement in absorbance due to the expansion of the solvent volume remains to be considered. Since over the usual limits of 25° to 100° the volume expansion is never more than 4%, an investigator may feel justified in disregarding the correction for manual calculations, especially with nucleoprotein denaturation where other factors may cause larger errors. The results of a minimum calculation would then provide a list of hyperchromicities and corresponding temperatures; these points would be plotted as a melting curve covering temperatures from ambient to well within the so called "plateau" region beyond the transition proper.

A coarse-grained derivative profile may also be plotted without great difficulty from manually collected data. A straightforward procedure giving values of $(\Delta H/\Delta T)$ is to let the numerator be the interpolated change in hyperchromicity over a two-degree interval extending from $T(i)$-1 to $T(i) + 1$, where $T(i)$ is any temperature at which an observation has been made. While calculation of this 2°C increment may often give a satisfactory description of the broad transitions characteristic of typical nucleoproteins, it is certainly inadequate for presenting the derivative profiles of pure DNA, where individual subtransitions frequently have breadths at half-height of about 0.35°C (15). If temperature increments in nucleoprotein data are closer than 1°C, it may be desirable to incorporate some simple smoothing procedure, such as a running average taken over three points at a time and plotted at the central position of the three, or the more elaborate procedure discussed later for computer-processed results. Following a manually performed smoothing by 3-point averaging or other simple method, it is an easy matter to graph the individual ΔH values, versus the temperature for the interval. This gives a derivative profile more accurately representing the experiment than a plot arbitrarily calculated for 2°C intervals.

The type of smoothing procedure employed should be reported. In all cases, honesty of data presentation suggests that both hyperchromicity plots and derivative profiles should be composed of individual points that are calculated and graphed at the temperature where observations actually were made. Possible exceptions to this principle would be a directly recorded (analog) plot of absorbance vs an ordinate that is linear in temperature, and the corresponding analog derivative of the rate of absorbance change with temperature (16, 17). However, digital rather than analog calculations appear more appropriate to the protracted time scale usually involved in thermal denaturation experiments.

The computation of the vast amount of data that can be collected by automated spectrophotometer systems is too laborious to perform without a computer. Once the raw data have been entered into a computer, it is then appropriate to carry out more refined calculations than those discussed above, as well as to take care of simple arithmetic associated with the particular form in which numbers have been recorded. Absorbance readings should be corrected for expansion relative to the solvent volume at 25°C. A correction formula for the expansion of dilute aqueous solutions, based on the density of pure water as a function of temperature, is the following: $V_T/V_{25°} = 1 + A(\Delta T) + B(\Delta T)^2 + C(\Delta T)^3$, where $\Delta T = T - 25°$, $V_T =$ volume at a temperature of $T°C$; $A = 2.497 \times 10^{-4}$ per degree, $B = 4.939 \times 10^{-6}$ degree^{-2}, $C = 1.456 \times 10^{-8}$ degree^{-3}. Different parameters apply for other media, as is indicated by the following constants for a 3.6 M urea solution: $A = 4.417 \times 10^{-4}$ per degree, $B = -3.747 \times 10^{-7}$ degree^{-2}, $C = 3.958 \times 10^{-8}$ degree^{-3}. In the case of pure DNA, failure to apply the expansion correction makes the hyperchromicity curve look more nearly plateau-like in the posttransitional region, while in reality it has a gentle positive slope. Accurate hyperchromicity values can then be calculated by dividing each corrected absorbance by the absorbance at ambient temperature, where the latter might be taken as the average of the three or so smallest readings. In most cases, the computer can be utilized also to plot the values of hyperchromicity vs temperature.

A good estimate of dH/dT may be derived from the parameters associated with the least-squares fit of a string of 5 or more hyperchromicity values to a quadratic polynomial in temperature: $H(i) = A + B \cdot T_i + C \cdot T_i^2$, where i is the observation number in a series of n successive data pairs at different temperatures (T_i), and A, B, and C are constant to be determined by a least-squares fitting routine.[4] The derivative is then given by: $(dH/dT)_c$

[4] A cubic calculation is preferable for pure DNA.

$= B + 2C \cdot T_c$, where c is the central observation in an odd number, n, of successive data pairs. The evaluation of a derivative at the central point of a group of n observations must be repeated for all temperatures where the data are sufficient. Two advantages accrue from this type of calculation. First, a significant smoothing is achieved which can be increased or decreased, as desired, by increasing or decreasing the value of n. Second, the reliability of the data in a derivative profile may be assessed by means of the standard errors calculated at each point from the residuals obtained in the fitting procedure (12).

Finally, the derivative profile can be normalized to the expected maximum hyperchromicity of pure DNA at the wavelength chosen for measurements, or to some purely arbitrary value. This is done by multiplying each value of the quantity (dH/dT), from the actual experiment, by the ratio $(H_f - 1)_{\text{theoretical}} : (H_f - 1)_{\text{actual}}$; where H_f is the hyperchromicity at the final temperature. The best combination of full data presentation and maximum utility probably is achieved by displaying a derivative profile normalized as above, together with an expansion-corrected hyperchromicity plot that has *not* been normalized.

A quantitative evaluation of the areas of subcomponents under a derivative profile can be undertaken by means of an analog curve resolver, such as that manufactured by Du Pont Instruments (Wilmington, Delaware). In this procedure, the general shape of all subcomponents to be fitted is assumed when the curve resolver is calibrated and the specific parameters of fit and number of components are chosen by trial and error. The experimental curve is then duplicated as the sum of the fitted components. The most common shape assumed for individual subcomponents is a Gaussian curve, which has the three adjustable parameters, height, breadth (twice the standard deviation), and position. With the Du Pont analog computer, the area of any subcomponents can be read from a panel meter. Unfortunately, such manually performed fitting procedures usually include a high degree of arbitrariness, except for especially well defined profiles. Furthermore, only in special cases is there a clear justification for assuming a Gaussian form for the subcomponents. Nonetheless, when the proper conditions can be satisfied, quantitative curve resolution adds a new dimension to the analysis of denaturation profiles.

VI.   Analysis of Profiles

The most important single improvement in thermal denaturation technique in recent years has been the adoption of derivative profiles to display the results of melting experiments. This type of presentation, introduced apparently independently by investigators in several different laboratories, is now in such common use that it frequently supplants the more traditional melting curves altogether. No new information is generated by the derivative profile. However, the fact that subcomponents of a complex transition can be identified as peaks at characteristic temperatures greatly facilitates the comparison of corresponding regions from one profile to another, enhances the viewer's ability to recognize irregularities signaling the presence of minor components, and allows quantitative comparisons. All these factors contribute to making thermal denaturation a very useful technique for the qualitative characterization of nucleoproteins. When an analog curve resolver is used, individual subcomponents of a derivative profile can be measured so that profiles may be quantitated in terms of the fractional change of hyperchromicity within all resolvable subtransitions at the modal temperatures of the transitions.

It is apparent that many useful nucleoprotein denaturation experiments should be regarded as qualitative or semiquantitative, rather than quantitative, in nature. Given this usual type of experiment, what can be deduced from it? One type of behavior that can be interpreted with relative ease occurs when successive additions of a ligand to DNA cause the T_m to increase, but do not result in the separation of a second, stabilized peak. From this, one could conclude that the ligand binds apparently at random to native DNA and probably involves a dynamic equilibrium. Examples of this type include addition of Ca^{2+}, Mg^{2+}, and the oligoamine spermidine. Another recognizable situation occurs when successive, subequivalent additions of ligand leave a portion of the DNA apparently unaffected in T_m, while the remainder is stabilized to a constant, elevated T_m. If the decrease in the area of the DNA peak corresponds to the increase of area under the stabilized peak and the change is proportional to the amount of ligand added, it is usually assumed that the binding of ligand to native DNA is cooperative, probably as a result of ligand–ligand interactions. Most histones, for instance, show evidence of cooperative binding (18), and in general one might anticipate that this type of binding should occur with basic proteins having two or more hydrophobic regions. The occurrence of two peaks should be analyzed with care, however, since the theoretical calculations of McGhee (19) show that noncooperative binding also may result in bimodal profiles. A final circumstance that is worthy of attention, even though it has been disregarded in this discussion up to now, is the case where the binding of a protein (either basic or acidic in composition), destabilizes the DNA so that the nucleoprotein complex melts at a *lower* temperature than does pure DNA. This is understandable in thermodynamic terms as the consequence of a much stronger binding of the ligand to single-stranded DNA than to the native double-stranded nucleic acids. An example of a basic protein with this property is pancreatic ribonuclease (20). In all likelihood, the capacity of specific proteins to destabilize DNA is biologically significant, as suggested by the discovery of a variety of "unwinding" or "melting" proteins (21–23). An unwinding protein from calf thymus has been investigated particularly well in thermal denaturation studies by Herrick and Alberts (24).

In principle, one might expect that there should be many opportunities to apply quantitative denaturation data to obtain a detailed thermodynamic analysis of DNA–polypeptide interactions. However, several factors have limited the realization of these possibilities. For instance, the opportunities to derive valid quantitative data are reduced by the practical problems of turbidity development, uncertain knowledge of the shape of individual derivative peaks for a nucleoprotein containing DNA with a broad melting transition (e.g., bovine DNA), and the possibility that equilibrium denaturation rates for nucleoproteins are very slow. In addition, the extreme complexity of a natural material such as chromatin presents a formidable hurdle to be surmounted before information of a very fundamental nature can be obtained. When one contemplates the difficulty of completing a full thermodynamic analysis on chromatin, taking into account each type of protein–DNA interaction for all proteins together with the effect of protein–protein interactions as a function of temperature, the task appears beyond hope of accomplishment. Furthermore, until recently the incentives for making careful measurements were reduced by the lack of a theory dealing with the many aspects of the thermal denaturation of even a simple ligand–nucleic acid complex.

Two major qualifications should be added to the above comments that might simplify some of the theoretical difficulties involved in analyzing denaturation experiments conducted on chromatin. One is the general observation that the major features of thermal denaturation profiles appear to be controlled by the histones, rather than by the highly variable nonhistone proteins. Furthermore, there is growing evidence that four of the five major histones interact with DNA as multihistone aggregates, rather than associating independently (25). The second simplifying qualification is that the interaction of histone clusters with DNA, just referred to, is believed to form chromatin subunits, or nu bodies (26). These subunits of chromatin appear to be organized essentially independently of the distribution of base sequences in the DNA (27,28). In this case, the basic structure of chromatin should be highly repetitious, as suggested by Noll (29), and therefore, from a theoretical standpoint, the major structural features of natural nucleoproteins might be reducible to the behavior of an equivalent simple system that could be analyzed by thermal denaturation techniques to obtain apparent binding constants. Pragmatically, it is already apparent from results such as those shown in Fig. 4 that thermal denaturation is highly useful for the characterization of chromatin preparations. At the same time, these results are an indication that the view of a perfectly repetitious structure for chromatin must not be completely valid and that the truth probably lies somewhere between this and the concept of an unfathomable structure.

One of the fundamental questions about the interpretation of thermal denaturation profiles is whether the observations bear any relevance to the characteristics of the system at room temperature. Another way of putting it, is that one would like to know whether the patterns observed provide information that can be related to the previous treatment of the nucleoprotein examined, or whether denaturation profiles are determined strictly by the equilibrium thermodynamic parameters of the components as a function of temperature. From experiments with model nucleohistones there is evidence that in some cases the manner of formation of the complex may greatly influence the thermal denaturation profile (9,30). This can obviously be very useful for the purposes of characterization; at the same time it is another indication that thermal denaturation experiments, as usually conducted, are not at thermodynamic equilibrium. Other information relevant to the same question is that some degree of reorganization of histones on DNA is to be expected within a nucleohistone during the process of thermal denaturation. Upon standing at low temperature, the components of chromatin can undergo rearrangements in the presence of urea or added polynucleotides (31); also, just the mild shearing involved in normal preparations of chromatin from nuclei causes the detailed structure of chromatin to be modified (32,33). It seems inevitable, therefore, that many elements of structure will be lost during thermal denaturation with no input of information being made to the recorded profile. Unfortunately, no general rule can be given about whether or not significant "history" effects will be found in a given denaturation experiment.

It is beyond the scope of this chapter to discuss the details of thermodynamic theory concerning the interactions of ligands with nucleic acids and their influence on helix–coil transitions. While it is always tempting to speculate that the amount by which the T_m of DNA is elevated as a result of complex formation should be related to the binding strength of the ligand, the complexities alluded to earlier would appear to preclude such calculations for the majority of current experiments. However, a number of experiments already conducted do deal with simple peptides and uncomplicated polynucleotides and thus should be reasonable candidates for such quantitative investigations. When an equilibrium can be demonstrated and the interacting system is well characterized, information about interaction energies should be obtainable from denaturation experiments by the application of proper theory. The theory of the influence of ligands on the thermal denaturation of polynucleotides has been considered by Crothers (34), McGhee and von Hippel (35), Schellman (36), Herrick and Alberts (24), and McGhee (19). McGhee (19) has illustrated his fairly general theory with data from experiments on the binding of netropsin to poly$[d(AT)]$. Herrick and Alberts (24) applied their calculations to the analysis of the destabilization of naturally occurring DNA by a nonspecifically bound native protein. The success achieved in these cases suggests that the theoretical analysis of thermal denaturation data will be attempted much more often in the future.

ACKNOWLEDGMENTS

I would like to thank Dr. Hermann Bultmann for helpful comments on the manuscript. Methods described here were developed with the support of grants from the Robert A. Welch Foundation (G-290), the U.S. Energy Research and Development Administration [AT-(40-1)-2832], and the National Cancer Institute, DHEW (CA-16672).

REFERENCES

1. Inman, R. B., and Baldwin, R. L. *J. Mol. Biol.* **8**, 452. (1964).
2. Dove, W. F., and Davidson, N. *J. Mol. Biol.* **5**, 467. (1962).
3. Schildkraut, C., and Lifson, S. *Biopolymers* **3**, 195. (1965).
4. Shih, T. Y., and Bonner, J. *J. Mol. Biol.* **48**, 469. (1970).
5. Li, H. J., and Bonner, J. *Biochemistry* **10**, 1461. (1971).
6. Ansevin, A. T., and Brown, B. W. *Biochemistry* **10**, 1133. (1971).
7. Ohlenbusch, H. H., Olivera, B. M., Tuan, D., and Davidson, N. *J. Mol. Biol.* **25**, 299. (1967).
8. Lewin, S., and Pepper, D. S. *Arch. Biochem. Biophys.* **112**, 243. (1965).
9. Ansevin, A. T., Hnilica, L. S., Spelsberg, T. C., and Kehm, S. L. *Biochemistry* **10**, 4793. (1971).
10. Hnilica, L. S. "The Structure and Biological Function of Histones." Chem. Rubber Publ. Co., Cleveland, Ohio. (1972).
11. Van, N. T., and Ansevin, A. T. *Biochim. Biophys. Acta* **299**, 367. (1973).
12. Ansevin, A. T., Vizard, D. L., Brown, B. W., and McConathy, J. *Biopolymers* **15**, 153. (1976).
13. Taun, D. Y. H., and Bonner, J. *J. Mol. Biol.* **45**, 59. (1969).
14. Ansevin, A. T., Macdonald, K. K., Smith, C. E., and Hnilica, L. S. *J. Biol. Chem.* **250**, 281. (1975).
15. Vizard, D. L., and Ansevin, A. T. *Biochemistry* **15**, 741. (1976).
16. Yabuki, S., Fuke, M., and Wada, A. *J. Biochem.* (*Tokyo*) **69**, 191. (1971).
17. Näslund, P., Liljesvan, B., and Abrahamsson, L. *Anal. Biochem.* **57**, 211. (1974).
18. Rubin, R. L., and Moudrianakis, E. N. *J. Mol. Biol.* **67**, 361. (1972).
19. McGhee, J. D. *Biopolymers* **5**, 345 (1976).
20. Felsenfeld, G., Sandeen, G., and von Hippel, P. H. *Proc. Natl. Acad. Sci. U.S.A.* **50**, 644. (1963).
21. Alberts, B. M., and Frey, L. *Nature* (*London*) **227**, 1313. (1970).
22. Alberts, B., Frey, L., and Delius, H. *J. Mol. Biol.* **68**, 139. (1972).
23. Herrick, G., and Alberts, B. *J. Biol. Chem.* **251**, 2124. (1976).

24. Herrick, G., and Alberts, B. *J. Biol. Chem.* **251**, 2133. (1976).

25. Kornberg, R. *Science* **183**, 868. (1974).

26. Olins, A. L., and Olins, D. E. *Science* **184**, 330. (1974).

27. Axel, R., Cedar, H., and Felsenfeld, G. *Biochemistry* **14**, 2489. (1975).

28. Lacy, E., and Axel, R. *Proc. Natl. Acad. Sci. U.S.A.* **72**, 3978. (1975).

29. Noll, M. *Nature (London)* **251**, 249. (1974).

30. Yu, S. S., Li, H. J. and Shih, T. Y. *Biochemistry,* **15**, 2027. (1976).

31. Ilyin, Y. V., Varshavsky, A. Y., Mickelsaar, U. N., and Georgiev, G. P. *Eur. J. Biochem.* **22**, 235. (1971).

32. Noll, M., Thomas, J. O., and Kornberg, R. D. *Science* **187**, 1203. (1975).

33. Sollner-Webb, B., and Felsenfeld, G. *Biochemistry* **14**, 2915. (1975).

34. Crothers, D. M. *Biopolymers* **10**, 2147. (1971).

35. McGhee, J. D., and von Hippel, P. H. *J. Mol. Biol.* **86**, 469. (1974).

36. Schellman, J. A. *Biopolymers* **15**, 999. (1976).

Chapter 21

Assaying Histone Interactions by Sedimentation Equilibrium and Other Methods

DENNIS E. ROARK

Departments of Biology and Physics,
Maharishi International University,
Fairfield, Iowa

I. Introduction

The role of histones in determining chromatin structure must involve a combination of histone–DNA and histone–histone interactions. Histones have at least two interaction sites: a positively charged region to interact with the DNA, and a region to interact with another histone. Thus, the histones may serve as "adaptors" linking two DNA regions and restricting the conformational freedom of the chromatin.

Ultracentrifugal analyses offer unique advantages in elucidating histone interactions in solution. The early role of the analytical ultracentrifuge, the determination of molecular weights, has been substantially replaced by the sodium dodecyl sulfate (SDS)–polyacrylamide gel electrophoresis molecular weight methods of Weber and Osborn (*1*). Improved techniques, however, permit a new role for the ultracentrifuge: the investigation of macromolecular associative interactions. This review will discuss some of these techniques that may be usefully applied to histones. In sedimentation-equilibrium (SE) experiments of favorable systems, one may estimate the molecular weight and concentrations of all macromolecular species, and determine the stoichiometry and thermodynamics of reversible associations. The SE experiments are uncomplicated by nonequilibrium and hydrodynamic factors that occur in the analysis of transport experiments (sedimentation-velocity, gel permeation chromatography, and electrophoretic mobility). The SE experiments, however, may be incapable of resolving a complex system of many sedimenting species. The limitations of these experiments and modern strategies of data analysis are discussed in Section IV.

Elegant studies of histone interactions may be performed by the SDS–polyacrylamide gel electrophoresis technique after the interacting histones have been corss-linked with a bifunctional reagent. The cross-linking may occur either while the histones are part of the chromatin, or after histone fractions have been isolated in solution. Cross-linking studies are useful as a sensitive and simple means of identifying histone interactions. The magnitude of the interactions may then be measured by a thermodynamic technique, such as SE. The cross-linking technique, particularly when applied to chromatin as a whole, does not actually indicate macromolecular interaction, but only proximity. Two histones may both be interacting with some common third component (e.g., DNA) and be proximal; and yet have no significant attractive interaction for each other.

II. Sedimentation Equilibrium Theory

We define the "reduced molecular weight" σ_i (2):

$$\sigma_i = \frac{M_i(1 - \bar{v}_i\rho)\,\omega^2}{RT} \qquad (1)$$

for sedimenting species i of molecular weight (MW) M_i, and partial specific volume $\bar{v}_i$, sedimenting in a solution of density ρ and temperature T at an angular velocity ω (R is the gas constant). The SE condition reflects the balance of sedimentation and diffusional forces:

$$\sigma_i = (1/RT)\left[d(\mu_i - \mu_i^\circ)/d\xi\right] \qquad (2)$$

where μ_i is the chemical potential of species i, and ξ is the radial parameter $r^2/2$. With the usual assumptions of ideal behavior, Eq. (2) reduces to the well-known exponential concentration distribution:

$$c(\xi) = \sum_i A_i e^{\sigma_i\xi} \qquad (3)$$

[$c(\xi)$ is the total macromolecular concentration on a weight/volume basis; and the amplitudes A_i reflect the relative amounts of each species.] The weight-average reduced MW σ_w, corresponding to the weight-average MW $M_{w'}$ is

$$\sigma_w = d\ln c/d\xi \qquad (4)$$

If the sedimenting species are charged, the equilibrium condition of Eq. (2) is accompanied by a similar relation for the low MW supporting electrolyte, and by the restriction of local electroneutrality. These conditions result in nonexponential concentration distributions and anomalously low apparent values of σ_w [defined by Eq. (4)]. This nonideality is the Donnan effect (3), but may be suppressed either by appropriate choice of pH to reduce the charge of the sedimenting species, or by the addition of sufficient low MW salt. For only moderate nonideality and a single sedimenting species, Eq. (3) becomes

$$c(\xi) = A \; e^{(\sigma\xi - \beta')} \tag{5}$$

where the nonideality factor is

$$\beta'(\xi) = (Z^2/M)\left[c(\xi)/m_3\right] \tag{6}$$

Z is the charge on the macroion, and m_3 is the molality of the univalent salt. Values of β' less than 0.1 generally yield an apparent MW within 5% of the true value.

Equation (4) is utilized by curve-fitting local regions of the concentration distribution to determine values of σ_w as a function of concentration or position. Also, additional local MW averages may be determined. The reduced Z-average MW,

$$\sigma_z = \Sigma \; c_i\sigma_i{}^2/\Sigma \; c_i\sigma_i \tag{7}$$

is estimated from previously determined values of σ_w:

$$\sigma_z = \left[d\ln(c \; \sigma_w)\right]/d\xi \tag{8}$$

Estimation of the reduced number-average MW is possible if the meniscus is essentially depleted of sedimenting species by high rotational speeds (2):

$$\sigma_n(r) = c(r)/\int_0^c \frac{d \; c(r)}{\sigma_w(r)} \tag{9}$$

Sophisticated computer programs are widely used to calculate these and other averages (4,5).

Estimation of true MWs requires an accurate knowledge of $\bar{v}$. Actually, the formulation given here is approximate, as Casassa and Eisenberg (6) have pointed out. In experiments involving dilute salt solutions, however, use of partial specific volumes is acceptable and generally easier than determination of "density increments." Values of $\bar{v}$ are often estimated from amino acid composition rather than by direct measurement (7,8). Van Holde (9) reviews this problem in his excellent discussion of sedimentation analysis. The value of $\bar{v}$ also depends on the Donnan effect. This influence is not suppressed by the addition of supporting electrolyte since its origin lies in the proper choice of an electrically neutral sedimenting thermodynamic component (3). Values of $\bar{v}$ for histones have been estimated from amino acid composition and corrected for Donnan effect: H1, 0.748 ml/gm; H2A, 0.734 ml/gm; H2B, 0.73 ml/gm; H3 and H4, 0.733 ml/gm.

III. Experimental Methods

For SE studies of histone–histone interactions, the Rayleigh interference optical system is the most useful and precise. Histone–polynucleotide interactions are best investigated with the UV scanning system, which permits separate determination of the protein and polynucleotide concentration distributions. A procedure of optical alignment for the interference system has been described by Richards et al. (10,11). The camera lens must be focused at the two-thirds plane of the cell to prevent gradient-dependent skewing of the pattern. Fringes at high gradients may be improved by masking the interference aperture to form a 1 mm width slit perpendicular to the light source slit. A Polaroid filter, mounted above the light source with its polarizing axis radially oriented, reduces the fringe blurring caused by cell window distortions at high speeds. Several laboratories use laser light sources to improve fringe quality (12,13). The interference fringe patterns are recorded on either Kodak II G or IV F spectroscopic plates, if the mercury light source or an argon laser source, respectively, is used, and are developed with Kodak HRP developer. The plates are read either manually with a microcomparator or by a computer-automated microcomparator (14). Five fringe positions are averaged at 50- to 100-um intervals along the photographic plate. Cell distortions and optical imperfections make blank-pattern corrections necessary (15).

Investigation of interactions by SE depends upon adequate spatial fractionation of sedimenting species (i.e., variation of the weight fractions across the cell). Thus, the high-speed, meniscus-depletion technique of Yphantis (2) offers obvious advantages. Choice of equilibrium speeds is important, but must often be made on the basis of experience and visual evidence. Generally, values of σ between 3 and 6 cm^{-2} are appropriate. In order to assess heterogeneity, SE experiments are performed simultaneously at three cell loading concentrations, ranging from approximately 0.3 to 2 mg/ml. Solution column lengths are 2.5 to 3 mm. The time required to essentially attain equilibrium can be reduced by overspeeding a calculated time (typically, 1 hour) at $2^{1/2}$ the equilibrium speed (*16*). Over-speeding is feasible because the overspeed time is only slightly dependent on the MWs, but depends, instead, on the frictional ratio.

IV. Analysis of Interacting Systems

A. Introduction

There are two general approaches to SE data analysis of associating systems. One can study the behavior of the various local MW averages: the manner in which they vary with concentration, radial position, or as a function of each other; or, one may perform direct curve fitting of the concentration distribution in terms of sums of exponentials. The local average approach is useful in indicating what MW species are present. With knowledge of the σ's of all sedimenting species, linear least-squares analysis of the concentration distribution [Eq. (3)] estimates the relative amounts of each species and the association constants. An SE experiment often cannot discriminate between two possible association models involving three or more species. It is possible to increase the resolving power of SE by simultaneous concentration distribution fits of the same sample at several equilibrium speeds. Additional techniques for studying interactions have been reviewed recently (*9*), including the elegant method of Steiner (*17,18*).

B. Detection of Heterogeneity

Both associating and nonassociating systems may display heterogeneity of sedimenting thermodynamic components. A self-associating system of histones is homogeneous, while a mixed association such as

$$H3 + H4 \rightleftharpoons (H3)(H4) \tag{10}$$

involves two thermodynamic components and is thus heterogeneous. For homogeneous interacting systems, the local MW averages are unique functions of local equilibrium concentration, in accordance with the law of mass action. In heterogeneous systems, on the other hand, sedimentation leads to a fractionation of thermodynamic components so that the local moments are also functions of position. A test for heterogeneity is the presentation of $\sigma_w(\xi)$ vs $c(\xi)$ for several initial loading concentrations. A particular local concentration will represent different radial positions.

Nonoverlap of these σ_w vs c curves, with higher loading concentration curves corresponding to lower values of σ_w, indicates heterogeneity.

C.   Local Molecular Weight Average Methods

A simple technique uses values of σ_w to determine the reduced MW σ_2 of an oligomer if the reduced MW σ_1 of the monomer is known. If α is the weight fraction of species 2, then, from the definition of σ_w, the concentration of species 2 is

$$c_2 = \alpha c \sim c(\sigma_w - \sigma_1) \tag{11}$$

Thus, σ_2 is given by

$$\sigma_2 = d\ln[c(\sigma_w - \sigma_1)]/d\xi \tag{12}$$

A presentation of $\ln[c(\sigma_w - \sigma_1)]$ vs $r^2/2$ is linear with a slope σ_2 if just two species are present.

If two unknown species are assumed to be present, a "two-species plot" estimates the MWs of both species (4). By expressing σ_n, σ_w, and σ_z in terms of σ_1, σ_2, and α, we may eliminate α between any two equations:

$$\sigma_k = -(\sigma_1\sigma_2)(1/\sigma_{k-1}) + (\sigma_1 + \sigma_2) \tag{13}$$

where σ_k and σ_{k-1} are either σ_w and σ_n, or σ_z and σ_w. Thus, presentation of σ_z vs $1/\sigma_w$ is linear if two species are present. The two-species plot, when extended, intersects the hyperbola $\sigma_z(1/\sigma_w) = 1$ at $\sigma_z = \sigma_1$ and σ_2, thereby permitting easy estimation of the two MWs. Upward curvature of this plot, or of that for Eq. (12), indicates the presence of additional species; while downward curvature indicates nonideality.

An extension of the two-species plot to three species is possible with a new set of moments, I_1 and I_2, that are analogous to σ_w and σ_z, except for the unusual property that they are independent of the weight fraction of the monomer species σ_1:

$$I_1 = \Sigma\, w_i\sigma_i/\Sigma\, w_i \tag{14a}$$

$$I_2 = \Sigma\, w_i\sigma_i^2/\Sigma\, w_i\sigma_i \tag{14b}$$

where, the weighting factors w_i are

$$w_i = c_i\,(1 - \sigma_1/\sigma_i) \tag{15}$$

(note that $w_1 = 0$.) If σ_1 and local values of σ_n and σ_w are known, I_1 and I_2

may be estimated by local curve fitting using expressions analogous to those used to estimate σ_w and σ_z [Eqs. (4) and (8)]:

$$I_1 = \frac{d\ln[c(1 - \sigma_1/\sigma_n)]}{d\xi} \tag{16a}$$

$$I_2 = \frac{d\ln[c(\sigma_w - \sigma_1)]}{d\xi} \tag{16b}$$

These moments are used in a two-species plot to yield estimates of σ_2 and σ_3:

$$I_2 = -(\sigma_2\sigma_3)(1/I_1) + (\sigma_2 + \sigma_3) \tag{17}$$

This three species technique is useful only if values of σ_1 and σ_n are sufficiently accurate and different.

D. Concentration Distribution Techniques

Least-squares fits of the observed concentration, $c(\xi)$, in terms of a sum of exponentials, Eq. (3), may be used to estimate the species present and their relative amounts. If the σ_i are chosen to correspond to a proposed association model, the least-squares technique solves a set of linear equations to determine the amplitudes, A_i, and the standard deviation of the fit. The correctness of the assumed model is evaluated on the basis of physical reasonableness of the amplitudes and small standard deviation of the fit. It is essential to test alternative models, since discrimination between models by the fits may not be possible. If the MWs are not known, one or more σ_i may be parameters determined by the fit. In such cases, the fitting techniques are nonlinear and are performed by more complex computer algorithms. Other fitting criteria are possible than the simple least-squares condition. Minimization of the sum of the absolute values of the residuals has some advantage, in principle, since it weights unreliable data less than in least-squares fits. In practice, estimated parameters are rarely significantly different for the two techniques. The simple "mini-max" technique minimizes the maximum residual, but is not recommended since the fits are dominated by the least reliable data.

Haschemeyer and Bowers (19) considered the uniqueness of concentration distribution fits and concluded that not more than five degrees of freedom can be analyzed in an SE experiment. If the MWs of all sedimenting species, and therefore the association model, are known, least-squares analysis can determine the association constants of many systems. If the MWs of none of the species are known, on the other hand, then, since each

species contributes two degrees of freedom, even the simplest mixed association cannot be uniquely analyzed. One often knows, as with histone interactions, the MWs of one or more species participating in the interaction; and
one can propose a limited number of interactions leading to other sedimenting species. It is then usually possible for fits to discriminate among the
proposed interactions. If the MWs of the species are similar, however,
discrimination on the basis of a single SE experiment may not be possible.
A pertinent example is the dissociation of the histone tetramer to a
dimer (20). Two modes of dissociation are possible:

$$(H3)_2(H4)_2 \rightleftharpoons 2(H3)(H4) \tag{18a}$$

$$(H3)_2(H4)_2 \rightleftharpoons (H3)_2 + (H4)_2 \tag{18b}$$

The mixed dimer, Eq. (18a), has an MW of 26,560; and the H3 and H4
dimers, Eq. (18b), have MWs of 22,480 and 30,640, respectively. No analysis
of a single SE experiment can demonstrate which model is correct. It is
possible, however, to perform a series of SE experiments at different speeds,
which are then analyzed by a simultaneous least-squares fit (16). This
"multispeed fit" significantly increases the resolving power of SE by requiring that data from the various speeds serve as mutual constraints on the fit,
thereby reducing the effective degrees of freedom.

Equation (3) is rewritten in terms of the mean species concentrations
c_{oi} (the initial loading concentrations for noninteracting systems) and a
position function $\Gamma_i(\xi, \sigma_i)$:

$$c(\xi) = \sum_i c_{oi} \Gamma_i(\xi, \sigma_i) \tag{19}$$

We assume a mixed association, for example,

$$A + B \rightleftharpoons C \tag{20}$$

and define the "total mean component concentrations" of A and B. Component A, for instance, is present as species A and as a weight fraction of
species C:

$$c_{at} \equiv c_{oa} + (\sigma_a/\sigma_c)c_{oc} \tag{21a}$$

$$c_{bt} \equiv c_{ob} + (\sigma_b/\sigma_c)c_{oc} \tag{21b}$$

Equations (21) are combined with Eq. (19) to eliminate c_{oa} and c_{ob}. Although,
c_{oa}, c_{ob}, and c_{oc} depend upon rotational speed (in accordance with mass
action), the "total mean component concentrations" are speed invariant.
A single fit is performed over three speeds of data to determine c_{at}, c_{bt},

and three values of c_{oc}. Multispeed fits assuming the models of Eqs. (18) indicate that the tetramer dissociates to the mixed dimer (H3) (H4).

V.   Other Methods

A.   Histone Cross-Linking

Many agents have been used to identify chromatin organization: formaldehyde (21–23), dimethyl suberimidate (24–26), carbodiimide (27), tetranitromethane (28), and methylmercaptobutyrimidate (29). The results of these studies have generally been consistent with SE analysis and have identified the histone interactions: H3–H4, H2A–H2B, and H2B–H4.

The electrophoretic mobility of a polypeptide in an SDS gel depends upon the proper unfolding of the polypeptide chain. For this reason, disulfide bonds are reduced prior to electrophoresis. The cross-linking reagent should, in principle, prevent proper unfolding of the polypeptides and cause anomalous migration. In practice, linked dimers and tetramers are still easily recognized on the basis of their apparent MWs. The greater the number of cross-links, however, the less ideal will be the mobility. Complexes believed to be octamers, for instance, are substantially cross-linked and may not demonstrate the same mobility as a single polypeptide of the same MW. Such highly cross-linked products should be purified and studied by SE or by cleavage of the cross-links and identification of the subunits. SDS gel analysis of cross-linked dimers and tetramers may exhibit slight nonideal mobility and make risky attempts to identify which monomer units compose the oligomers: the MWs of the monomer units of histones, for example, may differ by only 10%.

B.   Spectroscopic Methods

D'Anna and Isenberg (30–32) used fluorescence anisotropy and circular dichroism measurements to detect histone interactions. They observed salt-induced effects on the tyrosine fluorescence and percentage of α-helix of the separate histones. Histone interactions interfere with these salt perturbations, thus permitting investigation of the associations. This method revealed the 1:1 interactions of H2B–H4 and of H2A–H2B.

Lilley et al. (33) performed nuclear magnetic resonance studies as a function of histone concentration. Concentration-dependent effects indicated associative interactions, but not stoichiometries.

VI.   Histone Interactions

A.   Cross-Associations

Histones are capable of self-associations and a variety of cross-interactions. Histones H3 and H4 interact to form the mixed dimer (H3)(H4) and tetramer $(H3)_2(H4)_2$ (*16,20,24,32,34–37*). At low ionic strength or pH, the dimer dissociates to separate histone monomers. The tetramer has a helix content typical of globular proteins (28%), but an anomalously high frictional ratio. This apparent asymmetry may, instead, be due to the amino-terminal, positively charged regions extending freely into solution (*35,38*). These regions in chromatin are probably more α-helical and interact electrostatically, with DNA. Histones H2A and H2B also reversibly interact to form a dimer (*31,35,39–41*). There is no evidence of tetramer.

D'Anna and Isenberg (*30*) reported an H2B–H4 interaction. If histone is dissociated by salt extraction at moderate pH, little H2B–H4 forms. The individual histones may be prepared without denaturation (*35,42*) by chromatography on Bio-Gel P-60 in 0.1 *M* NaCl–0.02 *M* HCl. If equimolar amounts of histones H2B and H4 are mixed in this solvent, and then dialyzed into 0.2 *M* NaCl–0.02 *M* NaOAc (pH 5) an H2B–H4 complex forms. Sedimentation equilibrium studies indicate the presence of the mixed dimer equilibrium with small amounts of the tetramer and separate monomers. The existence of an H2B–H4 dimer may explain the often apparent greater molar amount of H4 than of H3. Histone H1 does not appear to interact with other histones.

Sperling and Bustin (*41*) performed SE experiments on all pairs of H2A, H2B, H3, and H4. Unfortunately, their analysis is partially erroneous. The comparatively low speed employed ($\sigma \cong 0.4$ cm^{-2} for a 25,000 MW species) causes only slight fractionation of sedimenting species with MWs differing by only a factor of 2. Therefore, it is not possible to estimate the MWs of the separate species from the limiting slopes of ln c vs $r^2/2$ at meniscus and base.

B.   Self-Associations

All histones except H1 and H5 are capable of significant self-associations; and, even H1 associates weakly to a dimer (*35,41,43*). Histone H4 associates readily to form large fibrous aggregates visible by electron microscopy (*37*). This self-association can be suppressed by low ionic strength or pH. Diggle *et al.* (*43*) found that pH > 2 and ionic strengths greater than 0.1, H4 forms a gel. This extreme aggregation is probably due to denaturation occurring during preparation (*35*). Sedimentation equilibrium experiments on H4, prepared as described above (42), demonstrate MW aggregates from 100,000 to 20,000,000, with no insoluble or gel fraction.

Histone H2A associates to dimers, tetramers, and octamers (35). In the concentration range 0.1–1 mg/ml (ionic strength 0.2 and pH 5), the predominant species is the tetramer. Our recent SE studies under the same conditions indicate that H2B associates to dimers and, weakly, to tetramers. This is consistent with Diggle et al. (43). Fraction H3 forms high MW aggregates (41,43).

The self-assembly properties of the histones are essentially suppressed by the cross-interactions H3–H4, H2B–H4, and H2A–H2B. Indeed, the interaction strength of the three pairs of mixed dimers are sufficiently strong that monomer histones (except H1 and H5) or self-aggregates may not occur in chromatin.

ACKNOWLEDGMENTS

Unpublished work described here was supported in part by USPHS, National Institutes of Health Grant GM 18456. I wish to thank Richard L. Engle for his technical assistance.

REFERENCES

1. Weber, K., and Osborn, M. In "The Proteins" (H. Neurath and R. L. Hill, eds.), 3rd ed., Vol. 1, p. 179. Academic Press, New York. (1975).
2. Yphantis, D. A. Biochemistry 3, 297. (1964).
3. Roark, D. E., and Yphantis, D. A. Biochemistry 10, 3241. (1971).
4. Roark, D. E., and Yphantis, D. A. Ann. N.Y. Acad. Sci. 164, Artic 1, 245. (1969).
5. Teller, D. C., Horbett, T. A., Richards, E. G., and Schachman, H. K. Ann. N.Y. Acad. Sci. 164, Artic. 1, 66. (1969).
6. Casassa, E. F., and Eisenberg, H. Adv. Protein Chem. 19, 287. (1964).
7. Cohn, E. J., and Edsall, J. T. "Proteins, Amino Acids, and Peptides." Van Nostrand-Reinhold, Princeton, New Jersey. (1943).
8. Haschemeyer, R. H., and Haschemeyer, A. E. V. In "Proteins," p. 162. Wiley, New York. (1973).
9. Van Holde, K. E. In "The Proteins" (H. Neurath and R. L. Hill, eds.), 3rd ed., Vol. 1, p. 225. Academic Press, New York (1975).
10. Richards, E. G., Teller, D., and Schachman, H. K. Anal. Biochem. 41, 189. (1971a).
11. Richards, E. G., Teller, D., Hoagland, V. D., Jr., Haschemeyer, R. H., and Shcachman, H. K. Anal. Biochem. 41, 215. (1971).
12. Paul, C. H., and Yphantis, D. A. Anal. Biochem. 48, 588 and 605. (1972).
13. Williams, R. C., Jr. Anal. Biochem. 48, 164. (1972).
14. Carlisle, R. M., Patterson, J. I. H., and Roark, D. E. Anal. Biochem. 61, 248. (1974).
15. Ansevin, A. T., Roark, D. E., and Yphantis, D. A. Anal. Biochem. 34, 237. (1970).
16. Roark, D. E. Biophys. Chem. 5, 185 (1976).
17. Steiner, R. F. Arch. Biochem. Biophys. 39, 333. (1952).
18. Steiner, R. F. Arch. Biochem. Biophys. 49, 400. (1954).
19. Haschemeyer, R. H., and Bowers, W. F. Biochemistry 9, 435. (1970).
20. Roark, D. E., Geoghegan, T. E., and Keller, G. H. Biochem. Biophys. Res. Commun. 59, 542. (1974).

21. Chalkley, R., and Hunter, C. *Proc. Natl. Acad. Sci. U.S.A.* **72**, 1304. (1975).
22. Hyde, J. E., and Walker, I. O. *FEBS Lett.* **50**, 150. (1975).
23. Van Lente, F., Jackson, J. F., and Weintraub, H. *Cell* **5**, 45. (1975).
24. Kornberg, R. D., and Thomas, J. O. *Science* **184**, 865. (1974).
25. Thomas, J. O., and Kornberg, R. D. *Proc. Natl. Acad. Sci. U.S.A.* **72**, 2626. (1975).
26. Chalkley, R. *Biochem. Biophys. Res. Commun.* **64**, 587. (1975).
27. Bonner, W. M., and Pollard, H. B. *Biochem. Biophys. Res. Commun.* **64**, 282. (1975).
28. Martinson, H. G., and McCarthy, B. J. *Biochemistry* **14**, 1073. (1975).
29. Hardison, R. Eichner, M. E., and Chalkley, R. *Nucleic Acids Res.* **2**, 1751. (1975).
30. D'Anna, J. A., and Isenberg, I. *Biochemistry* **12**, 1035. (1973).
31. D'Anna, J. A., Jr., and Isenberg, I. *Biochemistry* **13**, 2098. (1974).
32. D'Anna, J. A., Jr., and Isenber, I. (1974). *Biochemistry* **13**, 4992.
33. Lilley, D. M. J., Howarth, O. W., Clark, V. M., Pardon, J. F., and Richards, B. M. *Biochemistry* **14**, 4590. (1975).
34. Geoghegan, T. E., Keller, G. H., and Roark, D. E. *Fed. Proc., Fed. Am. Soc. Exp. Biol.* **33**, 1598. (1974).
35. Roark, D. E., Geoghegan, T. E., Keller, G. H., Matter, K. V., and Engle, R. L. *Biochemistry* (in press). (1976).
36. D'Anna, J. A., Jr., and Isenberg, I. *Biochem. Biophys. Res. Commun.* **61**, 343. (1974).
37. Sperling, R., and Bustin, M. *Proc. Natl. Acad. Sci. U.S.A.* **71**, 4625.
38. Moss, T., Cary, P. D., Crane-Robinson, C., and Bradbury, E. M. *Biochemistry* **15**, 2261. (1976).
39. Skandrani, E., Mizon, J., Santiere, P., and Biserte, G. *Biochimie* **54**, 1267. (1972).
40. Kelley, R. I. *Biochem. Biophys. Res. Commun.* **54**, 1588. (1973).
41. Sperling, R., and Bustin, M. *Biochemistry* **14**, 3322. (1975).
42. Böhm E. L., Strickland, W., Strickland, M., Thwaits, B. H., van der Westhuyzen, D. R., and von Holt, C. *FEBS Lett.* **34**, 217. (1973).
43. Diggle, J. H., McVittie, J. D., and Peacock, A. R. *Eur. J. Biochem.* **56**, 173. (1975).

Chapter 22

The Study of Histone–Histone Associations by Chemical Cross-Linking

JEAN O. THOMAS

*Department of Biochemistry, University of Cambridge,
Cambridge, England*

ROGER D. KORNBERG

*Department of Biological Chemistry,
Harvard Medical School,
Boston, Massachusetts*

I. Introduction

Chemical cross-linking can be used to reveal both the pattern and the degree of association of polypeptides in a multisubunit structure. Limited cross-linking results in dimers, formed from neighboring polypeptides. Extensive cross-linking gives a series of higher molecular weight products, the largest of which comprises the total number of subunits in the structure. Both types of analysis have been applied to the histones with the use of a variety of cross-linking agents, such as formaldehyde (*1,2*) imidoesters (*3,4*), tetranitromethane (*5*), ultraviolet light (*6*), and dicyclohexylcarbodiimide (*7*). This article will be concerned primarily with the imidoesters, which have proved useful in our own work and whose reaction with proteins is well understood.

The procedures for cross-linking with imidoesters are straightforward, and success in their application to histones and chromatin is largely dependent on the resolving power of the methods used to identify the cross-linked products. Fractionation of the histones and cross-linked products is difficult because of their similar charges and molecular weights. Standard procedures for sodium dodecyl sulfate (SDS)–polyacrylamide gel electrophoresis do not give adequate separation of three of the four main types of histone, and acetic acid and urea-containing gels, which do resolve all the histones, are unsuitable because each type is further resolved into various acetylated and phosphorylated forms, giving a complex pattern even before the introduction of cross-links. Here we describe a modified procedure for SDS–gel electrophoresis that gives a clean separation of the main types of histone and also resolves at least five species of cross-linked dimer. We further describe the use of two-dimensional gels for analyzing the histone compositions of cross-linked dimers and higher oligomers.

II. Cross-Linking of Histones in Chromatin and in Free Solution

A. Reagents and Conditions

1. REAGENTS

The reagents used (Fig. 1) are either diimidoesters, introduced for cross-linking by Davies and Stark (8), or diesters of N-hydroxysuccinimide, introduced by Lomant and Fairbanks (9). Diimidoesters of different length are readily prepared as dihydrochlorides from the corresponding dinitriles (commercially available) by treatment with methanolic HCl (10,11). Methyl 3-mercaptopropionimidate, used for introduction of cleavable cross-links, may be synthesized as described (12) and has a melting point of 78°C. The higher homolog, methyl 3-mercaptobutyrimidate, and also dithiobis(succinimidyl propionate), Lomant's reagent, are available from Pierce.

2. BUFFERS

All the reagents described react with proteins through nucleophilic attack by amino groups at alkaline pH. This precludes the use of amino-containing buffers (e.g., Tris), and suitable buffers used instead are, at pH 8, triethanolamine-HCl (pK_a triethanolamine, 7.8), phosphate, or N-ethylmorpholine-acetic acid (pK_a N-ethylmorpholine, 7.7); and at pH 9, sodium borate (pK_a boric acid, 9.2).

HN   NH
|| ||
H₃COC(CH₂)₆COCH₃

Dimethyl
suberimidate

NH
||
HSCH₂CH₂COCH₃

Methyl 3-mer-
captopropionimidate

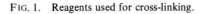

Lomant's reagent

FIG. 1.   Reagents used for cross-linking.

3. REACTION

Solutions of reagent and protein are combined and allowed to react at 20°–25°C for the required time. The solutions of imidoesters are prepared in the buffer used for the reaction, whereas Lomant's reagent, which is much less soluble in water, is dissolved in dimethylformamide (at a high concentration, e.g., 50 mg/ml, to minimize the final concentration of dimethylformamide in the reaction mixture). Both types of reagent are susceptible to hydrolysis, so solutions must be prepared fresh just before use.

Since the object is to form cross-links within, but not between, oligomers, the concentration of protein in the reaction mixture is kept fairly low (0.1–2 mg/ml). The reagent is also used at a low concentration, 1 mg/ml for dimethyl suberimidate hydrochloride and Lomant's reagent, which corresponds to about 4 and 5 mM, respectively. Use of a high reagent concentration would increase the rate of monofunctional reaction with the protein but not necessarily the rate of bifunctional reaction since the number of remaining free amino groups would be reduced.

4. TERMINATION OF THE REACTION

There are three convenient ways of terminating the cross-linking reaction: (a) Lowering the pH to about 5. For reactions carried out in buffer at pH 8 or 9, $I = 0.1$, this may be achieved by addition of 0.5 volume of 3 M sodium acetate (pH 5). Amino groups become protonated, and hence inactivated, and the reagent is rapidly hydrolyzed (13). The sample may be desalted by dialysis against 0.5 mM phenylmethylsulfonyl fluoride–0.2 mM NaEDTA and freeze-dried for analysis in gels. (b) Precipitating the protein with an equal volume of 50% trichloroactic acid at 2°C. After 15 minutes the protein precipitate is collected by centrifugation, washed 2–3 times with acetone, and dried *in vacuo*. This is a quicker method than (a) of preparing samples for analysis in SDS gels. (c) Adding an excess of amino groups (e.g., 0.1 volume of 2 M ammonium bicarbonate) to consume remaining reagent. After a reaction time of about 5 minutes at room temperature, the solution may be used, for example for physical studies of the cross-linked protein, or samples may be prepared for gel analysis by dialysis and freeze-drying as described in (a), or by precipitation with trichloroacetic acid as in (b).

5. SIDE REACTION OF IMIDOESTERS

The normal reaction of an alkyl amine with an imidoester, for example, a protein with a methyl imidoester (Fig. 2), proceeds via the formation of a tetrahedral intermediate (14) and its breakdown by loss of methanol to give a protein N-alkyl amidine (13). However, the intermediate can also break down by loss of ammonia to generate a protein N-alkyl imidate which may

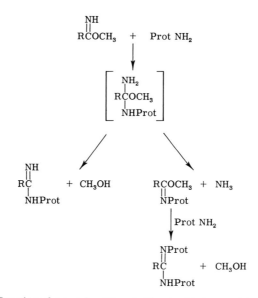

FIG. 2.   Reaction of a protein with an imidoester. Prot = protein, R = alkyl.

further react with protein to form a di(protein N-alkyl) amidine. Thus, even with an apparently monofunctional reagent, bifunctional reaction is possible and a cross-linked product is formed. This side reaction (15) is negligible at pH 9–10 but can be significant at a lower pH. For example, after treatment of chromatin at pH 8, $I = 0.1$, with methyl acetimidate (a monofunctional imidoester), cross-linked histone species are observed (J. O. Thomas, unpublished). This side reaction does not affect conclusions drawn from experiments with dimethyl suberimidate and other bifunctional imidoesters, but it means that reagents such as methyl 3-mercaptopropionimidate may introduce stable cross-links, in addition to ones that can be cleaved by thiol reagents (see Section II,C).

B.   Examples of Cross-Linking with Dimethyl Suberimidate

1.   THE TETRAMER $(H3)_2(H4)_2$

A typical cross-linking experiment for the tetramer from calf thymus was as follows (3). An ammonium sulfate precipitate prepared as described (3), was dissolved at a protein concentration of about 2 mg/ml [determined by the method of Lowry et al. (16)] in 50 mM sodium acetate–50 mM sodium bisulfite (pH 5.0) and dialyzed against 70 mM sodium phosphate–10 mM sodium bisulfite (pH 8.0) ($I = 0.25$). The protein concentration was adjusted to 0.5

mg/ml by dilution with the dialysis buffer and dimethyl suberimidate (freshly dissolved in the same buffer at 0°C and used within 30 seconds of dissolution) was added to a final concentration of 1 mg/ml. The mixture was kept at room temperature for 3 hours (by which time the reagent was largely consumed and termination of the reaction not necessary) and then dialyzed and freeze-dried as described above.

The cross-linked products were analyzed in SDS–polyacrylamide tube gels using the phosphate buffer system of Shapiro *et al.* (*17*) and Weber and Osborn (*18*) (see Section III, A, 1). The gel pattern (Fig. 3) comprised a band containing unresolved monomers and bands corresponding to all the oligomers expected as intermediates in cross-linking of an $\alpha_2\beta_2$ tetramer, namely the dimers α_2, β_2, and $\alpha\beta$; the trimers $\alpha_2\beta$ and $\alpha\beta_2$; and the tetramer $\alpha_2\beta_2$. These assignments may be confirmed by constructing a semilog plot of the molecular weight of the oligomers (calculated from the amino acid sequences of the individual histones) vs mobility measured from the top of the gel. The resulting straight line is evidence that the set of molecular weights is internally consistent. (The absolute values are not readily checked by running the usual set of proteins, e.g., bovine serum albumin, aldolase, as molecular weight standards, since histones migrate anomalously in SDS gels owing to their high positive charge.)

The effects of pH on the cross-linking are as expected for a reaction involving nucleophilic attack by an amino group. The positions of the bands in

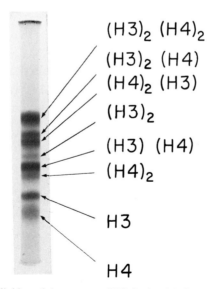

(H3)$_2$ (H4)$_2$
(H3)$_2$ (H4)
(H4)$_2$ (H3)
(H3)$_2$
(H3) (H4)
(H4)$_2$
H3
H4

FIG. 3. Cross-linking of the tetramer (H3)$_2$(H4)$_2$ with dimethyl suberimidate.

a gel are the same in experiments at pH 7, 8, and 9 [90 mM sodium phosphate–10 mM sodium bisulfite (pH 7), $I = 0.2$; 70 mM sodium phosphate–10 mM sodium bisulfite (pH 8), $I = 0.25$, 0.1 M sodium borate–0.15 M NaCl–10 mM sodium bisulfite (pH 9), $I = 0.25$], but the relative intensities shift toward the higher molecular weight bands as the pH is raised.

The pattern of cross-linking at pH 8 is independent of concentration over the range tested (0.1–2 mg/ml), indicating that the tetramer does not readily aggregate or dissociate.

2. HISTONES IN 2 M SODIUM CHLORIDE

Dissociation of chromatin at high ionic strength and pH results in an octamer of histones which is believed to represent the "core" of the repeat unit of chromatin structure (19). This octamer can be obtained free of H1 and nonhistones by treatment of the chromatin with 0.6 M NaCl before dissociation, for example, as described by Oudet et al. (20). DNA can be removed from the octamer by centrifugation.

For cross-linking, octamer at 2 mg/ml in buffer at pH 9, $I = 2$ (1.95 M NaCl–137 mM sodium borate–0.5 mM phenylmethylsulfonyl chloride) is brought rapidly to 23°C and treated with dimethyl suberimidate exactly as described above for the tetramer. Cross-linking is complete after about 1 hour at 23°C. The solution becomes turbid if cross-linking is so extensive that octamers are cross-linked to one another, and it is important to terminate the reaction before this occurs, which may be done in any one of the ways described above (Section II,A,4).

Pure octamer, free of DNA and H1, gives a single band in SDS-polyacrylamide gels after cross-linking for 45–60 minutes. Analysis of the gel pattern at shorter times of reaction shows intermediates in cross-linking consisting of dimer, trimer, etc., up to heptamer and octamer (20a); this identifies the band seen after complete reaction as the octamer.

If H1 is not removed from chromatin before dissociation in 2 M salt, the products of cross-linking give two bands in an SDS gel. One corresponds to the octamer and the other to monomeric H1; apparently the octamer and H1 do not interact under these conditions. With increasing time of cross-linking the band due to H1 broadens, probably because of both monofunctional reaction and intramolecular cross-linking which give rise to a heterogeneous population of molecules.

The results do not depend on whether the DNA is removed before cross-linking. If, however, the protein concentration is reduced below 0.1 mg/ml then additional bands due to hexamer and dimer (which runs slightly ahead of H1 in the gel) are seen. At much lower concentrations there is a further band due to tetramer (21). The appearance of these bands has been attributed to dissociation of the octamer.

3.   CHROMATIN

Cross-linking of chromatin has been carried out in 70 mM sodium phosphate (pH 8) $I = 0.2$; 274 mM triethanolamine–HCl (pH 8) $I = 0.1$; and 274 mM sodium borate (pH 9) $I = 0.1$ [or 100 mM sodium borate–60 mM NaCl (pH 9) $I = 0.1$] (19). A typical procedure is as follows. Chromatin containing DNA of weight average size 1600 base pairs, prepared in 0.2 mM NaEDTA by the nuclease method (22), is diluted approximately 30-fold in buffer to give a solution having an absorbance at 260 nm of 1 and then treated with dimethyl suberimidate at 23°C as described above for the tetramer. For reaction times of the order of 3 hours, the solution is dialyzed against 0.2 mM phenylmethylsulfonyl fluoride–0.1 mM NaEDTA (pH 7) and freeze-dried. For shorter times the reaction is terminated as described above (Section II,A,4). The products of cross-linking are analyzed in SDS gels using either of the systems described below (Section III) depending on what range of oligomer molecular weights is of interest.

C.   Introduction of Cleavable Cross-Links

Monofunctional reaction of methyl 3-mercaptopropionimidate (Fig. 1) with amino groups of histones, followed by oxidation of the newly introduced thiol groups, results in cross-links that differ from those formed with dimethyl suberimidate only in replacement of a –CH$_2$–CH$_2$–moiety by –S–S– (21). The cross-linked products can be resolved and their component histones identified by a two-dimensional SDS–polyacrylamide gel electrophoresis procedure involving cleavage of the cross-links by treatment of the gel with a thiol reagent between dimensions (see Section III,B).

1.   DISULFIDE CROSS-LINKING PROCEDURE

To a solution of histones or chromatin in buffer (for example, histone octamer at 1 mg/ml in buffer at pH 9, $I = 2$, or chromatin at 0.1 mg/ml in buffer at pH 8 or 9, $I = 0.1$, as described in Sections II,B,2 and 3) is added dithiothreitol in buffer to a final concentration of 10 mM, followed by 0.1 volume of methyl 3-mercaptopropionimidate freshly dissolved at 11 mg/ml in buffer at 0°C. The mixture is allowed to react at 23°C for 45–60 minutes, and then hydrogen peroxide (30% w/v) is added to a final concentration of 20 mM. The mixture is kept for 10 minutes at 23°C and at 1 minute intervals during this period 20 μl samples of the reaction mixture are tested with 1 ml of 5,5'-dithiobis(2,4-dinitrobenzoic acid) [0.2 mg/ml in 0.1 M Tris-HCl (pH 8)] to follow the loss of thiol groups, which is complete after 3–4 minutes. The solution is then treated with 2 μl of catalase solution (BDH Ltd., diluted 10-fold with water), kept for 5 minutes at 23°C, transferred to ice, mixed with 0.2 volume of 3 M sodium acetate (pH 5), and finally dialyzed against 0.2 mM phenylmethylsulfonyl fluoride–0.1 mM NaEDTA (pH 7) and freeze-dried. For analysis in SDS gels, the cross-linked material is dissolved in "sample buffer" (*18,23*) from which 2-mercaptoethanol is omitted.

2. DISCUSSION

The following modifications of the procedure have been investigated and found not to affect the band pattern seen in SDS gels: (a) omission of dithiothreitol during reaction with the imidoester; (b) dialysis to remove dithiothreitol and excess methyl 3-mercaptopropionimidate before oxidation with hydrogen peroxide (this causes slight precipitation and should therefore be avoided); (c) alkylation with iodoacetamide of any thiol groups remaining after oxidation. Thiol groups that are not converted by oxidation with hydrogen peroxide into disulfide bonds could catalyze disulfide interchange. If this were to occur during preparation of the cross-linked products for gel electrophoresis, artifactual histone–histone associations could arise.

It seems, however, that no thiol groups survive oxidation, as judged from a lack of any reaction with 5,5′-dithiobis(2,4-dinitrobenzoic acid) at pH 8; this is probably a consequence of performing the hydrogen peroxide treatment without removing dithiothreitol or excess methyl 3-mercaptopropionimidate. Under these conditions any thiol groups that are not converted into histone–histone disulfide cross-links are instead converted into mixed disulfides with excess reagent.

When the peroxide oxidation step is omitted, and dithiothreitol is also omitted from the first stage in the procedure, quite extensive cross-linking still occurs, as judged from bands up to hexamer seen in SDS–polyacrylamide gels. This is likely to be due to spontaneous formation of disulfide cross-links under the conditions of reaction with the imidoester at pH 9.

The peroxide oxidation step must be avoided in the cross-linking of proteins that, unlike histones, contain tryptophan. In such cases, the alternative procedure of oxidizing the thiol groups before reaction of the imidoester with the protein [as, for example, in the use of dimethyl bisdithiopropionimidate (*24*)] is advisable.

III. SDS–Polyacrylamide Gel Electrophoresis of Histones and Cross-Linked Products

A. One Dimension

The products of cross-linking are resolved in one of two types of gel, depending on the molecular-weight range of interest. A "low-resolution" gel, containing 5% or 7.5% polyacrylamide and a homogeneous buffer system, is used for examining the distribution of products among oligomers up to 12-mer, 13-mer, and beyond. A "high-resolution" gel, containing 18% polyacrylamide and a discontinuous buffer system, is used to separate the four main histones into well resolved bands, and also for resolution of cross-linked dimers; higher oligomers are not as conveniently observed with this gel system, probably owing to broadening of the bands by the effect of partial unfolding mentioned below (Section III,A,1).

1. Low Resolution

SDS–5% polyacrylamide gels are prepared in tubes (8cm × 0.5 cm gel) and 7.5% gels as slabs (0.15 cm × 15 cm × 15 cm gel, with 1 cm wide sample wells), using the phosphate buffer system of Shapiro *et al.* (*17*) and Weber and Osborn (*18*). The gels are run until bromophenol blue, used as tracking dye, is about 1 cm from the bottom, which takes about 3.5 hour at a constant current of 7 mA/gel for tubes and 12 hours at 55–60 mA for slabs. In each case about 30 μg of cross-linked protein in about 25 μl of "sample buffer" (*18*) provides a suitable loading. Tube gels are fixed for at least 2–3 hours, and slab gels for at least an hour, in methanol:acetic acid:water (5:1:5) before staining in 0.1% Coomassie Brilliant Blue in the same solvent for 30 minutes. Destaining is carried out by diffusion in 7.5% acetic acid–5% methanol (v/v); the first change of the destain solution is not made until several hours after staining. Staining and destaining are carried out at 30°C for tube gels and at room temperature for slab gels.

The mobility of a cross-linked product may be greater than expected, due to intramolecular cross-linking which prevents complete unfolding of the molecule in SDS. In addition, the bands obtained from products of cross-linking are broader than those given by unmodified proteins, due to the variety of possible intramolecular cross-links and thus the heterogeneity of the partially unfolded molecules.

2. High Resolution

A 15 cm-long slab gel is used for the qualitative assessment of monomer histones and cross-linked dimers (four dimer bands are obtained); a 30 cm-long gel is required to separate the monomer histones sufficiently for quantitation and gives further resolution of the dimers (five bands). The gels (0.15 cm thick, 15 cm wide) contain 18% polyacrylamide, are overlaid with a 3% stacking gel, and are run in a discontinuous buffer system with SDS. The procedure is that of Laemmli (*23*) with the following modifications (*19*), the first two of which are known to be essential: the concentration of Tris buffer in the separating gel is raised to 0.75 M; the ratio N,N'-methylene bisacrylamide: acrylamide is lowered to 0.15: 30; and the electrode buffer is 50 mM Tris–0.38 M glycine–0.1% (w/v) SDS.

The ingredients for a 15 cm-long gel are as follows:

Acrylamide, 30% (w/v)–bisacrylamide, 0.15%	30 ml
Tris-HCl, 3 M, pH 8.8	12.5 ml
Water	6.25 ml
N, N, N', N'-Tetramethylethylenediamine	0.01 ml
SDS, 10% (w/v)	0.5 ml
Ammonium persulfate, 10% (w/v)	0.5 ml

A mixture of the first four components is degassed under aspirator vacuum, the remaining components are added, and the solution is poured into a gel mold to a depth of 15 cm, followed by water-saturated sec-butanol to a further depth of about 1 cm. When the gel has set and cooled (polymerization is exothermic), the butanol is replaced with 0.1% SDS and the gel is left for about an hour. The SDS is then removed and replaced with a stacking-gel solution of the following composition:

Acrylamide, 10% (w/v)–bisacrylamide, 0.5%	6 ml
Water	8.8 ml
Tris-HCl, 0.5 M, pH 6.8	4.8 ml
N, N, N', N'-Tetramethylethylenediamine	0.01 ml
SDS, 10% (w/v)	0.2 ml
Ammonium persulfate, 10% (w/v)	0.2 ml

A mixture of the first four components is degassed, the remaining components are added, the solution is poured to fill the gel mold (a further depth of 3–5 cm), and a well-forming comb is inserted. As soon as the stacking gel has set (after about 15 minutes), electrophoresis is begun. The electrode buffer is prepared from a 5-fold concentrated stock solution consisting of 30.2 gm of Tris base, 144 gm of glycine, and 50 ml of 10% (w/v) SDS per liter. Samples are applied in "sample buffer" (*23*) to which bromophenol blue is added as tracking dye. The gel is run until the dye reaches the end, which takes about 6 hours at a constant current of 30 mA for 15 cm-long gels and about 15 hours at 4 W for 30 cm gels. Fixing, staining, and destaining are carried out as described above (Section III, A, 1) for phosphate-buffered slabs.

For quantitation of monomer histones, a fully destained 30 cm gel is scanned with a Joyce-Loebl microdensitometer. The peaks in the trace (Fig. 4) are sufficiently well resolved that peak heights can be used as a measure of the relative amounts of protein in the bands.

B.   Two Dimensions

The component histones of an oligomer cross-linked through disulfide bonds are determined after electrophoresis as described above, by soaking

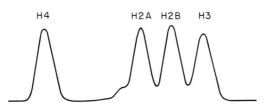

FIG. 4. Monomer region of a high-resolution sodium dodecyl sulfate–polyacrylamide gel of the histones.

the gel in 2-mercaptoethanol and subjecting it to electrophoresis in a second dimension at right angles to the first. Either a low- or high-resolution gel is run in the first dimension (the choice depending on the molecular weight range of interest, as described above), and a high-resolution gel is invariably used for separating the histones in the second dimension.

1. LOW RESOLUTION IN THE FIRST DIMENSION, HIGH IN THE SECOND

This gel system has been used (21) to analyze the dimer, hexamer, and octamer formed by cross-linking histones in 2 M sodium chloride (see Section II, C, 1). Electrophoresis in the first dimension was carried out in a tube (9 cm × 0.4 cm gel) as described in Section III, A, 1, except that 2-mercaptoethanol was omitted from the sample buffer. The gel was then soaked for 1 hour in 0.12 M Tris-HCl (pH 6.8)–0.1% (w/v) SDS–1.4 M 2-mercaptoethanol and placed on a slab for electrophoresis in a second dimension at right angles to the first. The slab was 0.4 cm thick and consisted (from bottom to top) of separating gel (12 cm), stacking gel (1.5 cm), both as described above (Section III, A, 2), and 1% (w/v) agarose (2 cm) containing 0.12 M Tris-HCl (pH 6.8)–0.1% SDS–1.4 M 2-mercaptoethanol (24); about 1 hour was allowed for the agarose layer to set. The assembly of first and second dimension gels was sealed under a layer of 1% (w/v) agarose containing 0.12 M Tris-HCl (pH 6.8)–0.1% SDS. Electrophoresis was for about 16 hours at a constant current of 35 mA. Fixing, staining, and destaining were as described in Section III, A, 1.

2. HIGH RESOLUTION IN BOTH DIMENSIONS

This sytem has been used (21) to analyze the dimers formed by cross-linking histones in chromatin (see Section II, C, 1). Electrophoresis in the first dimension was carried out on 280 μg of protein in a 30 cm-long slab gel, prepared as described (Section III, A, 2) except that the sample well was 2.5 cm wide.

The dimer region, which was well resolved from monomers and trimers, and constant in position in different runs (usually the middle third of the gel), was cut from the unstained gel and soaked and assembled for electrophoresis in a second dimension as described (Section III, B, 1), except that the slab was 24 rather than 12 cm long. Electrophoresis was for 24 hours at a constant current of 35 mA, and fixing, staining, and destaining were as described in Section III, A, 1.

REFERENCES

1. Van Lente, F., Jackson, J. F., and Weintraub, H. *Cell* 5, 45 (1975).
2. Chalkley, R., and Hunter, C. *Proc. Natl. Acad. Sci. U.S.A.* 72, 1304. (1975).
3. Kornberg, R. D., and Thomas, J. O. *Science* 184, 865. (1974).
4. Ilyin, Y. V., Bayev, A. A., Jr., Zhuze, A. L., and Varshavsky, A. J. *Mol. Biol. Rep.* 1, 343. (1974).
5. Martinson, H. G., and McCarthy, B. J. *Biochemistry* 14, 1073. (1975).
6. Martinson, H. G., Shetlar, M. D., and McCarthy, B. J. *Biochemistry* 15, 2002. (1976).
7. Bonner, W. M., and Pollard, H. B. *Biochem. Biophys. Res. Commun.* 64, 282. (1975).
8. Davies, G. E., and Stark, G. R. *Proc. Natl. Acad. Sci. U.S.A.* 66, 651. (1970).
9. Lomant, A. J., and Fairbanks, G. *J. Mol. Biol.* 104, 243. (1976).
10. Pinner, A. "Die Imidoäther und ihre Derivate." Oppenheim, Berlin. (1892).
11. McElvain, S. M., and Schroeder, J. P. *J. Am. Chem. Soc.* 71, 40. (1949).
12. Perham, R. N., and Thomas, J. O. *J. Mol. Biol.* 62, 415. (1971).
13. Hunter, M. J., and Ludwig, M. L. *J. Am. Chem. Soc.* 84, 3491. (1962).
14. Hand, E. S., and Jencks, W. P. *J. Am. Chem. Soc.* 84, 3505. (1962).
15. Browne, D. T., and Kent, S. B. H. *Biochem. Biophys. Res. Commun.* 67, 126. (1975).
16. Lowry, O. H., Rosenbrough, N. J., Farr, A. L., and Randall, R. J. *J. Biol. Chem.* 193, 265. (1951).
17. Shapiro, A. L., Viñuela, E., and Maizel, J. V., Jr. *Biochem. Biophys. Res. Commun.* 28, 815. (1967).
18. Weber, K., and Osborn, M. *J. Biol. Chem.* 244, 4406. (1969).
19. Thomas, J. O., and Kornberg, R. D. *Proc. Natl. Acad. Sci. U.S.A.* 72, 2626. (1975).
20. Oudet, P., Gross-Bellard, M., and Chambon, P. *Cell* 4, 281. (1975).
20a. Thomas, J. O. *In* "Molecular Biology of the Mammalian Genetic Apparatus" (P.O.P. Ts'o, ed.), Part I, p. 199. Elsevier/North-Holland Biomedical Press (1977).
21. Thomas, J. O., and Kornberg, R. D. *FEBS Lett.* 58, 353. (1975).
22. Noll, M., Thomas, J. O., and Kornberg, R. D. *Science* 187, 1203. (1975).
23. Laemmli, U. K. *Nature (London)* 227, 680. (1970).
24. Wang, K., and Richards, F. M. *Isr. J. Chem.* 12, 375. (1974).

SUBJECT INDEX

A

Absorption spectroscopy, of intercalated DNA, 356–357

Acetic acid fixatives, for chromosomes, 230–231

Acriflavine-Feulgen staining, of DNA, 251–253

Aldehyde fixation, of nucleohistones, 229–230

Antinonhistone antibody, animals used in production of, 124

Anti-RGG
iodination of, 146
purification of, 144–146

Artifacts, in circular dichroism, 347–348

B

Balbiani rings, 1

N-Bromosuccinimide, histone cleavage by, 195–197

Buoyant density-gradient sedimentation of chromatin, 23–39

C

Cameras, for neutron scattering, 320–322

Cells, preparation of, for flow-microfluorometry, 248–259

Chickens, use in antinonhistone antibody production, 124

Chromatin
buoyant density-gradient sedimentation of, 23–39
chemical composition of, 13
circular dichroism analysis of, 327–349
cleavage of, by restriction endonucleases, 52–53
core particles of, 71–72, 81–82
characterization of, 87–94
DNase II fractionation of, 11–16
ECTHAM-cellulose chromatography of, 4, 5–6, 10
electron microscopy of, 229–246
fibers of, metal shadowing of, 238
fixation and staining of, 251–259
flow-microfluorometric analysis of, 247–276

fractionation of, 1–103
criteria for, 3–6
methods for, 6–16
interphase type, 240–241
isolation of, 42
laser microbeam studies on, 277–294
negative staining of, 237–238
neutron scattering studies of, 295–325
nuclease digestions of, 42–43, 72–81
phase partition of, 11
repeating unit of, 71
sequence complexity of, 5
sheared, fractionation of, 2
spacer of, 72
spreading on Langmuir trough, 238–242
staining of, 258–259
structure of intercalating agents as probes for, 351–384
subunits of, isolation and characterization of, 69–103
sucrose density gradient centrifugation of, 4, 6–9
template-active, 3
structure of, 16–19
template-active and template-inactive portions of, 1–21
thermal denaturation of, 255
thermal denaturation analysis of, 385–415
thermal elution chromatography of, 10–11
ν-bodies of, visualization of, 61–68

Chromatin gels, isolation and digestion of, 79–80

Chromosome(s)
electron microscopy of, 229–246
eukaryotic, dissection of, 41–54
fixation and squashing of, 157–161
immunochemical analysis of, 105–167
laser microbeam studies on, 282–288
mitotic, 242
organization studies on, 282–285
proteins of, immunofluorescent techniques for, 151–167
stability of, laser microbeam studies of, 286–287

Chymotrypsin, histone cleavage by, 200–201

CONTENTS OF PREVIOUS VOLUMES

Volume I

Volume VII

Volume IX

Volume XI

Volume XIV

Volume XV

Volume XVI

Volume XVII

A
B
C   8
D   9
E   0
F   1
G   2
H   3
I   4
J   5